Classification and inventory of the world's wetlands

Advances in vegetation science 16

The titles published in this series are listed at the end of this volume.

Classification and inventory of the world's wetlands

Edited by

C. MAX FINLAYSON AND A. G. VAN DER VALK

Reprinted from Vegetatio, volume 118

Kluwer Academic Publishers

Dordrecht / Boston / London

Library of Congress Cataloging-in-Publication Data

*GB
621
C57
1995*

A C.I.P. Catalogue record for this book is available from the Library of Congress.

ISBN 0-7923-3714-X

Published by Kluwer Academic Publishers,
P.O. Box 17, 3300 AA Dordrecht, The Netherlands.

Kluwer Academic Publishers incorporates
the publishing programmes of
D. Reidel, Martinus Nijhoff, Dr W. Junk and MTP Press.

Sold and distributed in the U.S.A. and Canada
by Kluwer Academic Publishers,
101 Philip Drive, Norwell, MA 02061, U.S.A.

In all other countries, sold and distributed
by Kluwer Academic Publishers Group,
P.O. Box 322, 3300 AH Dordrecht, The Netherlands.

Printed on acid-free paper

Printed in the Netherlands

Vegetatio **118**: 1–2, 1995.

Editorial

The IV International Wetlands Conference was held in Columbus, Ohio, USA, in 1992 as part of the celebration of the 500th anniversary of the discovery of the new world. The general theme of the conference was 'Wetlands – Old World and New World'. In keeping with this theme, a symposium was organized to examine wetland classification systems and the status of wetland inventories in both the old and new world. Another objective of this symposium was to try to establish a framework for developing an international classification system that would eventually be used for a global inventory of wetlands.

Currently, the only international wetland classification system in use is the one that was developed by the Ramsar Convention (Scott 1989). Contracting parties to the Ramsar Convention are required to compile an inventory of wetlands as part of a process of developing and implementing a national wetland policy for the wise use of wetlands in their territories. This classification system was loosely adapted from the American system of Cowardin *et al.* (1979). The Ramsar Convention classification is outlined in this volume by Scott & Jones. A detailed description of the American system is given in Cowardin & Golet, and the status of the American inventory of wetlands that uses this system in Wilen & Bates. Other national and regional attempts to classify and inventory wetlands were also presented at the symposium: Europe (Huges, Pakarinen); Asia (Lu, Gopal & Sah); South America (Naranjo); Africa (Taylor *et al.*); Australia (Pressey & Adam); and Canada (Zoltai & Vitt). A new approach to wetland classification that is based primarily on hydrogeomorphology is presented by Semeniuk & Semeniuk. A new approach for determining changes in the status of wetlands is presented by Novitzki.

An inventory of wetlands is generally regarded as an essential prerequisite for developing wetland conservation and management plans at the national, regional or international level. With the exception of the United States, Canada, Australia and some European countries, national wetland inventories tend to consist of little more than lists of major wetlands and should be considered as only very preliminary. Consequently, they are of little use in establishing rates of wetland loss or degradation. Because good information on wetland location, area, and status are needed as a basis of effective conservation and management efforts, these preliminary inventories are of little use to policy makers at either the national or international level. Guidelines are urgently needed for conducting a minimal wetland inventory, including the kinds of information needed and how this information is to be collected and stored. These guidelines would enable a global inventory of wetlands to proceed piecemeal by standardizing local, national, and regional inventories.

The following resolution was passed at the IV International Wetlands Conference:

Whereas:
An internationally accepted wetlands classification system and the national inventories based on it have many benefits, including

1. a universally acceptable terminology for use in scientific research, international conservation projects and wetland management projects;
2. a framework for implementing international legal instruments for wetland conservation; and
3. a framework for disseminating information on wetlands to planners, policy makers, managers, and decision makers in international government and non-government agencies.

Be it resolved that:
The Conference Participants encourage the development and adoption of an international wetland classification system and guidelines for national wetland inventories as a contribution to achieving the goals of the Ramsar Convention.

A number of the participants in this symposium believe that we may not be far from having an international wetland classification system. They also believe, however, that we are a long way from having a detailed global inventory of wetlands. In keeping with the resolution passed at the Conference, we propose that an international committee under the auspices of an international agency, such as IWRB, the Ramsar Bureau or IUCN, needs to be established to develop an international wetland classification system and practical guidelines for carrying out national wetland inventories.

C. M. Finlayson

A. G. van der Valk

References

Cowardin, L. M., Carter, V., Golet, F. C. & LaRoe, E. T. 1979. Classification of wetlands and deepwater habitats of the United States. U.S. Fish and Wildlife Service FWS/OBS 79/31.

Scott, D. A. 1989. Design of wetland data sheets for database on Ramsar sites. Photocopied report to Ramsar Bureau, Gland, Switzerland.

Acknowledgments

This volume contains the proceedings of a sympo sium 'Classification and inventory of the world's wetlands' held at the INTECOL sponsored IV International Wetland Conference. The conference was held in Columbus, Ohio, USA, in September 1992. Support for this symposium was provided by the Institute for Wetland and Waterfowl Research, a joint project of Ducks Unlimited (DU) in the United States, Canada and Mexico. One of IWWR's major missions is to enhance the communication of the latest information on wetlands and waterfowl biology and conservation. Additional support for this symposium was provided by a travel grant from the Society of Wetland Scientists. We would like to thank Bill Mitsch and his staff, who organized the IV International Wetland Conference, for their help with this symposium both before and during the Conference.

Vegetatio **118**: 3–16, 1995.

Classification and inventory of wetlands: A global overview

D.A. Scott & T.A. Jones

International Waterfowl and Wetlands Research Bureau, Slimbridge, Gloucestershire GL2 7BX, UK

Key words: Ramsar Convention, Wetland classification, Wetland inventory

Abstract

Classification of wetlands is extremely problematical, definition of the term wetland being a difficult and controversial starting point. Although considerable effort has gone into the development of national and regional wetland classifications, the only attempt at establishing a global system has been under the auspices of the Ramsar Convention on Wetlands of International Importance. In view of the fact that the Ramsar Convention has 70 Contracting Parties world-wide, it is suggested that the Convention's definition and classification system should be adopted generally for international purposes. Much of the world has been covered by preliminary wetland inventories, but there is an urgent need to extend coverage to those areas not yet included. It is essential that all inventory projects give adequate attention to meeting the real information needs of agencies and individuals which have an impact on the conservation and wise use of wetlands. Attention should also be given to providing for wide dissemination and regular updating of information and establishment of procedures for monitoring ecological change at the sites identified.

Introduction

The classification and inventory of wetlands is fraught with difficulty. In fact, what is a wetland? Some parts of the world include land which may be completely dry for years, but which may nevertheless, support internationally important wetlands after period of exceptional rainfall. Where should the line be drawn between coastal wetlands and wholly marine systems? Should entirely man-made wetlands be given the same status as natural or semi-natural habitats? How should natural vegetational succession be covered? The delimitation of a wetland site is equally problematical; the impossibility of separating a wetland from its hydrological support system means that it is necessary to consider factors operating throughout a catchment, including sub-surface features, both upstream and downstream of the area under consideration.

Many published accounts of wetlands (especially national and local inventories; e.g. Environmental Problems Foundation of Turkey 1989) have largely avoided addressing such problems by classifying wetlands geographically and by using local terminology in descriptive text. Unfortunately, the attractive simplicity of this approach begins to break down when applied at an international scale. Although knowledge of the locations of the world's most important wetlands has taken great strides forward in the last fifteen years (e.g. Scott & Carbonell 1986; Whigham *et al.* 1993) there are still substantial areas of the globe which remain relatively uncharted. Even in areas which have been covered by preliminary wetland inventories, we lack, all too frequently, even a basic understanding of their hydrology, limnology and ecology.

Compilers of wetland inventories need to examine carefully whether they are gathering the data sets that are actually required for furthering the conservation and wise use of the sites which have been identified. The pressures which are leading to the degradation and destruction of many of the world's most important wetlands have an increasingly strong international dimension (Dugan 1993). It is, therefore, imperative that information concerning the locations and values of these sites is readily available to and understood by all those involved in formulating and implementing policies which affect them. Finally, wetland inventories of all kinds must be regularly reviewed and updated if they are not to become items of historical interest only.

International wetland classification

Many countries have national or regional wetland terminology that is not understood internationally. For example, how many wetland scientists could confidently and correctly assign a geographical region and accurate meaning to all of the following wetland types, each of which is used commonly in one or more parts of the world:

vlei	rybník
lochan	valle
turfmoor	hammock
rhyne	turlough
qa	jheel

All of these terms have precise meanings which can rarely be translated into another language through use of a single word. Usually, a short phrase is required, employing internationally understood terms such as 'lagoon' or 'floodplain', together with qualifying statements covering, for example, seasonality, vegetation, salinity and human impacts.

An increasing number of countries have established some kind of national wetland classification, usually in association with the development of national wetland inventories (e.g. Silvius *et al.* 1987). By definition, these national classifications tend to focus on the unique characteristics of a country's wetlands and are therefore of limited use for international applications. The evolution of a national wetland classification is determined by many factors, but the geographical location of a country immediately determines that some wetland types common elsewhere in the world are likely to be excluded (to take an extreme example, there are no mangroves in Canada, but equally, there are no tundra wetlands in Indonesia). However, in spite of the large and increasing volume of international wetland research and conservation activities, there have been few attempts to produce international wetland classifications.

The purposes and values of international wetland classifications can be summarised as follows:

A. To provde readily understood terminology for use in scientific research and conservation projects with an international dimension.
B. To provide a framework for implementing international legal instruments for wetland conservation.
C. To assist international dissemination of information to as many relevant individuals and organisations as possible.

Whilst recognizing the technical and scientific benefits of establishing certain common standards and terms for describing wetlands, the rich cultural and linguistic heritage of wetland areas must not be overlooked. It would surely be a case of 'killing the goose that laid the golden egg' if, in our quest for harmonisation and synthesis we jeopardised the continuing existence of uniqueness and diversity, or risked alienating the local people responsible for day-to-day stewardship of the world's wetlands.

Before elaborating a wetland classification, it is necessary to adopt a definition of the term 'wetland'. Internationally, the most widely used and accepted definition is the one provided by the Convention on Wetlands of International Importance, usually referred to as the 'Ramsar' Convention after the Iranian town of Ramsar where the treaty was adopted in 1971 (Matthews 1993). Almost 80 countries, from all regions of the world, are now Contracting Parties to the Convention (see Table 1) and have therefore accepted the following definition for international purposes:

> '... wetlands are areas of marsh, fen, peatland or water, whether natural of artificial, permanent or temporary, with water that is static or flowing, fresh, brackish or salt, including areas of marine water the depth of which at low tide does not exceed six metres.'

The rationale behind the very broad Ramsar definition stemmed from a desire to embrace all the 'wetland' habitats of migratory water birds; the full title of the treaty is, after all, *Convention on Wetlands of International Importance especially as Waterfowl Habitat* (Matthews 1993). Hence, the inclusion of areas of marine water less than six metres deep at low tide, which, at northern latitudes, are often important wintering habitats for loons (divers), grebes and sea ducks; and the inclusion of man-made wetlands such as reservoirs and seasonally flooded agricultural land, which are often important habitats for ducks, geese, cranes and shorebirds. Inevitably, however, a definition as broad as this has created problems. All areas of rice cultivation would

Table 1. Contracting parties to the Ramsar Convention – July 1993.

	Country	Date Convention came into force	Date Paris Protocol applied	Date Regina amendments accepted[1]	Number of wetlands designated	Area of wetlands (hectares)
1.	Australia	21.12.75	12.08.83	25.07.90	40	4,510,468
2.	Finland	21.12.75	15.05.84	27.03.90	11	101,343
3.	Norway	21.12.75	3.12.82	20.01.89	14	16,256
4.	Sweden	21.12.75	3.05.84	6.04.89	30	382,750
5.	South Africa	21.12.75	26.05.83	14.02.92	12	228,344
6.	Islamic Rep. Iran	21.12.75	29.04.86		18	1,357,550
7.	Greece	21.12.75	2.06.88	22.05.92	11	107,400
8.	Bulgaria	24.01.76	27.02.86	21.06.90	4	2,097
9.	UK	5.05.76	19.04.84	27.06.90	63	274,883
10.	Switzerland	16.05.76	30.05.84	9.06.89	8	7,049
11.	Germany	26.06.76	13.01.83	21.06.90	31	672,852
12.	Pakistan	23.11.76	13.08.85	20.09.88	9	20,990
13.	New Zealand	13.12.76	9.02.87	07.07.93	5	38,099
14.	Russian Federation	11.02.77	11.02.92	11.02.92	3	1,168,000
15.	Italy	14.04.77	27.07.87		46	56,950
16.	Jordan	10.05.77	15.03.84		1	7,372
17.	Yugoslavia	28.07.77			2	18,094
18.	Senegal	11.11.77	15.05.85		4	99,720
19.	Denmark	2.01.78	3.12.82		38	1,832,968
20.	Poland	22.03.78	8.02.84		5	7,141
21.	Iceland	2.04.78	11.06.86	18.06.93	2	57,500
22.	Hungary	11.08.79	28.08.86	20.09.90	13	114,862
23.	Netherlands	23.09.80	12.10.83	19.11.91	21	314,928
24.	Japan	17.10.80	26.06.87	2.06.88	9	83,454
25.	Morocco	20.10.80	3.10.85		4	10,580
26.	Tunisia	24.03.81	15.05.87	26.01.93	1	12,600
27.	Portugal	24.03.81	18.12.84		2	30,563
28.	Canada	15.05.81	2.06.83	8.11.88	31	13,020,203
29.	Chile	27.11.81	14.02.85		1	4,877
30.	India	1.02.82	9.03.84		6	192,973
31.	Spain	4.09.82	27.05.87		26	122,418
32.	Mauritania	22.02.83	31.05.89		1	1,173,000
33.	Austria	16.04.83	18.12.92	18.12.93	7	102,541
34.	Algeria	4.03.84			2	4,900
35.	Uruguay	22.09.84			1	435,000
36.	Ireland	15.03.85	15.11.84	28.08.90	21	13,035
37.	Suriname	22.11.85			1	12,000
38.	Belgium	4.07.86			6	7,945
39.	Mexico	4.11.86	4.07.86	2.11.92	1	47,840
40.	France	1.12.86	1.12.86		8	425,585
41.	USA	18.04.87	18.12.86		12	1,194,001
42.	Gabon	30.04.87	30.12.86		3	1,080,000
43.	Niger	30.08.87	30.04.87		1	220,000
44.	Mali	25.09.87	25.05.87		3	162,000
45.	Nepal	17.04.88	17.12.87		1	17,500
46.	Ghana	22.06.88	22.02.88		6	178,410

Table 1. Continued.

47.	Uganda	4.07.88	4.03.88		1	15,000
48.	Egypt	9.09.88	9.09.88		2	105,700
49.	Venezuela	23.11.88	23.11.88		1	9,968
50.	Viet Nam	20.01.89	20.09.88		1	12,000
51.	Malta	30.01.89	30.09.88		1	11
52.	Guinea-Bissau	14.05.90	14.05.90		1	39,098
53.	Kenya	5.10.90	5.06.90		1	18,800
54.	Chad	13.10.90	13.06.90		1	195,000
55.	Sri Lanka	15.10.90	15.06.90		1	6,216
56.	Guatemala	26.10.90	26.06.90		1	48,372
57.	Bolivia	27.10.90	27.06.90		1	5.240
58.	Burkina Faso	27.10.90	27.06.90		3	299,200
59.	Panama	26.11.90	26.11.90		2	97,179
60.	Ecuador	7.01.91	7.09.90		2	90,137
61.	Croatia	25.06.91			4	80,455
62.	Slovenia	25.06.91			1	650
63.	Romania	21.09.91	21.05.91		1	647,000
64.	Liechtenstein	6.12.91	6.08.91	6.08.91	1	101
65.	Zambia	28.12.91	28.08.91		2	333,000
66.	Peru	30.03.92	30.03.92		3	2,415,691
67.	Costa Rica	27.04.92	27.12.91		3	30,269
68.	China	31.07.92	31.03.92		6	586,870
69.	Indonesia	8.08.92	8.04.92	8.04.92	1	162,700
70.	Argentina	4.09.92	4.05.92		3	82,474
71.	Bangladesh	21.09.92	21.05.92	21.05.92	1	59,600
72.	Czech Republic	1.01.93	1.01.93		4	18,109
73.	Slovak Republic	1.01.93	1.01.93		7	25,519
74.	Guinea	18.03.93	18.11.92		6	264,109
75.	Trinidad & Tobago	21.04.93	21.12.92	21.12.92	1	6,234
76.	Papua New Guinea	16.07.93	16.03.93		1	590,000
77.	Brazil	24.09.93	24.05.93		2	168,400
78.	Armenia	9.10.93	9.06.93		2	51,976
79.	Honduras	23.10.93	23.06.93		1	8,500
	former USSR				9	1,819,185
					623	38,202,706

[1] For an explanation of the Paris Protocol and Regina Amendments, see Matthews 1993.
Information supplied by the Ramsar Database.

technically qualify as wetlands, though most such areas are of scarcely any conservation value. Similarly, a large part of the world's coral reefs and sea-grass beds qualify as wetlands. If coral reefs are to be included, perhaps the definition should embrace all such systems, rather than only those above the six metres limit.

At a national level, many countries have adopted narrower definitions; for example, some countries do not consider large rivers or water storage reservoirs as wetlands. However, the Ramsar definition is increasingly providing the basis for both national and international inventories, as more and more countries ratify the Convention and, in doing so, accept the definition for at least international purposes.

One of the first international wetland classifications was employed by Scott (1980) in *A Preliminary Inventory of Wetlands of International Importance for Waterfowl in West Europe and Northwest Africa.* Correspondents in each country were asked to complete a simple datasheet for each site, indicating which of 25 habitat types was present within the site. The classification was based on work being undertaken in Paris on behalf of the European Community in relation to the then fledgling Community-wide Directive on the Con-

Table 2a. Classification of wetland type used in the Directory of Neotropical Wetlands (Scott & Carbonell 1986).

01	shallow sea bays and straits
02	estuaries, deltas
03	small offshore islands, islets
04	rocky sea coasts, sea cliffs
05	sea beaches (sand, pebbles)
06	intertidal mudflats, sandflats
07	coastal brackish and saline lagoons & marshes, salt pans
08	mangrove swamps, brackish forest
09	slow-flowing rivers, streams (lower perennial)
10	fast-flowing rivers, streams (upper perennial)
11	riverine lakes (including oxbows), riverine marshes
12	freshwater lakes and associated marshes (lacustrine)
13	freshwater ponds (< 8 ha), marshes, swamps (palustrine)
14	salt lakes, salars (inland systems)
15	reservoirs, dams
16	seasonally flooded grassland, savanna, palm savanna
17	rice paddies, flooded arable land, irrigated land
18	swamp forest, temporarily flooded forest
19	peat bogs, wet Andean meadows (bofedales), snow melt bogs

Table 2b. Classification of wetland type used in the Directory of Asian Wetlands (Scott 1989a).

01	shallow sea bays and straits (< 6 m depth at low tide)
02	estuaries, deltas
03	small offshore islands, islets
04	rocky sea coasts, sea cliffs
05	sea beaches (sand, pebbles)
06	intertidal mudflats, sandflats
07	mangrove swamps, brackish forest
08	coastal brackish and saline lagoons and marshes
09	salt pans
10	shrimp ponds, fish ponds
11	rivers, streams; slow-flowing (lower perennial)
12	rivers, streams; fast-flowing (upper perennial)
13	oxbow lakes, riverine marshes
14	freshwater lakes and associated marshes (lacustrine)
15	freshwater ponds (< 8 ha), marshes, swamps (palustrine)
16	salt lakes, saline marshes (inland drainage systems)
17	water storage reservoirs, dams
18	seasonally flooded grassland, savanna, palm savanna
19	rice paddies
20	flooded arable land, irrigated land
21	swamp forest, temporarily flooded forest
22	peat bogs

Source: Ramsar Convention Bureau (1990).

servation of Wild Birds. In addition to wetlands *per se*, this classification included certain dryland habitats which are commonly found in association with Western Palearctic wetlands.

A number of subsequent international wetland inventories have followed the simple type of classification described above. For example, the Directories of Neotropical Wetlands (Scott & Carbonell 1986) and Asian Wetlands (Scott 1989a) employed broadly similar systems, which are reproduced below in Table 2. The introductions to both of these Directories include the note that, 'Although more sophisticated wetland classification systems are available, the information was seldom adequate to permit a more detailed breakdown, and in any case for many of the enormous wetlands described in the Directory, a detailed classification of habitat types would be extremely cumbersome'.

Recognizing the limitations, in terms of both quantity and quality, of data available for many countries is fundamental to the construction of an international wetland classification. In the development of hierarchical classifications, there is always a temptation to focus debate on the most detailed (and hence, usual-

ly the most controversial) level rather than on broader generic categories. This can result in a classification which is partly (or even mostly) irrelevant to the level of information available from much of the world.

During the late 1980s, the Contracting Parties to the Ramsar Convention recognised the need for establishing a database to hold information on those wetlands designated under Article 2.1 of the treaty for the Ramsar *List of Wetlands of International Importance*. In connection with setting up a database, the Contracting Parties also charged the Ramsar Bureau with establishing a wetland information sheet and classification of wetland type aimed at standardising the data gathered for each Ramsar site.

In 1990, as a result of this initiative, the Fourth Meeting of the Conference of the Contracting Parties adopted a Recommendation approving an information sheet and hierarchical classification of wetland type (Scott 1989b) based loosely on the *Classification of Wetlands and Deepwater Habitats of the United States* (Cowardin *et al.* 1979). The US classification is divided into systems, sub-systems, classes and sub-classes, together with a series of modifiers concerning water regime, water chemistry (salinity, pH) and soil. The basic unit of the hierarchy is the system, of which five types are distinguished (marine, estuarine, riverine, lacustrine and palustrine). Both the US classification and the version adopted for use with the Ramsar Convention are reproduced here as Tables 3 and 4, respectively.

A number of authors have suggested the use of highly simplified groupings of basic wetland types for use in general information and education materials. For example, Dugan (1990) has suggested that it is possible to reduce the more detailed groupings of the Ramsar classification to, 'seven landscape units which are wetlands, or where wetlands form an important component, and which therefore define the planning framework for wetland conservation'. These units are estuaries, open coasts, floodplains, freshwater marshes, lakes, peatlands and swamp forest.

The Ramsar database, which is maintained at Slimbridge, UK by IWRB, has been in operation for approximately 4 years, during which time habitat information received from Contracting Parties concerning their designated Ramsar sites has been coded and entered into a dBaseIV system. The habitat information is stored in conjunction with a wide range of other site data, from geographical coordinates to landuse. When updated material has been provided by all Contracting Parties, use of the Ramsar classification will make it possible to analyse the Convention's coverage of the principal wetland types, thereby allowing the identification of gaps for immediate attention. The Ramsar classification is, like all classifications, a compromise. However, experience to date suggests that the Ramsar system is workable and readily understood and we suggest that it should be used as the basis for appropriate international projects in the future.

One on-going mathematical classification and inventory project has recently been completed; the *Directory of Important Wetlands in Australia* (Usback & James 1993) was published in June 1993. It was compiled using the Ramsar classification system with minor modifications to provide specifically for wetland types which it is important to distinguish in a national context. A number of other national or regional classifications of wetland type have been elaborated; countries covered by recent publications include Canada (National Wetlands Working Group 1987), Greece (Heliotis 1988), Indonesia (Silvius et at. 1987) and South Africa (Walmsley & Boomker 1988). Accounts of several such projects will be presented later in this volume but it is worth looking at some of the contrasting classifications that have been adopted. For exam-

ple, the Canadian classification includes five wetland classes and 70 wetland forms, of which 18 are types of bog and 17 are types of fen; while the Indonesian classification has broken down forested wetlands into six types of mangrove forest and eight types of freshwater swamp forest.

In July 1992, the European Community published the official version of a Community-wide Directive obliging the twelve Member States to undertake measures which will conserve certain scarce or threatened habitats and species, as specified in Annexes to the so-called *Habitats Directive* (Official Journal of the European Communities 1992). Annex 1 to the Directive lists 'Natural habitat types of Community interest whose conservation requires the designation of Special Areas of Conservation'. The classification of habitats used is that developed during the 1980s under the Community's *CORINE* biotopes project. The CORINE classification is hierarchical with a strong phytosociological element and is extremely detailed. Thus, Annex 1 of the Habitats Directive (which, as indicated above, includes only those habitats thought to be in need of special conservation measures) includes at least 50 specific wetland habitat types which fall within the Ramsar definition. Although it is an international classification, it is strictly concerned with the territory of the European Community and is much too elaborate for effective world-wide application.

International wetland inventories

In the course of developing an effective wetland conservation programme, one of the first steps is the compilation of a basic inventory of wetlands (covering at least the more important and/or vulnerable sites) in the relevant geographical area. One expression of the burgeoning interest in wetlands in recent years has been the proliferation of inventory projects. Such inventories may:

- aid identification of priorities for future action in research, protection and management;
- establish the basis for monitoring the conservation status of wetlands;
- facilitate local, national and international comparisons between sites;
- promote increased awareness of/interest in key wetland sites on the part of politicians, government officials, land use planners, students and scientists.

Table 3. Hierarchy of wetlands and deepwater habitats in the U.S. wetland classification (Cowardin *et al.*, 1979), showing systems, subsystems and classes. The Palustrine system does not include deepwater habitats.

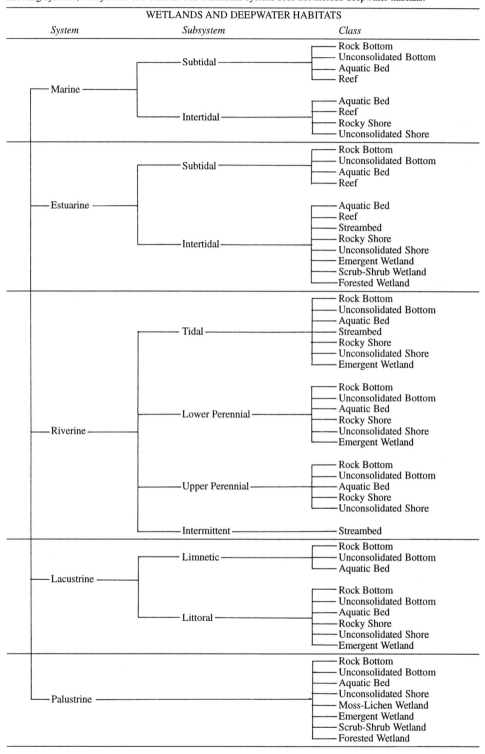

WETLANDS AND DEEPWATER HABITATS

System	Subsystem	Class
Marine	Subtidal	Rock Bottom / Unconsolidated Bottom / Aquatic Bed / Reef
	Intertidal	Aquatic Bed / Reef / Rocky Shore / Unconsolidated Shore
Estuarine	Subtidal	Rock Bottom / Unconsolidated Bottom / Aquatic Bed / Reef
	Intertidal	Aquatic Bed / Reef / Streambed / Rocky Shore / Unconsolidated Shore / Emergent Wetland / Scrub-Shrub Wetland / Forested Wetland
Riverine	Tidal	Rock Bottom / Unconsolidated Bottom / Aquatic Bed / Streambed / Rocky Shore / Unconsolidated Shore / Emergent Wetland
	Lower Perennial	Rock Bottom / Unconsolidated Bottom / Aquatic Bed / Rocky Shore / Unconsolidated Shore / Emergent Wetland
	Upper Perennial	Rock Bottom / Unconsolidated Bottom / Aquatic Bed / Rocky Shore / Unconsolidated Shore
	Intermittent	Streambed
Lacustrine	Limnetic	Rock Bottom / Unconsolidated Bottom / Aquatic Bed
	Littoral	Rock Bottom / Unconsolidated Bottom / Aquatic Bed / Rocky Shore / Unconsolidated Shore / Emergent Wetland
Palustrine		Rock Bottom / Unconsolidated Bottom / Aquatic Bed / Unconsolidated Shore / Moss-Lichen Wetland / Emergent Wetland / Scrub-Shrub Wetland / Forested Wetland

10

Table 4. Wetland classification used by the Ramsar Convention Bureau.

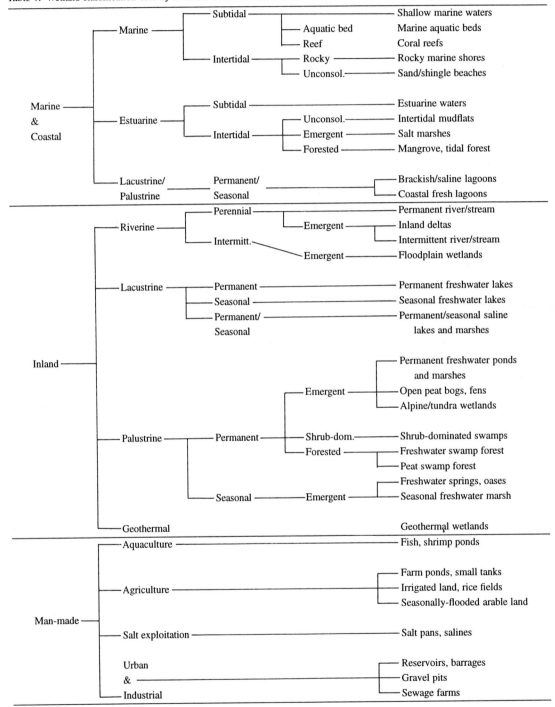

National wetland inventories date back at least as far as the early 1950s and a number of countries, particularly in Europe and North America, have now produced some kind of inventory of their wetland resources. These vary greatly in scope and depth of treatment, from simple lists of major water bodies, to comprehensive descriptions of all wetland resources in the country concerned.

Nowhere has this inventory work been taken to greater technical lengths than in the United States. A major project was initiated by the US Fish & Wildlife Service in 1974 'to provide needed data and information so that decision-makers can make informed decisions about wetland resources, knowing how many, of what type, are where, as well as what they functionally contribute'. Scientists responsible for implementing the inventory experimented with various types of remote sensing, and finally decided to utilize high altitude stereoscopic aerial photography. At that time, satellite imagery could not consistently identify and classify wetlands to the degree of accuracy required. However, it seems likely that the more recent generations of satellites with enhanced capabilities will be able to do this, and will certainly be useful in monitoring changes in wetland resources.

Comprehensive inventories of this type are extremely expensive and time consuming. The U.S. Wetlands Inventory has already cost tens of millions of dollars over the last 18 years and will require further massive investment prior to completion. It is clearly impossible for all but the richest countries to carry out inventories of this kind; indeed, the time scale for such detailed work could mean that internationally important wetlands become degraded or destroyed during the course of inventory compilation. If we are to conserve and make wise use of the most important wetlands in global terms, it is necessary to consider the results which have been obtained from simpler methodologies.

During the late 1950s, the International Society of Limnologists (SIL) decided to prepare a worldwide list of lakes and rivers whose protection was considered to be particularly desirable. The International Biological Programme eventually took over the project, aptly named Aqua, and published its list of sites in 1971 (Luther & Rzoska 1971). Meanwhile, IUCN had embarked on the compilation of a list of marshes, bogs and other wetlands of international importance, primarily as waterfowl habitat, in Europe and North Africa. This list, known as the MAR List (Olney 1965), was deliberately restricted to about 200 sites, since it

was felt that governments and conservation bodies at that time would not be able to cope with many more sites.

With the rapid advances in knowledge of wetlands in the 1960s and early 1970s, the Aqua and MAR lists rapidly became out of date. Table 5 makes a comparison between the number of wetlands in each country included in the MAR list, and those which, almost 30 years on, have been designated under the Ramsar Convention. It is clear that substantial progress has been made in Western Europe, but the situation in Eastern Europe and North Africa gives cause for some concern.

With the Ramsar Convention coming into force in 1975, there arose a need for widely accepted criteria for the identification of sites of international importance. Provisional criteria were approved at a Wetlands Conference at Heiligenhafen, Germany, in 1974, and later refined at the Conference of the Contracting Parties to the Convention (most recently at Montreux, Switzerland, in 1990; see Table 6). These criteria pertain to: (a) the representative character or uniqueness of the sites; (b) the value of the sites for threatened or endemic species of animals and plants and for maintenance of biodiversity; and (c) the importance of the sites for populations of waterfowl.

Under the terms of the Ramsar Convention, Contracting Parties are required to designate sites for inclusion in the Convention List of Wetlands of International Importance. Once criteria for site selection had been developed, it became possible to compile basic lists, or 'shadow lists', of all those sites which might be eligible for designation as Ramsar sites. The first of these such lists was compiled for IUCN and the United Nations Environment Programme (UNEP), and covered the Western Palearctic Region. This *Directory of Western Palearctic Wetlands* (Carp 1980) was based on both the MAR and Aqua lists, and combined sites of ornithological interest with sites of limnological and botanical interest. It covers forty-four countries in Europe, North Africa and the Middle East, and lists almost 900 sites.

At about the same time, the International Waterfowl and Wetlands Research Bureau (IWRB) was pulling together the extensive information derived from wetland surveys and waterfowl counts in western Europe and north-west Africa. Detailed counts of wildfowl, shorebirds, and in many cases also pelicans, herons and other water birds, were available for most wetlands in western Europe, as were some estimates of the total size of the populations, making it possible to

Table 5. Comparison of wetland sites listed by the MAR project and those subsequently designated under the Ramsar Convention.

Country	No. of sites on MAR list published 1965	No. of Ramsar sites designated by July 1993
Albania	0	Not a Ramsar Party
Algeria	5	2
Andorra	0	Not a Ramsar Party
Armenia	0	2
Austria	3	7
Azerbaijan	1	Not a Ramsar Party
Belarus	0	Not a Ramsar Party
Belgium	2	6
Bulgaria	4	4
Croatia	1	4
Czech Republic	3	4
Denmark	4	3
Estonia	2	Not a Ramsar Party
Finland	3	11
France	21	8
Georgia	0	Not a Ramsar Party
Germany	16	31
Greece	7	11
Hungary	6	13
Iceland	0	2
Italy	7	46
Latvia	1	Not a Ramsar Party
Liechtenstein	0	1
Lithuania	1	Not a Ramsar Party
Luxembourg	0	Not a Ramsar Party
Malta	0	1
Moldova	0	Not a Ramsar Party
Monaco	0	Not a Ramsar Party
Morocco	7	4
Netherlands	10	21
Norway	7	14
Poland	15	5
Portugal	4	2
Romania	5	1
Russia	4	3
Slovak Republic	2	7
Slovenia	9	1
Spain	10	26
Sweden	17	30
Switzerland	7	8
Tunisia	8	1
Turkey	8	Not a Ramsar Party
Ukraine	0	Not a Ramsar Party
United Kingdom	20	62
Yugoslavia	4	2

Table 6. Ramsar criteria for listing wetlands of international significance.

A wetland qualifies for designation as a Ramsar site if:

1. The site is:

(a) a particularly good representative example of a natural or near-natural wetland, characteristic of the appropriate biogeographical region; or

(b) a particularly good representative example of a natural or near-natural wetland, common to more than one biogeographical region; or

(c) a particularly good representative example of a wetland, which plays a substantial hydrological, biological or ecological role in the natural functioning of a major river basin or coastal system, especially where it is located in a trans-border position; or

(d) an example of a specific type of wetland, rare or unusual in the appropriate biogeographical region.

and/or

2. The site:

(a) supports an appreciable assemblage of rare, vulnerable or endangered species or subspecies of plant or animal, or an appreciable number of individuals of any one or more of these species; or

(b) is of special value for maintaining the genetic and ecological diversity of a region because of the quality and peculiarities of its flora and fauna; or

(c) is of special value as the habitat of plants or animals at a critical stage of their biological cycles.

and/or

3. The site:

(a) is of special value for one or more endemic plant or animal species or communities; or

(b) regularly supports 20,000 waterfowl; or

(c) regularly supports substantial numbers of individuals from particular groups of waterfowl, indicative of wetland values, productivity or diversity; or

(d) where data are available, regularly supports 1% of the individuals in a population of one species or subspecies of waterfowl.

apply objectively the Ramsar criteria concerning water birds and thereby determine which wetlands qualified for inclusion on the Ramsar List on the basis of numbers of birds present. IWRB published this assessment in *A Preliminary Inventory of Wetlands in West Europe and Northwest Africa* (Scott 1980), which describes over 500 wetlands in 22 countries. This work has subsequently been updated and expanded by a joint ICBP/IWRB project, the final report of which is entitled *Important Bird Areas in Europe* (Grimmett & Jones 1989) and contains brief accounts of around 1200 wetland sites important for birds. The project covered all of Europe, including Greenland, Russia as far east as the Ural mountans, and the whole of Turkey.

Wetland inventory and monitoring in the Mediterranean region will be extended in the next few years with the implementation of the MEDWET initiative; a major wetland conservation project being funded by the European Community. IWRB has been given the responsibility to test a wetland inventory methodology which can be employed throughout the Mediterranean Basin. MEDWET involves close cooperation between Governments and NGOs, with the Bureau of the Ramsar Convention playing an important coordinating role.

Since the development of initial inventories of European wetlands in the late '70s and early '80s, most of the globe has been covered by some kind of wetland inventory. In Africa, the lead was taken by limnologists who began in the 1970s to compile information on wetlands of limnological interest for publication in handbook form. This work was subsequently taken over by UNEP and IUCN, and expanded into a full-scale directory of major wetlands in the Afrotropical Realm, covering about thirty-five countries in Africa south of the Sahara (Hughes & Hughes 1992), although it would be misleading to suggest that this work forms a comprehensive inventory of internationally important wetlands. Another important publication on the wetlands and shallow water bodies of Africa sponsored by ORSTOM (Burgis & Symoens 1987) provides detailed information on many of Africa's largest and best-known wetlands, but has many gaps in coverage, and is therefore also of limited value; much work remains to be done.

In the Neotropical Region, the lead was taken by IWRB. At a meeting in Edmonton, Canada, in 1982, IWRB launched an ambitious Neotropical Wetlands Project which, amongst other things, included the compilation of an inventory of wetlands of international importance in South and Central America and the Caribbean. The project was jointly sponsored by ICBP and IUCN and was funded by a variety of organizations, notably WWF-US, the US Fish and Wildlife Service and the Canadian Wildlife Service. The resulting *Directory of Neotropical Wetlands* summarises information received from over 280 contributors, and describes 730 wetlands covering 118 million hectares in forty-five countires (Scott & Carbonell 1986).

The lead was taken in South and East Asia by the International Council for Bird Preservation (ICBP) at its 10th Asian Continental Section Meeting in Sri Lanka in 1984. This meeting focused on the wetlands and waterfowl of southern and eastern Asia and concluded that the identification of important sites in the region was a priority task. The reports presented at the meeting were published in the form of a preliminary inventory which listed 488 wetlands of importance for waterfowl in 21 countries (Karpowicz 1985). In late 1985, IWRB and IUCN joined ICBP in a three-year project – the Asian Wetlands Inventory – to compile a comprehensive inventory of wetlands of international importance in southern and eastern Asia. This inventory, funded by WWF, was similar in general approach to the Neotropical Inventory, but much broader in scope in that it gave consideration to all the natural functions and values of wetlands, and was therefore less specifically oriented towards the values of wetlands for wildlife.

The Asian Wetlands Inventory covered all twenty-four countries from Pakistan to China, Japan, Indonesia and Papua New Guinea. Over 500 individuals and organisations participated in the project, and in most countries, national coordinators were appointed and wetland working groups or committees set up. The final report of the project, entitled *A Directory of Asian Wetlands* describes a total of 947 wetlands covering over 73 million hectares (Scott 1989). The Directory was published by the World Conservation Monitoring Centre, and has been distributed free of charge to all major participants.

The *Directory of Neotropical Wetlands* and the *Directory of Asian Wetlands* vividly demonstrate the value of international wetland inventories in establishing priorities. They constitute overviews of the wetland situation throughout large regions of the globe, and provide valuable information on the total area of wetlands of international importance, the number of wetlands enjoying some legal protection, and the total area under protection. They provide us with an excellent basis for planning future research, enabling us to identify areas in urgent need of basic surveying work or more detailed study. They also provide a considerable amount of information on the principal threats to wetlands; in fact, some threat was reported at over 80% of the sites in both regions, and no less than 50% of the sites were considered to be under moderate to serious threat (Scott & Carbonell 1985; Scott & Poole

14

1989). The directories reveal the great regional variations that exist both in the extent of protection and in the degree of threat, and can thus be used to identify those wetland ecosystems which are least well represented in networks of protected areas and which are in most urgent need of attention.

Two other wetland inventories of this type have very recently been completed and published; one covering Oceania and the other covering Australia. *A Directory of Wetlands in Oceania* (Scott 1993), funded jointly by IWRB, the Asian Wetland Bureau (AWB), the South Pacific Regional Environment Program (SPREP) and the Ramsar Bureau, was initiated in September 1989. The report describes the principal wetland ecosystems in 24 island nations and territories in the Pacific. A companion volume dealing with the internationally important wetlands of New Zealand, is being prepared by the New Zealand Department of Conservation. *A Directory of Important Wetlands in Australia* was compiled under the auspices of the Australian Nature Conservation Agency (formerly Australian National Parks and Wildlife Service).

Excluding Antarctica, which has few wetlands in the conventional sense, the only other major regions of the globe which have not as yet been covered by national or international wetland inventories are the Asian part of Russia, together with other Central Asian Republics of the CIS; and the Middle East. IWRB is currently coordinating the project development phase of an inventory of the Baltic Republics, CIS and Georgia. Parts of the Middle East were incorporated in the UNEP/IUCN Directory of Western Palearctic Wetlands (Carp 1980), but very little information was given for most of the listed sites, and the Arabian Peninsula was excluded. A project to remedy this situation is currently being elaborated (Scott 1992).*

With increasing coverage of regional wetland inventories came attempts to compile global accounts of particular ecosystems. For example, the Working Group on Mangrove Ecosystems of IUCN's Commission on Ecology has produced a report on the global status of mangroves (Saenger *et al.* 1983), while the Scientific Committee on Oceanic Research (SCOR) is conducting a 'Biosphere Inventory Report' of mangroves around the world. UNEP and IUCN have also sponsored the compilation of a world inventory of coral reefs, many of which fall under the definition of wetlands contained in the Ramsar Convention (UNEP/IUCN 1988). On a much broader scale, the International Society of Ecologists (INTECOL) has been working for some years on the preparation of

a major publication on the world's wetlands (Whigham *et al.* 1993); IWRB has recently produced a general account of the world's principal wetlands (Finlayson & Moser 1991), and IUCN has taken the lead in developing a global wetlands atlas (Dugan 1993). In another recent initiative, the UK Department of the Environment commissioned the first phase of a study of wetlands in the UK Dependent Territories aimed at reviewing the potential for Ramsar site designations (Hepburn *et al.* 1992).

Obviously, wetland inventories of this type, useful though they may be, are only 'snap shots' of the situation at the time of their compilation. Within a very few years, they become so out of date as to become almost useless for conservation planning. It is essential that the information, once collected and centralized, be updated as new information becomes available. Unfortunately, a proposal by the World Conservation Monitoring Centre to establish just such a global wetland database to serve as a central clearing house for information has so far failed to attract the necessary funding. As a consequence, the information gathered during the various regional inventories remains scattered between a variety of international and regional conservation bodies, and almost no coordinated updating of information has been possible, except with respect to wetlands designated for the Ramsar List.

Information on Ramsar sites is currently maintained in a database by IWRB on behalf of the Ramsar Convention Bureau. In the past, the World Conservation Monitoring Centre has updated site data at regular intervals for publication in conjunction with each Conference of the Contracting Parties. The most recent edition of the *Directory of Wetlands of International Importance*, prepared by WCMC for the Fourth Conference of the Contracting Parties in Montreux, Switzerland, in 1990, contains information on all 465 sites listed by the 52 countries which were parties to the convention at that time (Ramsar Convention Bureau 1990). As of July 1993, there were 623 Ramsar sites; completely revised texts on each (approved by the Contracting Parties concerned) were published in June 1993 at the Fifth Meeting of the Conference of the Contracting Parties (Jones 1993).

There also needs to be careful consideration of the data sets gathered; for example, few of the existing international publications contain easily analysed information on wetland functions, nor has there been any systematic collection of quantifiable data for use in monitoring ecological change in wetlands. The latter point provided the focus for debate in one of the workshops at IWRB's Board Meeting in Florida (St. Petersburg Beach, 16–17 November 1992), the results of which were also conveyed to the Contracting Parties to the Ramsar Convention at their triennial meeting in Kushiro, Japan, in June 1993.

By the time that the entire world has been covered by preliminary inventories, it seems likely that we will have identified over 5,000 wetlands as being of 'international importance' for nature conservation. If we are able to take full advantage of these inventories and the wealth of information which they have generated, it is essential that the information be centralized and standardized for easy access and updating. Otherwise there is a real danger that much of the original information will be lost or become so out of date as to be almost useless, in which case we will find ourselves having to repeat the inventories all over again, almost from scratch.

Conclusion

Definition of wetland. A globally accepted definition is desirable. The Ramsar definition has been accepted in principle by 79 Contracting Parties and has been used in many international inventory projects. It will also be used in compilation of the forthcoming inventory of Middle Eastern wetlands. The Ramsar definition cannot be changed except by a complex and time-consuming legal procedure to amend the Convention text. We therefore recommend use of the Ramsar definition for international purposes, although the Ramsar Bureau could be encouraged to develop guidelines for interpretation of the definition – especially in relation to man-made wetlands and marine ecosystems.

Global classification of wetlands. There is a need for a simple global classification system, and in spite of its inevitable shortcomings, much progress has already been made with the Ramsar classification. We believe that there is little to be gained in terms of wetland conservation by working on the development of an entirely new system and therefore advocate the utilisaton of the Ramsar system for use in all international fora.

Regional, national and local classifications. These can and should be as detailed as is necessary or feasible. However, for ease of international exchange and transfer of information on key sites, it is preferable if the broader categories in such classifications are compatible with the Ramsar hierarchy.

Global coverage of wetland inventories. There is an urgent need to complete global coverage of preliminary wetland inventories. International wetland organisations should aim to produce before the year 2000, an inventory of all the world's wetlands which qualify for designation under the Ramsar Convention. This inventory should be compiled on the basis of maximising technical objectivity and should not be controlled solely by political considerations. Governments are as free as they have always been to produce their own inventories.

Coverage of national wetland inventories. All countries that have not already done so, should be encouraged to produce their own, detailed national wetland inventories. These should cover wetlands of national or local importance as well as the sites already identified by international projects.

Follow-up to inventory projects. All wetland inventories should provide scope for:

– regular updating
– functional analysis
– monitoring ecological change and wetland loss
– provision of information most useful to wetland conservation
– wide dissemination of inventory results.

Location and accessibility of wetland inventory data. There is an urgent need for original data gathered under international inventory projects to be centralised at a location that will:

– provide for networking with other databases
– permit ready access to data
– facilitate updating of information
– publicise/promote the existence of such data sets.

* The Middle East Wetland Inventory Project is now (1995) nearing completion.

References

Burgis, M. J. & Symoens, J. J. (Eds.). 1987. African Wetlands and Shallow Water Bodies. ORSTOM, Paris, France.

Carp, E. 1980. A Directory of Western Palearctic Wetlands. UNEP, Nairobi, Kenya & IUCN, Gland, Switzerland.

Cowardin, L. M., Carter, V., Golet, F. C. & LaRoe, E. T. 1979. Classification of Wetlands and Deepwater Habitats of the United States. 140 pp.

Dugan, P. J. 1990. Wetland Conservation – a Review of Current Issues and Required Action. IUCN, Gland, Switzerland. 100 pp.

16

Dugan, P. J. (Ed.). 1993. Wetlands Under Threat. Mitchell Beazley, London.

Environmental Problems Foundation of Turkey. 1989. Wetlands of Turkey. EPFT, Ankara, Turkey. 178 pp.

Finlayson, M. & Moser, M. 1991. Wetlands. IWRB. Facts-on-File, Oxford & New York.

Grimmett, R. F. A. & Jones, T. A. 1989. Important Bird Areas in Europe. ICBP Technical Publication No. 9. ICBP, Cambridge, U.K.

Heliotis, F. D. 1988. An inventory and review of the wetland resources of Greece. Wetlands, Vol. 8. 18 pp.

Hepburn, I., Oldfield, S. & Thompson, K. (in press). UK Dependent Territories – Ramsar Study Stage 1. Report to UK Department of the Environment. IWRB/NGO Forum for Dependent Territories.

Hughes, R. H. & Hughes, J. S. 1992. A Directory of African Wetlands. IUCN, Gland, Switzerland/UNEP, Nairobi, Kenya/WCMC, Cambridge, U.K.

Jones, T. A. (compiler). 1993. A Directory of Wetlands of International Importance, Volumes I-IV. Ramsar Convention Bureau, Gland, Switzerland.

Karpowicz, Z. 1985. Wetlands in East Asia – A Preliminary Review and Inventory. ICBP Study Report No. 6. ICBP, Cambridge, U.K.

Luther, H. & Rzoska, J. 1971. Project Aqua; a source book of inland waters proposed for conservation. IBP Handbook No. 21. Blackwell Scientific Publications, Oxford and Edinburgh.

Matthews, G. V. T. 1993. The Ramsar Convention; its History and Development. Ramsar Convention Bureau, Gland, Switzerland.

National Wetlands Working Group, Canada Committee on Ecological Land Classification. 1987. The Canadian Wetland Classification System. Ecological Land Classification Series No. 21. Lands Conservation Branch, Canadian Wildlife Service, Environment Canada. 20 pp.

Official Journal of the European Communities. 1992. L 206, Vol. 35. 54 pp.

Olney, P. (Ed.). 1965. Project MAR. List of European and North African Wetlands of International Importance. IUCN New Series 5. IUCN, Morges, Switzerland.

Patten, B. C. 1990. Wetlands and Shallow Continental Water Bodies: Volume 1 – Natural and Human Relationships. SPB Academic Publishing.

Ramsar Convention Bureau. 1990. Directory of Wetlands of International Importance. Ramsar Convention Bureau, Gland, Switzerland.

Saenger, P., Hegerl, E. J. & Davie, J. D. S. 1983. Global Status of Mangrove Ecosystems. IUCN Commission on Ecology Papers No. 3. IUCN, Gland, Switzerland.

Scott, D. A. 1980. A Preliminary Inventory of Wetlands of International Importance for Waterfowl in West Europe and Northwest Africa. IWRB Special Publication No. 2. IWRB, Slimbridge, U.K.

Scott, D. A. & Carbonell, M. 1985. The IWRB/ICBP Neotropical Wetlands Project: a report on the completion of 'A Directory of Neotropical Wetlands'. In: Scott, D. A., Smart, M. & Carbonell, M. (Eds) Report of the XXXI Annual Meeting of IWRB, Paracas, Peru, 10–16 February 1985: 51–65. IWRB, Slimbridge, U.K.

Scott, D. A. & Carbonell, M. (Eds). 1986. A Directory of Neotropical Wetlands. IUCN, Gland, Switzerland. 713 pp.

Scott, D. A. (Ed.) 1989a. A Directory of Asian Wetlands. IUCN, Gland, Switzerland, and Cambridge, U.K. 1198 pp.

Scott, D. A. 1989b. Design of Wetland Data Sheet for Database on Ramsar Sites. Photocopied report to Ramsar Bureau, Gland, Switzerland.

Scott, D. A. 1992. Inventory of Wetlands of the Middle East: Project Proposal. Unpub. report.

Scott, D. A. (Ed.). 1993. A Directory of Wetlands in Oceania. IWRB, Slimbridge, UK & AWB, Kuala Lumpur, Malaysia. 461 pp.

Scott, D. A. & Poole, C. M. 1989. A Status Overview of Asian Wetlands. Asian Wetland Bureau Publication No. 53. AWB, Kuala Lumpur, Malaysia.

Silvius. M. J., Djuharsa, E., Taufik, A. W., Steeman, A. P. J. M., Berczy, E. T. (compilers). 1987. The Indonesian Wetland Inventory – a preliminary compilation of information on wetlands in Indonesia, Volume I. PHPA-AWB/Interwader, Indonesia & EDWIN, The Netherlands. 169 pp.

Walmsley, R. D. & Boomker, E. A. (Eds). 1988. Inventory and Classification of Wetlands in South Africa. Proceedings of a Workshop. Occasional Report No. 34. Foundations for Research Development, CSIR, Pretoria.

Whigham, D. F., Dykyjkova, P. & Hejny, S. (Eds). 1993. Wetlands of the World. I. Inventory, ecology and management. Kluwer Academic Publishers. 768 pp.

UNEP/IUCN. 1988. Coral Reefs of the World. 3 Vols. UNEP Regional Seas Directories and Bibliographies. IUCN, Gland, Switzerland/Cambridge, U.K./UNEP, Nairobi, Kenya.

Usback, S. & James, R. (compilers). 1993. A Directory of Important Wetlands in Australia. Australian Nature Conservation Agency. Canberra ACT, Australia. 717 pp.

Vegetatio **118**: 17–28, 1995.

The current status of European wetland inventories and classifications

J.M.R. Hughes
Department of Geography, University of Reading, Whiteknights, Reading RG6 2AB, UK

Key words: Classification, Database, European wetlands, Inventory

Abstract

The status of wetland inventory and classification is considered for 44 European countries, as well as for the continent as a whole. Data and information were obtained from questionnaires compiled by the International Waterfowl and Wetland Research Bureau, the MedWet sub-project on inventory and monitoring, and the Ramsar Bureau. Nine European countries have national wetland inventories, and 32 have inventories of sites of international importance listed under the Ramsar Convention. There has been a trend in producing regional or continental inventories for wetlands that are important as waterfowl habitat. There is an urgent need to produce wetland inventories for all European countries. The Ramsar database takes into consideration hydrological and economic wetland values, as well as ecological ones. The Ramsar classification lists a total of 35 wetland types, and is sufficiently flexible that it could be used for classifying European wetlands at the national scale.

Introduction

The European continent is endowed with a great diversity of climates, hydrologies, and human, animal and plant life. Consequently, the range of wetland types reflects this regional diversity, and they may be found from the semi-arid Mediterranean (e.g. salt lakes), to temperate north-west Europe (e.g. freshwater marshes, permanent rivers), to sub-arctic Europe (e.g. glacial lakes, peatlands). The need to classify and inventory these wetlands is great, especially as many wetlands in Europe have been destroyed during the last 100 years by drainage for agriculture and pollution (Jones & Hughes 1993). Human pressures on the remaining wetlands are possibly decreasing (but still prevail in most areas), due to increasing conservation awareness and the establishment of the Ramsar Convention in 1971 (Hollis & Jones 1991). Inventories are important for the management and conservation of wetlands. They are based on an existing classification scheme and designed to meet the needs of particular user-groups. Inventories are thus a vital tool for determining the number and type of wetlands in Europe, at the national, regional and continental scale.

This paper summarizes the state of wetland classifications and inventories for Europe west of the Urals and for the continent as a whole (Fig. 1). Europe is considered to embrace the following 44 countries: Albania, Andorra, Armenia, Austria, Azerbaijan, Belgium, Belarus, Bosnia and Herzogovina, Bulgaria, Croatia, Cyprus, Czech Republic, Slovak Republic, Denmark, Estonia, Finland, France, Georgia, Germany, Greece, Hungary, Iceland, Ireland, Italy, Latvia, Liechtenstein, Lithuania, Luxembourg, Malta, Moldova, Netherlands, Norway, Poland, Portugal, Romania, Russian Federation, Slovenia, Spain, Sweden, Switzerland, Turkey, Ukraine, United Kingdom and Yugoslavia. Most of these countries (32) are contracting parties to the Ramsar Convention on Wetlands of International Importance (Fig. 1). The total area of designated Ramsar wetlands in Europe is 7 412 748 ha (as from 1st January 1994).

Regional inventories

At the regional scale, there are several inventories which have been produced for Europe. The inventories produced by the Ramsar Convention on Wetlands

18

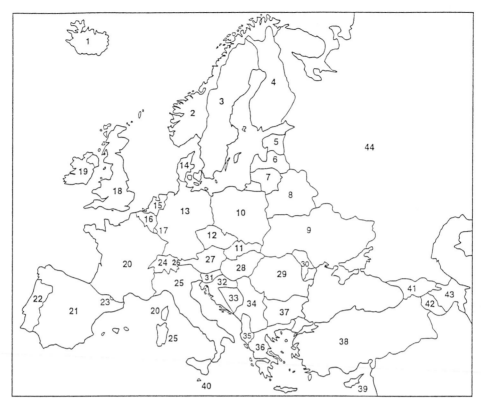

bold = contracting parties to Ramsar Convention (1.1.94)

1.	Iceland	13.	Germany	25.	Italy	37.	Bulgaria
2.	Norway	14.	Denmark	26.	Liechtenstein	38.	Turkey
3.	Sweden	15.	Netherlands	27.	Austria	39.	Cyprus
4.	Finland	16.	Belgium	28.	Hungary	40.	Malta
5.	Estonia	17.	Luxembourg	29.	Romania	41.	Georgia
6.	Latvia	18.	United Kingdom	30.	Moldova	42.	Armenia
7.	Lithuania	19.	Ireland	31.	Slovenia	43.	Azerbaijan
8.	Belarus	20.	France	32.	Croatia	44.	Russian
9.	Ukraine	21.	Spain	33.	Bosnia & Herzegovina		Federation
10.	Poland	22.	Portugal	34.	Yugoslavia		
11.	Slovak Republic	23.	Andorra	35.	Albania		
12.	Czech Republic	24.	Switzerland	36.	Greece		

Fig. 1. Countries included in the survey of European wetland classifications and inventories.

of International Importance for 32 European countries list a total of 411 wetlands that have been designated as internationally important, within a unified and systematic framework and database. There are 31 entries on the database sheet, including socio-economic, hydrological and physical, and ecological values, as well as a map (Scott 1989). The classification scheme produced by Ramsar for this inventory is divided into marine and coastal, inland and man-made wetlands. The three categories are sub-divided to give a total of 35 wetland types. This provides a relatively simple classification with wide applicability, which has been tested by many of the 80 contracting parties to the Convention (Ramsar Convention Bureau 1993). The Ramsar scheme is not aimed at producing an inventory for the whole of Europe, but rather an inventory of those sites designated as internationally important in each separate European country. Although there are very definite criteria for designating Ramsar wetlands, the reasons for including (or excluding) certain wetlands in these national inventories may be bureaucratically or politically linked. This is reflected in the variability of the number and area of designated wetlands (Table 1).

Table 1. Status of European wetland inventories.

Country	Contracting party to Ramsar Convention (with number and area of designated wetlands) (ha)		National wetland inventory	Other wetland inventory (by region, habitat or species)	Inventory proposed or underway
1 Iceland	x(2)	57,500		x	
2 Norway	x(14)	16,256		x	x
3 Sweden	x(30)	382,750		x	x
4 Finland	x(11)	101,343		x	
5 Estonia					x
6 Latvia				x	x
7 Lithuania	x(5)	50,541			x
8 Belarus					x
9 Ukraine					x
10 Poland	x(5)	7,141			
11 Slovak Rep.	x(7)	25,519			
12 Czech Rep.	x(9)	27,978	x		
13 Germany	x(31)	672,852		x	
14 Denmark	x(38)	1,832,968		x	
15 The Netherlands	x(21)	314,928		x	
16 Belgium	x(6)	7,945		x	
17 Luxembourg					
18 United Kingdom	x(67)	308,337		x	
19 Ireland	x(21)	13,035		x	
20 France	x(11)	641,585	x		
21 Spain	x(27)	124,058	x		
22 Portugal	x(2)	30,563	Preliminary	x	x
23 Andorra					
24 Switzerland	x(8)	7,049		x	
25 Italy	x(46)	56,950	x		
26 Leichtenstein	x(1)	101			
27 Austria	x(7)	102,541		x	
28 Hungary	x(13)	114,862			x
29 Romania	x(1)	647,000		x	x
30 Moldova					
31 Slovenia	x(1)	650			
32 Croatia	x(4)	80,455	Preliminary		
33 Bosnia & Herzegovina					
34 Yugoslavia	x(2)	18,094			
35 Albania					
36 Greece	x(11)	107,400	x		
37 Bulgaria	x(4)	2,097	x		
38 Turkey	not yet ratified		x		
39 Cyprus					
40 Malta	x(1)	11			
41 Georgia				x	x
42 Armenia	x(2)	492,239			x
43 Azerbaijan					x
44 Russian Federation	x(3)	1,168,000			x
	32(411)	7,412,748	9	15	14

Table 2. Summary of data contained in the national inventories of nine European countries.

National inventory	Ramsar scheme 31 data entries 35 wetland types No. of wetlands of International Importance	No. of wetlands in National inventory	No. of entries in the data base	No. of wetlands in the classification scheme
Czech Republic	9	820	8	16
France	11	14,000	17	14
Spain	27	1,500	17	31
Italy	46	103	14	18
Greece	11	381	17	8
Bulgaria	4	88	16	no particular scheme
Turkey	5 proposed but not yet ratified	75	6	no particular scheme
Preliminary inventory				
Portugal	2	49	5	no particular scheme
Croatia	4	30	4	10

In 1985, the CORINE Programme (Coordination of Information on the Environment) was established as an experimental project, and subsequently extended to provide a permanent information network to support the European Environment Agency. CORINE consists of projects covering subjects including air quality, soil, water, coastal erosion, land cover and biotopes. The Biotopes Project aims to 'identify and describe biotopes of major importance for nature conservation in the Community'. It is developing a standardised approach for the collection, storage and analysis of information on sites of European importance, including a hierarchical phyto-sociological classification system (including wetland habitats), which is currently being extended to include the entire Palearctic region. The database currently holds information on over 6,700 sites, which yield data on site location, landuse, species, habitat type, ecology, conservation and management. At the national level, coverage by the inventory is highly variable (Commission of the European Community 1991). This database is not specifically aimed at wetlands, and it could therefore be cumbersome to access.

There has been much recent attention given to the conservation of wetlands in the Mediterranean region, and this culminated in a conference held at Grado, Italy in 1991 on the management of Mediterranean wetlands. Later in 1991, MedWet was conceived as an EC funded initiative for the wise use and conservation of Mediterranean wetlands. As part of this project, the International Waterfowl and Wetlands Research Bureau (IWRB), in conjunction with the Portuguese Institute for the Conservation of Nature, is coordinating the development and testing of a wetland inventory methodology for use throughout the Mediterranean region, and resulting in continued close cooperation between NGOs and Governments. A standardised classification and datasheet is being developed which is taking into account compatibility with those adopted by the Ramsar Convention (Prentice & Tomas-Vives, in press).

Additionally, there is the internationally coordinated project for 'wetland inventories for some countries of the former USSR' (IWBR pers. comm.): Estonia, Latvia, Lithuania, Belarus, Ukraine, Georgia, Armenia, Azerbaijan and Russian Federation. This project is proceeding through the development of national wetland inventories which will follow technical and organisational guidelines which were agreed at a meeting in Minsk on 2–3 March 1993 (Minsk Declaration on Inventory of Wetlands of International and National Importance). This agreement will ensure that the national methodologies adopted are based upon the Ramsar Wetland Information Sheet and classification of wetland type, as adopted at the Fourth Meeting of the Conference of the Contracting parties in Montreux, Switzerland, in 1990.

In addition to these above databases, there are several wetland inventories at the continental scale which are based on waterfowl criteria. Project MAR (Olney 1965) lists 200 wetlands of international importance in Europe and North Africa. This list was the first international inventory of important wetlands and five data sheet categories were used: a) geographical coordinates; b) area of wetland; c) ecological category (eight major types); d) general description of the area; and e) ornithological description of the site.

Carp (1980) produced a directory of important wetlands in the Western Palearctic. 900 wetlands were listed in this inventory, and the sites were selected on the basis of the Ramsar criteria. 14 data categories were used in the inventory with 25 wetland-types in the classification. Both these inventories are biased towards wetlands as waterfowl habitat. This trend has pervaded wetland inventories in Europe until recently, when it became apparent that the conservation of wetlands should be based on hydrological and economic criteria as well as ecological ones.

Grimmett and Jones (1989) list important bird areas in Europe for 32 countries, including Europe west of the Urals. Of the 2444 sites listed, 1384 were important for waterfowl, and most of these were wetlands. The data sheet listed 15 categories for each site including protection status, land ownership and threats to the site. Scott (1980) produced an inventory of wetlands of international importance for waterfowl in West Europe and Northwest Africa. The data included in the inventory were mostly taken from the EEC Important Bird Areas Project (1979–81), and it lists 544 wetlands in 22 European countries.

An inventory of EEC rivers was produced by Haslam (1987), which is based on the distribution of aquatic macrophytes alone. This is an interesting divergence from wetland inventories based on waterfowl criteria, but obviously omits standing water wetlands. Menanteau (1991) is undertaking a satellite inventory of EEC coastal wetlands, but so far only France, Italy, Spain and Portugal have been completed. This example demonstrates how wetland inventories can be conducted using techniques other than ground surveys.

National inventories

Using unpublished data from a) the International Waterfowl and Wetlands Research Bureau (IWRB)

National Report Questionnaire which was circulated to all member countries in June 1992 (21 European countries responded), b) information from a report on the 'status of wetland inventory in the Mediterranean region' to be published by IWRB [Hecker & Tomàs-Vives (in press)], c) unpublished proceedings from IWRB workshops and working groups, and d) country reports to the Ramsar Convention, Kushiro Conference in June 1993 (Ramsar Convention Bureau 1993), it was possible to discover the extent and status of wetland classifications and inventories for most of Europe.

No information on inventories was available (or was found) for several countries, and for others the only information was whether the country was a contracting party to the Ramsar Convention (listed in Table 1) or whether it was participating in the project on 'wetland inventories for the countries of the former USSR' (discussed above). As a result twelve countries have been omitted from the following list: Andorra, Armenia, Austria, Azerbaijan, Bosnia and Herzegovina, Liechtenstein, Luxembourg, Malta, Moldova, Poland, Slovak Republic, Slovenia.

Iceland

The Icelandic Nature Conservancy Council has an inventory of areas of special importance for conservation which would include wetlands. There is no national inventory dealing with wetlands alone.

Norway

There are 14 wetlands of international importance designated under the Ramsar Convention. Eldoy (1990) summarizes the status of wetland conservation and management in Norway, and the country report to the Ramsar Convention (1993) states that by the end of 1992 there were 213 wetland nature reserves, 219 mire/bog reserves and 270 seabird reserves (which include coastal wetlands). An inventory of wetlands, mires and bogs is being undertaken (June 1993) for each of the 18 counties by the Directorate for Nature Management. The results of the county inventories will be amalgamated in 1995 to produce a national wetland inventory (O. Storkersen pers. comm.) but no information concerning database entries or classification scheme was provided.

Sweden

The Swedish Environment Protection Agency is currently coordinating a national wetland inventory (M. Lofroth, pers. comm.) This is being undertaken at the county level, with the regional County Boards carrying out the operative part of the inventory. To date 22 counties have completed inventories, one is on-going, and the county of Norrbotten will commence an inventory later in 1994. The amalgamated county inventories will produce a national inventory containing 26 786 wetlands. The classification scheme has divided wetlands into three main groups: mires (bogs, fens and mixed mires); shores (peat-forming or non-peat areas influenced by saline or brackish water); other wetlands (wet heaths, wet meadows, wet forests). Most of the data for the inventory come from aerial photographs, and from these a map of each wetland has been produced. Site visits have been carried out for wetlands of 'very high conservation value'. In general, only sites of 10 hectares or more are included in the inventory. The database scheme used is more complicated, and includes up to four levels of data for each wetland: 1) Site level (eg. mire complex) includes 20 information types; 2) Subsite level (eg. mire units) includes 11 information types; 3) Element level (eg. structures such as hummocks, palsas etc.) includes 10 information types; 4) Species levels (these are the only wetlands surveyed in the field) includes 4 information types. The inventory can provide information for individual wetlands on species richness and rareness (plants and animals), naturalness, representativeness, location, size and conservation status.

Finland

Finland has a national wetlands conservation programme (Haapanen 1981), but no national inventory was found. Ruuhujarvi (1983) has classified Finnish mires and described their regional distribution which provides useful groundwork for a wetland inventory.

Estonia

Estonia is a participant of the project for wetland inventories in the former USSR, and will be a contracting party to the Ramsar Convention with effect from 29th July 1994.

Latvia

Viksne (1993) has produced an inventory of 38 important bird areas in Latvia, the majority of which are bogs, lakes, marshes or swamp forests. There is a written account for each site (averaging about 200 words), including details of location, protection status, ecological and economic characteristics, and importance to birds. No particular classification scheme was used. Information about the biological values of wetlands is concentrated in the Latvian Institute of Biology (for lakes and ponds), and the Faculty of Biology of the Latvian University (for bogs) (J. Viksne pers. comm. via IWRB). Latvia is a participant of the project for wetland inventories of the former USSR.

Lithuania

World Wildlife Fund (WWF) is to fund an inventory which will be coordinated by the Lithuanian Institute of Ecology. The Danish National Environmental Research Unit is producing an inventory of Lithuanian marine areas, with the aim of identifing marine Ramsar sites.

Belarus

Belarus is a participant of the project to produce an inventory covering the area of the former USSR. To this end a methodology has been proposed using a Ramsar-based classification. Work on the national inventory has already started, with the formation of a national coordinating committee, and a list of sites for inclusion. The inventory will focus on 15 sites of international importance, but include approximately a further 20 sites of national importance.

Ukraine

A national wetland inventory is currently underway by the Ministry of Environmental Protection. It is anticipated that this inventory of all wetlands in Ukraine will be completed by 1995 (Wilson and Moser 1994). A list of wetlands of international significance in the south of Ukraine (Ministry of Education of Ukraine *et al.* 1993), gives detailed information on location, wetland type (but not using a specific classification scheme), dominant bird species, economic uses and threats, protection and administration, for 29 wetlands.

Czech Republic

A comprehensive inventory of wetlands has been produced (Hudec *et al.* 1993; Hudec, in press). The inventory is divided into four sections, covering wetlands of local importance (> 600); wetlands of regional importance (211); wetlands of national importance (includes almost all wetland sites protected as national nature reserves and sites listed in international inventories. Their numbers are included in the previous category); wetlands meeting the criteria of the Ramsar Convention (9). Each wetland is listed in a table, and these provide information on area, protection, wetland-type, location (district) and latitude. Sixteen wetland-types are listed in the classification used by the inventory: springs, streams, floodplain lake (backwater, oxbow, pool), floodplain forest, flooded or wet meadow, other aquatic or wetland biotopes, reedswamp/sedge marsh, mire (bog or fen), mountain lake, saltmarsh, canal, industrial settling pond, fishpond, fishpond system, reservoir, quarry, gravel or sand pit.

Germany

Germany has 31 wetlands listed as Ramsar sites. There is no national wetland inventory, but there are some inventories of biotopes (including wetlands) on the level of the Federal Lander (Ministry for the Enviroment, Nature Conservation and Nuclear Safety, pers. comm.). Individual classification and data sheet schemes operate by region. Studies of individual wetlands (e.g. Jacoby 1991) and censuses (e.g. Struwe & Nehls 1992) exist, as well as listings associated with the Ramsar Convention and bird reserves (Harengerd 1990; Mayr 1991; Rheinwald 1990).

Denmark

There are several ornithological inventories e.g. Ferdinand (1980), Dybbro (1985), but no national wetland inventory. Rates of wetland loss in Denmark are discussed in Waagepetersen (1986). The most recent report on the 27 wetlands listed under the Ramsar Convention was published as a national report by the Ministry of the Environment and the National Forest and Nature Agency (1993). A national waterfowl and seals monitoring programme has been initiated by the National Forest and Nature Agency together with the National Environmental Research Institute, to cover the combined wetland list of Ramsar sites and EEC Bird Protection Sites.

The Netherlands

Only various waterfowl inventories which list wetlands were found, e.g. Frikke (1991).

Belgium

There is no inventory for Belgium, but some wetlands are listed in an inventory of Areas of Special Conservation Value (Royal Institute for Natural Sciences, Belgium 1987) and by Van Vessem and Kuijken (1986). Additionally, an inventory is currently under way to list rivers and marshes in Flanders. This is a joint project by

the Institute of Nature Conservation and the University of Ghent.

United Kingdom

There is no published or officially recognised national wetland inventory. However, there are inventories of various wetland types e.g. estuaries (Buck 1993; Davidson et al. 1991), peatlands (Lindsay, in prep.) and saltmarshes (Burd 1989), and sites where waterbirds have been counted for many years under the National Waterfowl Counts and Birds of Estuaries Enquiry (e.g. Kirby et al. 1992). The Government's statutory nature conservation advisers, Joint Nature Conservancy Council (JNCC) have prepared a wetland register that meets the waterbird criteria for designation as Wetlands of International Importance under the Ramsar Convention and/or as Special Protection Areas (SPAs) under the European Community Directive on the Conservation of Wild Birds. This database of 5000 wetlands sites is continually reviewed and amended, and it uses the Ramsar data sheet and classification scheme (G. Boobyer, pers. comm.).

Ireland

An inventory of sites that meets the criteria for designation of Wetlands of International Importance under the Ramsar Convention and/or as SPAs under the European Community Directive on the Conservation of Wild Birds is currently being carried out by the Irish National Park and Wildlife Service, Department of the Environment for Northern Ireland, and JNCC.

France

There are two national inventories of French wetlands. Firstly, a national inventory of all natural areas of ecological, faunal and floral interest (ZNIEFF) has been published by the Secretariat de la Faune et de la Flore (SFF 1992; Barnaud & Richard 1993). The classification scheme (which has a total of 30 habitat-types) includes eight types of marine and coastal wetlands (near-coastal waters, bay, estuary or delta, lagoon, coastal lake, saltpan, salt marsh, coastal marsh) together with six inland wetland types (fast flowing river, slow flowing river, lake and reservoir, marsh and peatland, wet meadow, salt lake). The database for each site has 17 entries dealing with location (and map), landownership, conservation and protection, faunal and floral values (with comprehensive species lists), threats and disturbances, bibliography. To date a total of 14 000 sites have been surveyed, with wetlands of special biological interest covering an area of 31 000 km^2. Secondly, and with a different aim, Lierdeman and Mermet (1991, 1992) have produced a review of wetlands of national importance dealing with larger wetland sites of five types: Atlantic coastal (21 sites); Mediterranean coastal (8 sites); floodplains and rivers (23 sites); inland ponds and marshes (18 sites); peatlands and other upland wetlands (16 sites). The inventory includes 70 sites covering an area of 20 815 km^2 (excluding peatlands), 17 721 km^2 of which are inland. The aim of this inventory was to identify wetland sites of national importance based on ecological values with emphasis on their ornithology. There are numerous inventories of individual Department or regions e.g. inventory of the wetlands of Lorraine (Ministère de l'Environnement, Paris, 1987), Somme (Etienne 1990), the Camargue (Britton & Podlesjski 1981), and the Mediterranean (Barnaud & Richard, in press).

Spain

An inventory was undertaken between 1988 and 1990 by an interdisciplinary team from the Ministry of Public Works and Urbanisation (MOPU) and 1500 wetlands were listed (Casado *et al.* 1992; Ministerio de Obras Publicas y Transportes 1991; Montes 1990). The classification scheme used did not include artificial wetlands, so apart from a few exceptions (e.g. saltpans) the inventory covers natural systems only. Six main wetland-types are used in the classification scheme (montane, karstic, inland freshwater, inland saline, floodplain, coastal) and each is subdivided on the basis of origin and function to give a total of 31 categories. The database has 17 entries describing the location and surface area (with maps), hydrology, geology, flora and fauna, socio-economic values, conservation status (including wetland loss) and bibliography. The inventory has demonstrated that 60% of wetland area has been lost in the last four decades, primarily due to groundwater abstraction, leaving a total surface area of 1150 km^2. Montes and Bifani (1989) published an ecological and economic listing of Spanish wetlands, which has 125 wetlands belonging to 22 wetland types. An earlier inventory of wetlands as waterfowl habitat was produced in 1987 by the Sociedad Espanola de Ornitologia. Additionally, there exists an inventory for the Balearic Islands (Amengual Ramis 1991), and there are some regional wetland inventories e.g. Andalucia (Agencia de Medio Ambiente 1989), the Mediterranean (Montes & Bernués, in press).

Portugal

A preliminary inventory has been completed (Farinha & Trindade 1994) which is based on amalgamated data from existing international, national and regional inventories. The 49 wetlands listed in the inventory have been grouped into 16 catchments. For each site there is a map, and details on the location, threats, fauna (birds, mammals, reptiles and amphibians), conservation status and bibliography. No classification scheme was used, and wetlands were chosen by their faunal values. The development of standardised methodologies for wetland inventories in the Mediterranean region (MedWet project) is being undertaken jointly by IWRB and the Portuguese Institute for the Conservation of Nature (ICN). As a result, a more comprehensive inventory of Portuguese wetlands has commenced and is coordinated by the ICN (Costa, in press).

Switzerland

There are comprehensive inventories of Swiss wetlands which are valuable as waterfowl habitats (Leuzinger 1976; Marti & Schifferli 1987; Marti 1988) but there is a lack of information concerning the other functions and values of these sites, while other wetlands are excluded entirely. There are eight Ramsar sites of International Importance. A federal inventory of reserves for waterfowl and migratory birds is in progress.

Italy

An inventory of Italian wetlands was published by the Ministry of the Environment (Nature Conservation Service), Rome in 1990, listing the 46 Ramsar sites as well as wetlands of national importance. This uses the Ramsar classification and database scheme. In a second inventory of Italian wetlands, 103 sites have been listed by the Ministero dell'Ambiente (De Maria 1992). The classification scheme in this inventory divides wetlands into natural and artificial, with a total of 18 wetland-types. The database has 14 entries for each wetland dealing with the following information: location, surface area, wetland type; protection and conservation; administration and landownership; faunal and floral values; functions and uses; significance and importance of the wetland. Additionally, there is a separate volume containing maps for each of the sites (1:25 000).

Hungary

There is no national inventory for Hungary. In its reply to the IWRB National Report Questionnaire, it was stated that a national review of wetlands was prepared for IUCN, but no references or further information were provided. Various relevant research projects are being carried out in Hungary which will provide useful groundwork for a wetland inventory: listing of small size wetland areas by WWF Hungary and the Hungarian Ornithological Society; the creation of a new zonation system for Hungarian Protected Areas; and a basic Natural Status Survey according to IUCN prescriptions. These latter two projects are being undertaken by the National Authority for Nature Conservation (established in 1990 within the Ministry for the Environment and Regional Policy).

Romania

The Ministry of Waters, Forests and Protection of the Environment has responsibility for the conservation of wetlands in Romania (Wilson & Moser 1994). An inventory of wetlands (including detailed mapping) has been started by the Ministry to identify further sites which can be included on the Ramsar list. Additionally, a detailed inventory of the Important Bird Areas in Romania has been prepared.

Croatia

Four sites are listed as wetlands of international importance in Croatia (Kopacki Rit, Lonske Polje, Neretva Delta and Crna Mlaka). Grimmett & Jones (1989) describe 11 wetland sites which are important for birds. A preliminary national inventory has been produced (unpublished) by the Institute for Ornithology of the Croatian Academy of Arts and Sciences. Thirty wetlands are included, and the classification scheme has ten wetland types: inland freshwater lake; wet meadow, forest; flood meadows; fishpond; coastal freshwater lake; coastal saltwater lake; coastal mudflats; delta estuary; salting; and reservoir. The data sheets include information on the flora, fauna, threats and protection of each site. A more comprehensive inventory is planned. The status of wetlands in the Croatian Mediterranean is summarised by Muzinic (in press).

Yugoslavia

There is a list of 13 wetlands compiled by the Institute of Protection of Nature of Serbia and Montenegro.

Albania

Lake Karavastas, on the Adriatic coast, is expected to be Albania's first wetland site when it joins the Ramsar Convention in the near future. Albania does not have an inventory of wetlands, but four are listed in Grimmett and Jones (1989), and the same four are described by Gjiknuri and Peja (1992).

Greece

A Greek wetland inventory has been completed under the coordination of the Greek Biotope/Wetland Centre (Zalidis 1993). 381 sites have been listed (covering 201 267 ha), and there are 17 database entries for each. The database entries include information on location, wetland type, threats, species present, economic values and conservation measures taken. There are maps for about 50 of the sites, including all the Ramsar sites. The longterm goals of having such a comprehensive database are for monitoring, disseminating scientific information on, and protecting Greek wetlands. The classification scheme used with the inventory divides wetlands into eight types (delta, freshwater lake, saline lagoon, river, spring, marsh, constructed wetland, other). Additionally, several other reviews exist, which taken together, form a comprehensive listing of Greek wetlands (Tsiouris & Gerakis 1991; Heliotis 1988; Gerakis 1990). Psilovikos (1992) reviews the rates of wetland loss in Greece.

Bulgaria

The results of a national wetland inventory (involving 50 researchers and other experts) were published in the National Wetlands Plan (Ministère de l'Environnement 1993; Wilson & Moser 1994). Eighty-eight wetlands of national importance were identified, and were listed under three arbitrarily-defined wetland types: drained wetlands; natural lakes and marshes (the list included estuaries in this type); artificial lakes. For each wetland, there are 16 entries describing the wetland, and its ecological, hydrological and economic values, as well as conservation status and management. It is a very detailed inventory, which has helped to identify aims and priorities in Bulgarian wetland conservation.

Turkey

An inventory providing detailed information for some 75 wetlands was published in 1989 (and updated in 1993) by the Environmental Problems Foundation of Turkey with funding and assistance by the US Fish and Wildlife Service. The inventory is divided into six regions (Black Sea, Marmara, Mediterranean, Aegean, Central and Eastern Anatolia), with information on area, position, general description, ornithology and human usage for each wetland. A second inventory was published by the Society for the Protection of Nature, Turkey (DHKD) and the International Council for Bird Preservation (ICBP) (Ertan *et al.* 1989) which includes 61 wetlands and is biased towards wetlands as bird habitats. Turkey signed (but has not yet ratified) the Ramsar Convention in January 1994.

Cyprus

Not yet a signatory to the Ramsar Convention, but Cyprus has started the necessary procedures. Larnakas and Akrotiri Lakes are likely to be listed as wetlands of international importance. There are two marine inventories for Cyprus (Leontiades 1977; Demetropoulos 1989), and some wetlands are listed by Carp (1980).

Georgia

There is no wetland inventory for Georgia, and the country is not a contracting party to the Ramsar Convention. The Kolkheti lowlands hold the only significant wetlands in the Black Sea basin (encompassing rivers, lakes, coastal peat bogs and marshes), and it is hoped to produce an atlas of the area (110 maps) prepared by the Institute of Geography of the Georgian Academy of Sciences (Wilson & Moser 1994).

Russian Federation (west of Urals)

Russia is participating in the international project to carry out national wetland inventories in the countries of the former USSR. A proposal for a national inventory has been developed and work is planned to start in 1994. A first inventory of important wetlands was conducted in the 1970s and its results published (Skokova & Vinogradov 1986). The current inventory will seek to update this work.

Discussion

I have examined the status of European wetland classification and inventory, and several summary points can be made concerning inventories at the national scale (Table 1). There are nine European countries with comprehensive national inventories (Table 2), and fourteen with inventories that are in progress or proposed. Thus half the countries of Europe are engaged in making wetland inventories, and it is clear that an enormous coordination effort is required if a wetland inventory is to be produced for all European countries.

None of the nine countries have used the same database scheme, and thus the information that can be accessed for each inventory is highly variable. Uniformity of methodology allows for efficient comparison of results. Moreover, inventories have to reveal the extent and quality of the wetland resource for planning, monitoring, protection and management, and highlighting the values of the wetland.

All nine inventories used a different wetland classification scheme, and three countries used no classification scheme in particular. Again, it is impossible to compare or assess the wetland resource if different classifications are used to describe it. In 1990, at the Fourth Meeting of the Contracting Parties to the Ramsar Convention, a hierarchical classification was approved based on the US classification. This is a simple scheme with wide applicability that can incorporate additional wetland types or habitat categories that are appropriate to the country or region.

There is no central archive where information on European wetlands can be obtained. There is a need for a centralised repository of information to allow a coordinated overview of the status and trends of European wetland resources. The author found the task of amassing information on wetlands for 44 countries to be logistically difficult, and in some cases little or no information was found.

Acknowledgments

I am most grateful to Tim Jones for providing access to the IWRB archives, and to Pere Tomas-Vives, Mike Moser, Crawford Prentice and Paul Rose (all from IWRB) for their help and discussions. Max Finlayson provided some very useful advice on the structure of the paper.

References

Agencia de Medio Ambiente. 1989. Edudio de la Gestion Integrada de las Zonas Humedas Costeras en Andalucia. Direction General de Conservacion de la Naturaleza (AMA) con la colaboracion del Grupo de Expertos Comunitarios en Zonas Humedas por encargo de la DGXI de la CEC.

Amengual Ramis, J. F. 1991. Inventario de las Zonas Humedas de Baleares. Serveis Forestals de Balears, SA.

Barnaud, G. & Richard, D. 1993. Inventaires des zones humides en France. In: Ramsar Convention Bureau (ed), Proceedings of the 5th Meeting of the Conference of the Contracting Parties, Kushiro, Japan. Ramsar Convention Bureau, Gland, Switzerland.

Barnaud, G. & Richard, D. France. In: Hecker, N. & Tomàs-Vives, P. (Eds), The status of wetland inventories in the Mediterranean region. MedWet project, ICN and IWRB. (in press).

Buck, A. L. 1993. An Inventory of UK Estuaries (7 Volumes). Joint Nature Conservancy Council, Peterborough, UK.

Burd, F. 1989. The Saltmarsh Survey of Great Britain. An Inventory of British Saltmarshes. Research and Survey in Nature Conservation No. 17. Nature Conservancy Council, Peterborough.

Carp, E. 1980. Wetlands of International Importance in the Western Palearctic. UNEP (Kenya) and WWF (Switzerland).

Casado, S., Florin, M., Molla, S. & Montes, C. 1992. Current Status of Spanish Wetlands. pp. 56–58. In: Finlayson, C. M., Hollis, G. E. & Davis, T. J. (eds) Managing Mediterranean Wetlands and their Birds. IWRB Special Publication No. 20, Slimbridge, UK.

Commission of the European Communities. 1991. CORINE Biotopes- the design, compilation and use of an inventory of sites of major importance for nature conservation in the European Community. EUR 13231, Office for Official Publications of the European Communities, Luxembourg.

Costa, L.T. et al. Portugal. In: Hecker, N. & Tomàs-Vives, P. (Eds), The status of wetland inventories in the Mediterranean region. MedWet project, ICN and IWRB. (in press).

Davidson, N. C., Laffoley, D. d'A., Doody, J. P., Way, L. S., Gordon, J., Key, R., Drake, C. M., Pienkowski, M. W., Mitchell, R. & Duff, K. L. 1991. Nature Conservation and Estuaries in Great Britain. NCC, Peterborough.

De Maria, G. (ed) 1992. Inventario delle Zone Umide del Territorio Italiano. Servizio Conservazione della Natura, Ministerio dell'Ambiente.

Demetropoulos, A. 1989. Annual Report on the Fisheries Department and the Cyprus Fisheries, 1989. Fisheries Department, Ministry of Agriculture and Natural Resources, Nicosia.

Dybbro, T. 1985. Status of Danish Bird Localities. Danish Ornithological Society, Copenhagen.

Eldoy, S. 1990. Wetland Conservation and Management in Norway. In: Finlayson, C. M. and Larsson, T. (eds) Wetland Management and Restoration. Swedish Environmental Protection Agency.

Environmental Problems Foundation of Turkey. 1989. Wetlands of Turkey. EPFT.

Ertan, A., Kilic, A. & Kasparek, M. 1989. Türkiye'nin Onemli Kus Alanlan. DHKD/ICBP, Istanbul.

Etienne, P. 1990. Inventaire des Zones Humides de la Somme. Association Picarde Chasseurs de Gibier d'Eau, F. D. C., Somme.

Farinha, J. C. & Trindade, A. 1994. Contribuicao para o Inventario e Caracterizacao Nacional de Zonas Humidas. Estudos de Biologia e Conservacao da Natureza. Instituto da Conservacao da Natureza, Lisboa.

Ferdinand, L. 1980. Birds in the Landscape. Danish Ornithological Society, Copenhagen.

Frikke, J. 1991. Breeding waders and wet grassland habitats in Denmark. Wader Study Group Bulletin 61: 42–49.

Gerakis, P. A. 1990. (ed.) Conservation and management of the wetlands in Greece. Proceedings of the Thessaloniki Workshop, 17–21 April 1989. WWF and Laboratory of Ecology, Faculty of Agriculture, Aristotelian University, Thessaloniki.

Gjiknuri, L. & Peja, N. 1992. Albanian Lagoons: Their importance and economic development. pp. 130–133. In: Finlayson, C. M., Hollis, G. E. and Davis, D. J. (eds), Managing Mediterranean Wetlands and their Birds. IWRB Special Publication No. 20, Slimbridge, UK.

Grimmett, R. F. A. & Jones, T. A. 1989. Important Bird Areas In Europe. ICBP and IWRB, Cambridge, UK.

Haapanen, A. 1981. Finnish National Wetland Conservation Program. Committee Report 32, 1–197.

Harengerd, M. 1990. Die Ramsar-Konvention, Bericht fur die Bundesrepublik Deutschland 1990. Ber. Dtsch. Sekt. IRV 29, 87–99.

Haslam, S. M. 1987. River Plants of Western Europe. Cambridge University Press.

Hecker, N. & Tomàs-Vives, P. (Eds), The status of wetland inventories in the Mediterranean region. MedWet project, ICN and IWRB. (in press).

Heliotis, F. 1988. An inventory and review of the wetland resources in Greece. Wetlands 8, 15–31.

Hollis, G. E. & Jones, T. A. 1991. Europe and the Mediterranean Basin pp 27–56. In: Finlayson, M. & Moser, M. (eds) Wetlands. IWRB, Slimbridge, UK. pp. 27–56.

Hudec, K. Wetlands of the Czech Republic. Ministry of the Environment, Czechoslovakia (in press).

Hudec, K., Husak, S., Janda, J. & Pellantova, J. 1993. Survey of Aquatic and Wetland Biotopes of the Czech Republic (Summary Report). Czech Ramsar Committee.

Jacoby, H. 1991. Errichtung und Sicherung schutzwurdiger Teile von Natur und Landschaft mit gesamtstaatlich reprasentativer Bedeutung; Beispiel: Wollmatinger Ried. Nt. und Landsch. 66, 567–572.

Jones, T. A. & Hughes, J. M. R. 1993. Wetland inventories and wetland loss studies: a European perspective. pp. 164–170. In: Moser, M., Prentice, R. C. & van Vessem, J. (eds), Waterfowl and Wetland Conservation in the 1990s. IWRB Special Publication No. 26. IWRB, Slimbridge, UK.

Kirby, J. S., Ferns, J. R., Waters, R. J. & Prys-Jones, R. P. 1991. Wildfowl and Wader Counts 1990–91. The Wildfowl and Wetlands Trust, Slimbridge, UK.

Leontiades, L. 1977. Report on Wetlands and Marine Parks in Cyprus. Prepared for the UNEP Expert Consultation on Mediterranean Marine Parks and Wetlands, Tunis, 12–14 January, 1977.

Leuzinger, H. 1976. Inventar der Schweizer Wasservogelgebiete von Internationaler und Nationaler Bedeutung. Orn. Beob. 73, 147–194.

Lindsay, R. A. National Peatland Resource Inventory. Joint Nature Conservancy Council, Peterborough, UK (in prep.).

Marti, C. 1988. Zones d'Importance Internationale pour les oiseaux d'eau en Suisse; Cartes commentées pour la première revision de l'inventaire, 1987. Station Ornithologique Suisse, Sempach.

Marti, C. & Schifferli, L. 1987. Inventar der Schweizer Wasservogelgebiete von Internationaler Bedeutung. Erste Revision 1986. Orn. Beob. 84, 11–47.

Mayr, C. 1991. Europaische Vogelschutzgebiete in der BRD-Entwicklungen seit 1991. Ber. Dtsch. Sekt. IRV 30, 35–54.

Menanteau, L. 1991. Zones Humides du Littoral de la Communauté Européenne Vues de l'Espace.

Ministère de l'Environnement, Bulgarie. 1993. Plan National d'Actions Prioritaires de Conservation des Zones Humides les plus Importantes de Bulgarie. Ministère de l'Environnement, Bulgarie.

Ministère de L'Environnement, Paris, 1987. Les Zones Humides de Lorraine. Délégation à la Qualité de la Vie, Ministère de l'Environnement, Paris.

Ministerio de Obras Publicas y Transportes, Madrid. 1991. Estudio de las Zonas Humedas de la Espana Peninsular: Inventario y Tipificacion. Documento de Sintesis.

Ministry of Education of Ukraine, Academy of Sciences of Ukraine, State Committee of Ukraine on Science and Technology. 1993. The list of the Wetlands of International Significance in the South of Ukraine. Melitopol, Ukraine.

Ministry of the Environment, Rome. 1990. Inventory of Italian Ramsar Sites. Ministry of the Environment, Nature Convention Services, Rome.

Ministry of the Environment and the National Forest and Nature Agency, Denmark. 1993. Danish report 1993 on the Ramsar Convention, Denmark and Greenland. International Conference on the Convention at Kushiro, Japan, June 1993.

Montes, C. 1990. Estudio de las Zonas Humedas de la Espana Peninsular: Inventario y Tipification. INITEC, Direccion General de Obras Hidraulicas. Ministerio de Obras Publicas y Urbanismo. Madrid.

Montes, C. & Bernués, M. Spain. In: Hecker, N. & Tomàs-Vives, P. (Eds), The status of wetland inventories in the Mediterranean region. MedWet project, ICN and IWRB. (in press).

Montes, C. & Bifani, P. 1989. An Ecological and Economic Analysis of the Current Status of Spanish Wetlands. Department of Ecology, Autonomous University of Madrid, Spain, Report for OECD.

Muzinić, J. Croatia. In: Hecker, N. & Tomàs-Vives, P. (Eds), The status of wetland inventories in the Mediterranean region. MedWet project, ICN and IWRB. (in press).

Olney, P. J. S. 1965. Project MAR list of European and North African Wetlands of International Importance. IUCN New Series No. 5, Morges, Switzerland.

Prentice, C. & Tomas-Vives, P. Inventory and monitoring of coastal wetlands in the Mediterranean: the development of a standardised approach. Proceedings of the 4th EUCC Congress, Marathon, 1993 (in press).

Psilovikos, A. 1992. The prospective situation for wetlands and waterfowl in Greece. pp. 53–55. In: Finlayson, C. M., Hollis, G. E. and Davis, T. J. (eds.) Managing Mediterranean Wetlands and their Birds. Proc. Symp., Grado, Italy, 1991. IWRB Special Publication No. 20. Slimbridge, UK.

Ramsar Convention Bureau. 1990. Directory of Wetlands of International Importance. Ramsar Convention Bureau, Gland, Switzerland.

Ramsar Convention Bureau. 1993. Convention on Wetlands of International Importance Especially as Waterfowl Habitat. Proceedings of the 5th Meeting of the Conference of the Contracting Parties, Kushiro, Japan. Volume 3. Ramsar Convention Bureau, Gland, Switzerland.

Rheinwald, G. 1990. Europaische Vogelschutzgebiete (IBA) in Deutschland. Ber. Dtsch. Skt. IRV 29, 19–42.

Royal Institute for Natural Sciences, Belgium. 1987. Inventaire des Zones de Grands Intérêts pour la Conservation des Oiseaux Sauvages dans la Communauté Européenne. RINS, Belgium.

Ruuhijarvi, R. 1983. The Finnish mire types and their regional distribution. In: Gore, A. J. P. (ed.) Mires: Swamp, Bog, Fen and Moor. B. Regional Studies. Ecosystems of the World, Vol 4B. Elsevier Science Publishers, Amsterdam. pp. 47–68.

Scott, D. A. 1989. Design of Wetland Data Sheet for Database of Ramsar Sites. Mimeographed report to Ramsar Convention Bureau, Gland, Switzerland.

Scott, D. A. 1980. A Preliminary Inventory of Wetlands of International Importance for Waterfowl in West Europe and Northwest Africa. IWRB, Slimbridge, UK.

Secretariat de la Faune et de la Flore. 1992. Listes Nationales et Régionales des Zones Naturelles d'Intérêt Ecologique, Faunistique et Floristique (ZNIEFF). Paris, France.

Skokova, N. N. & Vinogradov, V. G. 1986. Waterfowl Habitat Conservation (Russian). Agropromizdat, Moscow.

Sociedad Espanola de Ornitologia. 1987. Clasification de las Zonas Humedas Espanolas en Funcion de las Aves Acuaticas. SEO, Madrid, Vols 1 and 2.

Struwe, B. & Nehls, H.-W. 1992. Ergebnisse der Internationalen Wasservogelzahlung im Januar 1990 an der Deutschen Ostseekuste. Seevogel 13, 17–28.

Tsiouris, S. A. & Gerakis, P. A. 1991. Wetlands of Greece: values, alterations and conservation. WWF and Laboratory of Ecology, Faculty of Agriculture, Aristotelian University, Thessaloniki.

Van Vessem, J. & Kuijken, E. 1986. Contribution to the Determination of Special Protection Areas for Bird Conservation in Flanders. Ministry of the Flemish Community, Belgium.

Viksne, J. 1993. Important Bird Areas in Latvia. International Council for Bird Preservation, Cambridge, UK.

Waagepetersen, J. 1986. Restoration of Drained Lakes and Fjords in Denmark. Forest and Nature Agency, Copenhagen.

Wilson, A. M. & Moser, M. E. (eds.) [1994] Conservation of Black Sea Wetlands: a preliminary action plan. IWRB Special Publication No. 33, IWRB, Slimbridge, UK.

Zalidis, G. C. 1993. The National Wetlands Inventory for Greece: Prospects and progress. pp. 178–184. In: Moser, M., Prentice, R. C. & van Vessem, J. (eds), Waterfowl and Wetland Conservation in the 1990s. IWRB Special Publication No. 26, IWRB, Slimbridge, UK.

Vegetatio **118**: 29–38, 1995.

Classification of boreal mires in Finland and Scandinavia: A review

P. Pakarinen

Department of Ecology and Systematics, University of Helsinki, P.O. Box 4, Fabianink. 24, FIN-00014 Helsinki, Finland

Key words: Boreal wetlands, Fens, Peatlands, Plant ecology, *Sphagnum* bogs, Vegetation types

Abstract

Mires have been classified in northern Europe at two levels: (1) mire complexes are viewed as large landscape units with common features in hydrology, peat stratigraphy and general arrangement of surface patterns and of minerogenous vs. ombrogenous site conditions; (2) mire sites are considered as units of vegetation research and used in surveys for forestry and conservation. This paper reviews the development of site type classifications in Fennoscandia (Finland, Sweden, Norway), with a discussion on circumboreal classification and corresponding mire vegetation types in Canada. The scale of observation affects classifications: small plot size (0.25–1 m^2) has been used in Scandinavia to make detailed analyses of ecological and microtopographical variation in mostly treeless mire ecosystems, while larger sampling areas (up to 100–400 m^2) have been commonly employed in Finnish studies of forested peatlands. Besides conventional hierarchic classifications, boreal mires have been viewed as an open, multidimensional, non-hierarchic system which can be described and classified with factor, principal component or correspondence analyses. Fuzzy clustering is suggested as an alternative method of classification in mire studies where only selected environmental and vegetational parameters are measured or estimated.

Nomenclature: Lid, J. (1987) Norsk, svensk, finsk flora (vascular plants). Corley *et al.* (1981) Journal of Bryology 11: 609–689 (bryophytes)

Introduction

In Scandinavia (Norway, Sweden) and Finland, mires cover extensive areas, on average 10 to 30% and in some northern regions up to 50% of the terrain (Goodwillie 1980; Heikurainen 1960; Löfroth 1991). Bioclimatically these three countries belong largely to the boreal zone which continues as a wide, nearly continuous belt through northern parts of Eurasia and North America (Abrahamsen *et al.* 1977; Hämet-Ahti 1982; Tuhkanen 1984). The hemiboreal and south boreal zones of Fennoscandia contain probably the largest concentration of raised mire ecosystems in Europe while extensive aapa-mires, patterned mire complexes with wet flarks and hummock strings, prevail in the north boreal zone in Finland and Sweden (Ruuhijärvi 1983; Sjörs 1983). Palsa mires with peat mounds re-

present the northernmost type of boreal mire complexes in Fennoscandia between 69–70° N (Fig. 1).

Peatlands are defined as areas with a minimum peat depth of 30 cm (Finland: Heikurainen 1960) or 40 cm (Canada: Wells & Zoltai 1985), but the term mire also covers waterlogged areas with shallow or discontinuous peat layer e.g. in paludified forests or groundwater discharge areas. Other types of wetlands – lakes, freshwater and brackish-water marshes, wet meadows (cf. Okruszko 1979; Hollis & Jones 1991) – are not treated in this paper. According to the hydrological-geographical classification of Orme (1990), a large majority of boreal mires represent 'palustrine' wetlands.

Despite wetland losses, mainly as a result of drainage for agricultural or timber production, the total area of intact mires in Fennoscandia (Finland, Sweden and Norway) – altogether c. 14 mill. hectares – is still

Fig. 1. Selected Fennoscandian and Estonian mire ecosystems of international importance for conservation and research. The boundary between the temperate and boreal bioclimatic zones mainly according to Hämet-Ahti (1982) and Tuhkanen (1984), and the area of palsa mires according to Ruuhijärvi (1960) and Dierssen (1982). Mires designated as Ramsar sites (Finlayson & Moser, 1990) are shown with triangles. Key to locations – *Norway:* (1) Måmyra, (2) Havmyrene, (3) N. Kisselbergmosen, (4) Hedmarksvidda, (5) Dvergberg, (6) Aiddejavvre; – *Sweden:* (7) Store Mosse-Kävsjön, (8) Åkhultsmyr, (9) Komosse, (10) Muddus, (11) Sjaunja, (12) Tavvavuoma; – *Estonia:* (13) Nigula, (14) Kuresoo, (15) Muraka; – *Finland:* (16) Munasuo-Kananiemensuo, (17) Torronsuo, (18) Patvinsuo, (19) Martimo, (20) Riisitunturi, (21) Koitelaiskaira, (22) Sammuttijänkä.

large when compared to central Europe (Goodwillie 1980). The site type and vegetation type classifications of undrained boreal mires are the subject of this paper, and an attempt is made to relate recent numeri-cal and non-hierarchic approaches to the conventional classifications, and also to discuss terminology and compare boreal site type systems in Fennoscandia and Canada.

Classifications and peatland surveys in Nordic countries

Finnish studies

The original classification of Cajander (1913) for Finnish peatlands included 35 site types and four major groups: (1) *Weissmoore*, treeless, oligotrophic or mesotrophic *Sphagnum* mires, (2) *Braunmoore*, treeless, eutrophic 'brown-moss' mires, (3) *Reisermoore*, pine mires, and (4) *Bruchmoore*, hardwood-spruce mires. The basic features of the Cajanderian mire classification, which is a combined vegetation/site type system, have been in use in Finland with some modifications both in applied surveys and in basic botanical/ecological research.

Peatland surveys carried out in National Forest Inventories (see Heikurainen 1960) and by the Geological Survey of Finland (e.g. Maunu 1984; Pajunen 1989; Sten & Svahnbäck 1989; Tuittila 1983) contain information on mire site types and peat deposits which has been useful in conservation programs and in monitoring of man-made changes (e.g. Eurola *et al.* 1991). Especially for conservation, mires have been viewed also as mire complexes – landscape units which have a characteristic hydrology, peat stratigraphy and surface topography and consequently a predictable arrangement of site types especially in the center parts of a basin. Raised mires, aapa-mires and palsa mires have a regional occurrence (Fig. 1) which has been recently mapped in Finland over the whole country in the scale of 1 to 1 mill. (Atlas of Finland 1988).

The vegetation types of northern Finnish aapamires were studied by Ruuhijärvi (1960) and numerically analysed by Pakarinen & Ruuhijärvi (1978). Eurola and Kaakinen (1979) and Eurola *et al.* (1984) distinguished altogether c. 70 vegetation types and regrouped them into seven major classes where the combination site types *(Waldweissmoore)* were considered as a major transitional group between open and forested wetlands (treed fens in Fig. 2).

For peatland forestry, Heikurainen (1960, 1979) used 32 site types which are quite comparable with the Cajanderian classification and two-dimensional schemes were used to show relationships of site types along wetness and site fertility gradients. However, the advanced successional phases of drained peatland forests have been generally grouped only to six or seven site types (Heikurainen 1979; Laine 1989) reflecting a one-dimensional fertility series and the general loss

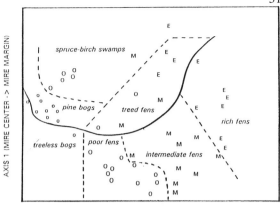

AXIS 2 (OMBROTROPHIC -> MINEROTROPHIC)

Fig. 2. A two-dimensional grouping of Finnish peatland vegetation types on the basis of detrended correspondence analysis (after Pakarinen 1985). Axis 1 indicates a gradient from mire center (with stagnant water and thick peat layer) to mire margin (with mobile surface waters and generally thin peat layer) while axis 2 represents a fertility gradient of trophic conditions: o = ombtrotophic, O = oligotrophic, M = mesotrophic, E = eutrophic.

of site type diversity and wet-dry variation characteristic of undrained wetlands.

Tuomikoski (1942) introduced correlation analysis as a method to form ecological species groups and strongly advocated the ordination approach in mire vegetation research. Havas (1961) utilized the correlation analysis to arrive at species groups and further used them in construct a phytosociological system of vegetation types of Finnish sloping mires. With the development and availability of statistical multivariate methods, ordination and classification of representative peatland data sets from southern and northern Finland were performed during late 1970s (Pakarinen 1976, 1979; Pakarinen & Ruuhijärvi 1978).

Scandinavian studies

Stenius in 1742 and Linnaeus in 1751 were according to Du Rietz (1957) the first authors to describe differences between fen and bog vegetation although with different terminology, and the bog/fen limit (*Mineralbodenwasserzeigergrenze*, Du Rietz 1954) is still recognized as the major dividing line in the Swedish site type system between ombrotrophic *mosse* and minerotrophic *kärr* vegetation (Löfroth 1991; Påhlsson 1984).

Swedish mire classifications are mostly included in monographs of specific mire complexes (Osvald 1923; Sjörs 1948; Malmer 1962). Major directions of variation according to which the described site types have been arranged, are the gradients wet – dry (hummock – mud-bottom), poor – rich (bog – poor fen – rich fen), and mire margin – mire expanse (Sjörs 1950, 1983; Malmer 1986).

In Norway, the Braun-Blanquet school has had some influence after Nordhagen (1943) developed a classification system for alpine vegetation with this method. More recently, Dierssen (1982) published an extensive and detailed phytosociological investigation of mire communities over a large area of northwestern Europe: Norway, Iceland, the Faeroer and the British Isles. In this study, the mire vegetation types (associations) were grouped into alliances, orders and classes, and subdivided into subassociations and variants; the resulting hierarchical classification was thus a six-level system where the exclusive fen plant limit and the increasing influence of minerotrophic water appeared at lower classification units and not in the major grouping.

The increasing use of numerical methods and statistically valid sampling methods is seen also in Scandinavian mire studies. In Sweden, principal component analysis was used by Tyler (1980) for species-rich fen vegetation and by Sonesson & Kvillner (1980) to study species-site relationships in subarctic mires near Abisko. The emphasis in recent Norwegian mire vegetation research has been in intensive studies of particular mire complexes and in the development of methods. Galten (1987) utilized clustering techniques while in the series of papers by Okland (1989, 1990), the major subject has been the evaluation of ordination techniques, particularly detrended correspondence analysis, for the identification of ecological gradients and for the construction of a framework of classification.

Discussion

Northern peatland complexes

The main mire complex types in Finland, southern raised mires and northern aapa-mires, show distinct differences in their peat stratigraphy and general arrangement of ombrotrophic vs. minerotrophic vegetation types (Fig. 3). Central parts of aapa-mires are usually characterized by a flark-string pattern and fen vegetation, while oligotrophic-ombrotrophic *Sphagnum* bogs are typically found in the margins or locally on peat ridges (high strings). There are also examples of mixed mire complexes (Palkinsuo in Fig. 3), southern aapa-mires (part of Kananiemensuo, site 16

in Fig. 1), or northern raised bogs (Atlas of Finland 1988).

Aapa-mires have been described also in Norway ('*Aapamoor*' or '*Strangmoor*' in Dierssen 1982), in central and northern Sweden ('mixed mires' by Sjörs 1983; patterned fens by Foster and Fritz 1987) as well as in northwestern Russia (Abramova *et al.* 1974), but according to Botch & Masing (1983) not in Siberia. Minerotrophic patterned wetland complexes resembling European aapa-mires have been recorded in boreal and subarctic North America both near the coast and in the interior of the continent: e.g. in Newfoundland (Wells 1981), Labrador (Foster & King 1984), Quebec ('structured fen', Gerardin & Grondin 1987), northern Minnesota (Glaser *et al.* 1981), Great Slave Lake area (Pakarinen & Talbot 1976), and the boreal and subarctic Canada in general ('ribbed fens', Zoltai & Pollett 1983; Zoltai *et al.* 1988).

Palsa mires and related hummocky mires occur in continental climate near the forest-tundra transition: locally in northern Fennoscandia (Fig. 1) and more extensively in northern Russia/Siberia (Botch & Masing 1983) as well as in northern Canada ('palsa bogs', Zoltai et al. 1988).

Bogs and fens

Especially in Scandinavia the term 'bog' has been mainly reserved for ombrotrophic mire communities where any ground-water influence and fen indicators are absent. Occasionally the term bog has been used in wider sense, covering all *Sphagnum* rich peatland vegetation (Jeglum *et al.* 1974; see also discussion in Malmer 1962), or even referring to all northern mires (Lugo *et al.* 1990).

Due to regional differences in climatic factors (see discussion in Gorham 1956), a similar community in highly oceanic conditions can be called 'bog', but in a more continental area 'fen': compare e.g. *Menyanthes trifoliata* flarks in Scotland (Dierssen 1982) with those in northern Finland (Ruuhijärvi 1960). Similarly, ombrogenous blanket bogs of the oceanic flank of Ireland (with *Sphagnum subsecundum* and *Schoenus nigricans*, Hammond 1979) can be compared with *Schoenus* fens of Sweden and Estonia (Tyler 1981) or with mesotrophic fens in the Finnish classification (Eurola *et al.* 1984). In North America, Glaser (1992) concluded that a large number of vascular species growing on maritime bogs were limited to fens or mineral upland in continental regions.

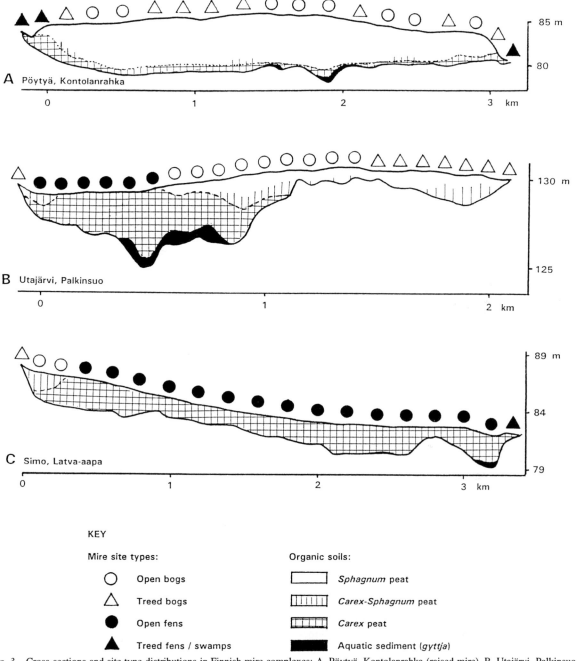

Fig. 3. Cross-sections and site type distributions in Finnish mire complexes: A. Pöytyä, Kontolanrahka (raised mire), B. Utajärvi, Palkinsuo (mixed mire), C. Simo, Latva–aapa (aapa-mire). Scale: vertical – m above sea level, horizontal – km. (modified from Tuittila 1983; Pajunen 1989; Maunu 1984).

The nutrient-ecological importance of poor fen/rich fen limit (i.e. the *Weissmoor/Braunmoor* limit of Cajander 1913) has been recently emphasized by several authors (e.g. Gorham & Janssens 1992; Malmer 1986; Malmer *et al.* 1992). Thus, bogs *sensu lato* can include ombrotrophic and weakly minerotrophic *Sphagnum* mires.

Drainage and conservation

Extensive drainage of peatlands in Finland has resulted in successional site types and in an increase of
treed/forested wetlands and generally in a decrease of
landscape diversity when drained wetlands gradually
change to forest ecosystems. According to Eurola *et
al.* (1991), virgin mire vegetation accounted in late
1980s for only 26% of all peatlands in southern and
central Finland while recently drained areas covered
14% and transitional drainage areas 49% of peatland
area. The same study showed that drainage was most
frequent in forested or semi-forested marginal parts
of mire complexes and thus the proportion of wet site
types, such as flark fens, had increased in the aapa-mire
region. Among the diversity of Finnish mire vegetation, meso-eutrophic site types in particular – fern-rich
spruce mires, spring-influenced spruce mires, black
alder swamps, paludified herb-rich hardwood-spruce
forests, rich birch fens and spring fens – have become
extremely rare and endangered and consequently have
had high priority in recent national and local conservation plans. Besides eutrophic lakes and coastal marshes, a few large mire complexes in Finland and Sweden
(see Fig. 1) have been designated as Ramsar sites (Hollis & Jones 1991).

The scale

In vegetation research and particularly in vegetation
mapping the scale of observation is an essential factor.
Classifications and regional comparisons are certainly
different, if the unit of analysis is 1 m^2 or less vs. a
100—1000 m^2 plot (or more) as discussed by Masing
(1974, 1984).

Small plot size (generally 0.25 to 1 m^2) has
been typical of Scandinavian phytosociology, and this
has resulted in fine-scale classification systems with
numerous site types. The classical study of Komosse in
southern Sweden by Osvald (1923) contained descriptions of over 100 'associations' which were grouped
into c. ten 'formations'. With a similar method – using
small plots and forming vegetation types as combinations of single dominants – Waren (1926) described
mire communities in Finland, and the use of dominants in site type names is still apparent in the recent
working group proposal for a general classification of
Nordic mire vegetation (Påhlsson 1984) where c. 100
types or subtypes are listed.

Mosaic structure of mire sites – characterized by
uneven surface topography and alternation or patch-
iness of specific microsites (hummocks, hollows,

flarks) – is a common characteristic of boreal wetlands – and for the estimation of the proportion of the
microtopographic community types in each site, systematic sampling schemes have been developed, e.g.
Heikurainen (1951). In forested mire vegetation, the
sample plot size for vegetation description has been
relatively large: in Finland 100 m^2 (Ruuhijärvi 1960)
and in Canada 100–400 m^2 (Jeglum 1991).

The Braun-Blanquet system

The phytosociological (Braun-Blanquet) system of
hierarchic classification has been used quite extensively in Britain by nature conservation authorities
(Rodwell 1991) and to some extent also in Canada
(e.g. Gauthier 1980; Wells 1981) and in Russia (Botch
1986; Botch & Smagin 1993). In Norway criticism has
been presented against the traditional Central European
phytosociology (Okland 1990), and national surveys
performed in Norway have utilized mainly a hydro-
topographical classification complemented with information on the occurrence of specific ecological species
groups characterizing a particular site or *synsite* (i.e. a
combination of sites, cf. Moen 1985).

In bogs with a pronounced surface relief the species
composition may completely change between adjacent
hummocks and hollows, and in the Braun-Blanquet
classification, these microsites are differentiated at
high level: the hummock vegetation belongs to the
class *Oxycocco – Sphagnetea* and the depressions to
Scheuchzerio – Caricetea nigrae (Dierssen 1982; see
discussion in Malmer 1986). Description of association complexes, i.e. estimating the cover of community types present in an area of up to 500 m^2, has been
proposed by Dierssen & Dierssen (1985) for regional
studies of mires. Interpretation of the synoptic tables
of such studies requires, however, a detailed knowledge of the phytosociological nomenclature of various
hierarchic units.

Multidimensional classifications

In his theoretically oriented study of eastern Finnish
spruce mire vegetation, Tuomikoski (1942) discussed
in detail the classification approaches and concluded
that a natural vegetation type system should reflect the
multidimensional nature of ecological and phytosociological variation in boreal mires, and the indicator
value of species groups rather than single species
should be used as the basis of classification.

A factor analytical classification was presented for
boreal mire vegetation in North Finland by Pakarinen

Table 1. Correspondence of selected mire site types in North European (Fennoscandian) and Canadian vegetation classifications. Studies used for comparison: Finland (Eurola *et al.* 1984, Pakarinen & Ruuhijärvi 1978), Norway/NW Europe (Dierssen 1982), Newfoundland (Wells 1981), S Quebec (Gauthier 1980), Quebec-Labrador (Foster & King 1984), N Ontario (Jeglum 1991), Prince Rupert area/British Columbia (Vitt *et al.* 1990). Names of the major site type groups (I–V) translated from Påhlsson (1984).

N Europe (Fennoscandia)	Canada
I. Treed bogs *(skogsmosse)*	
Sphagnum fuscum bog (Eurola *et al.* 1984); *Empetro – Sphagnetum fusci* (Dierssen 1982)	Treed ombrotrophic bog (T1/*Sphagnum fuscum*, Jeglum 1991); *Piceo marianae – Sphagnetum fusci* (Gauthier 1980)
Dwarf-shrub pine bog (Eurola *et al.* 1984); *Ledo – Sphagnetum fusci* (Dierssen 1982)	Treed ombrotrophic bog (T2/Other Sphagnum/Low shrub phase, Jeglum 1991); *Picea mariana – Sphagnum angustifolium* (Gauthier 1980)
II. Open bogs *(öppen mosse)*	
Sphagnum hollow bog (Pakarinen & Ruuhijärvi 1978); Flark-level bog (Eurola *et al.* (1984); *Caricetum limosae* (Dierssen 1982)	Ombrotrophic carpet – pool (Releve group 5, Vitt *et al.* 1990); *Sphagno majoris – Caricetum limosae* (Gauthier 1980)
Short-sedge intermediate-level bog (Eurola *et al.* 1984)	Ombrotrophic lawn (Releve group 3, Vitt *et al.* 1990); *Scirpo – Sphagnetum magellanici* (Wells 1981)
III. Treed fens *(skogskärr)*	
Tall-sedge pine fen (Eurola *et al.* 1984)	Intermediate treed fen (T5, Jeglum 1991); *Carex aquatilis – Sphagnum fuscum* (Gauthier 1980)
Rich pine fen (Eurola *et al.* 1984)	Rich treed fen (T6/*Sphagnum warnstorfii*, Jeglum 1991)
IV. Open fens *(öppet kärr)*	
Tall-sedge fen (Eurola *et al.* 1984); *Caricetum lasiocarpae, Caricetum rostratae* (Dierssen 1982)	*Caricetum lasiocarpae, Caricetum rostratae* (Gauthier 1980); *Carex rostrata* nodum (Foster & King 1984)
Campylium stellatum fen (Eurola *et al.* 1984); *Caricion davallianae/Drepanoclado revolventis – Trichophoretum cespitosi* (Dierssen 1982)	*Carex lasiocarpa – Campylium stellatum* (Gauthier 1980); *Potentillo – Campylietum stellati* (Wells 1981)
V. Paludified forests *(sumpskog)*	
Vaccinium myrtillus spruce swamp (Pakarinen & Ruuhijärvi 1978); True spruce mire (Eurola *et al.* 1984)	Transitional poor conifer swamp (T7, Jeglum 1991); *Abies balsamea – Sphagnum girgensohnii* nodum (Foster & King 1984)
Herb-rich spruce-birch swamp (Pakarinen & Ruuhijärvi 1978)	Transitional rich swamp (T10/*Alnus rugosa*, Jeglum 1991).

& Ruuhijärvi (1978), and the steps of factor analysis appeared to provide a theoretical and practical framework for a multidimensional classification and vegetational analysis of boreal mire sites in Finland: (1) the significant ecological gradients of characteristics are extracted by a few (up to six) factors; (2) corresponding ecological species groups are defined; (3) a site type is defined as a characteristic combination of species groups present, i.e. a specific combination of significant factor (component) loadings. A factor loading matrix can then be used to construct a multidimensional system or multiple-entry key of sites (Pakarinen 1985).

Okland (1990), based on his large data-set of systematically sampled microplots, outlined a multidimensional classification system of mire sites: mire vegetation was classified into 32 site types by a reticulate division of a four-dimensional ecological space.

Thus, a multidimensional classification can be based on direct or indirect ordination methods, including the factor analysis. On the other hand, fuzzy classifications recently discussed and introduced with phytosociological and soil/landscape-ecological examples (Burrough 1989; Feoli & Zuccarello 1988) may replace the traditional hierarchic techniques, especially in mire surveys where large sample plots and combined vegetation and soil data are collected.

Circumboreal comparisons

Sjörs (1963) discussed ecological gradients and species assemblages of Canadian bogs and fens in northern Ontario in comparison with Scandinavian mires and found a high degree of similarity in the species composition of the bottom layer (bryophytes and lichens) while the higher strata had less common species with corresponding European vegetation. More recently, Wheeler *et al.* (1983) studied the floristic similarities between the Red Lake Peatland, northern Minnesota, with other northern peatlands, finding that the similarity quotient between northern Minnesota and Fennoscandia (Sweden, Finland) was 16 to 21% for vascular plants and 39 to 60% for bryophytes. Ecology of circumboreal peat mosses *(Sphagnum)* has been discussed also in several other European and North American studies (e.g. Gauthier 1980; Horton *et al.* 1979; Pakarinen 1979; Tüxen *et al.* 1972; Vitt *et al.* 1990).

In Table 1, ecological and vegetational correspondence is shown between the major mire types described in Fennoscandia and Canada. The correspondence between site types has been determined primarily on the basis of the floristic composition and vegetation structure with additional information of such ecological characteristics as soil or water pH (cf. Jeglum 1991). There are remarkable similarities in the vegetation of boreal and subarctic peatlands of northern Europe and Canada, especially regarding the ground layer vegetation. The major groups used in Table 1 – treed bogs, open bogs, treed fens, open fens, and paludified forests – could form a basis for a general classification of boreal mires. In most cases, however, it would be desirable to refine such a classification with the information of the substrate quality (pH, ash content, peat depth), either by direct measurements or with the use of plant indicators.

References

Abrahamsen, J., Jacobsen, N., Dahl, E., Kalliola, R., Wilborg, L. & Påhlsson, L. 1977. Naturgeografisk regionindelning av Norden. Nordiska ministerrådet, NU B 34:1–130. Stockholm.

Abramova, T. G., Botch, M. S. & Galkina, E. A. (ed.) 1974. Tipy bolot S.S.S.R. i printsipy ikh klassifikatsii (Mire types of the USSR and principles of their classification, in Russian). Leningrad, 254 pp.

Atlas of Finland 1988. Biogeography, nature conservation. Mires 1:1000000 (map appendix 2). Folio 141–143. National Board of Survey & Geographical Society of Finland. Helsinki.

Botch, M. S. 1986. On the classification of mire vegetation (as exemplified by Sphagnum bogs of the north-west of RSFSR). Botanicheskii Zhurnal 71(9):1182–1192.

Botch, M. S. & Masing, V. V. 1983. Mire ecosystems in the U.S.S.R. pp. 95–152. In: Gore, A. J. P. (ed.) Mires: swamp, bog, fen and moor. Ecosystems of the world, Vo. 4B. Amsterdam.

Botch, M. S. & Smagin, V. A. 1993. Flora and vegetation of mires in north-west Russia and principles of their protection (In Russian with an English abstract). Proceedings of Komarov Botanical Institute (Russian Academy of Sciences), St. Petersburg, Russia. 225 pp.

Burrough, P. A. 1989. Fuzzy mathematical methods for soil survey and land evaluation. Journal of Soil Science 40:477–492.

Cajander, A. K. 1913. Studien über die Moore Finnlands. Acta Forestalia Fennica 2(3):1–208.

Dierssen, K. 1982. Die wichtigsten Pflanzengesellschaften der Moore NW-Europas. Conservatoire jardin botaniques, Geneve. 382 pp.

Dierssen, K. & Dierssen, B. 1985. Suggestions for a common approach in phytosociology for Scandinavian and Central European mire ecologists. Aquilo Ser. Botanica 21:33–44.

Du Rietz, G. E. 1954. Die Mineralbodenwasserzeigergrenze als Grundlage einer natürlichen Zweigliederung der nord- und mitteleuropäischen Moore. Vegetatio 6:571–585.

Du Rietz, G. E. 1957. Linnaeus as a phytogeographer. Vegetatio 7:161–168.

Eurola, S., Aapala, K., Kokko, A. & Nironen, M. 1991. Mire type statistics in the bog and southern aapa mire areas of Finland (60–66° N). Annales Botanici Fennici 28:15–36.

Eurola, S., Hicks, S. & Kaakinen, E. 1984. Key to Finnish mire types. pp. 11–117. In: Moore, P. D. (ed.) European mires. Academic Press, London.

Eurola, S. & Kaakinen, E. 1979. Ecological criteria of peatland zonation and the Finnish mire type system. pp. 20–32. Proceedings of the International Symposium on Classification of Peat and Peatlands, Hyytiälä, Finland, September 1979. International Peat Society, Helsinki.

Feoli, E. & Zuccarello, V. 1988. Syntaxonomy: a source of useful fuzzy sets for environmental analysis? Coenoses 3:141–147.

Finlayson, M. & Moser, M. (eds.) 1991. Wetlands. International Waterfowl and Wetland Research Bureau, Facts on File, Ltd., Oxford, 224 pp.

Foster, D. R. & Fritz, S. C. 1987. Mire development, pool formation and landscape processes on patterned fens in Dalarna, Central Sweden. Journal of Ecology 75:409–437.

Foster, D. R. & King, G. A. 1984. Landscape features, vegetation and developmental history of a patterned fen in south-eastern Labrador, Canada. Journal of Ecology 72:115–143.

Galten, L. 1987. Numerical analysis of mire vegetation at Åsenmyra, Engerdal, Central Southern Norway and comparison with traditional Fennoscandian paludicology. Nordic Journal of Botany 7:187–214.

Gauthier, R. 1980. La végétation des tourbières et les sphaignes du Parc des Laurentides, Quebec. Etudes Ecologiques 3:1–634. Laboratoire d'écologie forestière, Université Laval, Quebec.

Gerardin, V. & Grondin, P. 1987. Description and regionalization of peatland forms of Moyenne-et-Basse-Côte-Nord, Quebec. pp. 393–402. In: Rubec, C. D. A. & Overend, R. P. (eds.) Proceedings, Symposium 87 Wetlands/Peatlands, Edmonton.

Glaser, P. H. 1992. Raised bogs in eastern North America – regional controls for species richness and floristic assemblages. Journal of Ecology 80:535–554.

Glaser, P. H., Wheeler, G. A., Gorham, E. & Wright, H. E., Jr. 1981. The patterned mires of the Red Lake Peatland, northern Minnesota: vegetation, water chemistry and landforms. Journal of Ecology 69:575–599.

Gorham, E. 1956. On the chemical composition of some waters from the Moor House Nature Reserve. Journal of Ecology 44:375–382.

Gorham, E. & Janssens, J. A. 1992. Concepts of fen and bog re-examined in relation to bryophyte cover and the acidity of surface waters. Acta Societatis Botanicorum Poloniae 61:7–20.

Goodwillie, R. 1980. European peatlands. European Committee for the Conservation of Nature and Natural Resources, Council of Europe. 75 pp. Strasbourg.

Hämet-Ahti, L. 1982. The boreal zone and its biotic subdivision. Fennia 159:69–75.

Hammond, R. F. 1979. The peatlands of Ireland. Soil Survey Bulletin 35:1–58. An Foras Taluntais, Dublin.

Havas, P. 1961. Vegetation und Ökologie der ostfinnischen Hangmoore. Annales Botanici Societatis Vanamo 31(1):1–188.

Heikurainen, L. 1951. Ein Verfahren zur Analysierung der Moorvegetation. Silva Fennica 70:1–18.

Heikurainen, L. 1960. Metsäojitus ja sen perusteet. 378 pp. Porvoo-Helsinki.

Heikurainen, L. 1979. Peatland classification in Finland and its utilization for forestry. Proceedings of the International Symposium on Classification of Peat and Peatlands, Hyytiälä, Finland, Sept. 1979, pp. 135–146. International Peat Society.

Hollis, G. E. and Jones, T. A. 1991. Europe and the Mediterranean Basin. pp. 27–56. In: Finlayson, M. & Moser, M. (eds.) Wetlands. International Waterfowl and Wetlands Research Bureau, Oxford.

Horton, D. G., Vitt, D. H. & Slack, N. G. 1979. Habitats of circumboreal-subarctic sphagna: I. A quantitative analysis and review of species in the Caribou Mountains, northern Alberta. Canadian Journal of Botany 57:2283–2317.

Jeglum, J. K. 1991. Definition of trophic classes in wooded peatlands by means of vegetation types and plant indicators. Annales Botanici Fennici 28:175–192.

Jeglum, J. K., Boissonneau, A. N. & Haavisto, V. F. 1974. Toward a wetland classification for Ontario. Canadian Forestry Service, Department of the Environment, Information Report O-X-215, 54 pp. + Appendix. Sault Ste. Marie.

Jones, R. K., Pierpoint, G., Wickware, G. M. & Jeglum, J. K. 1983. A classification and ordination of forest ecosystems in the Great Clay-belt of northeastern Ontario. pp. 83–96. In: Wein, R. W., Riewe, R. R. & Methven, I. R. (eds.) Resources and dynamics of the Boreal Zone. Association of Canadian Universities for Northern Studies, Ottawa.

Laine, J. 1989. Metsäojitettujen soiden luokittelu (Summary: Classification of peatlands drained for forestry). Suo 40:37–51. Finnish Peatland Society, Helsinki.

Löfroth, M., 1991. Våtmarkerna och deras betydelse. Naturvårdsverket Rapport 3824:1–93. Solna.

Lugo, A. E., Brinson, M. & Brown, S. (eds.) 1990. Forested wetlands. Ecosystems of the World 15, 527 pp. Amsterdam–Oxford–New York–Tokyo.

Malmer, N. 1962. Studies on mire vegetation in the Archaean area of southwestern Götaland (South Sweden). I. Vegetation and habitat conditions on the Åkhult mire. Opera Botanica 7(1):1–322.

Malmer, N. 1986. Vegetational gradients in relation to environmental conditions in northwestern European mires. Canadian Journal of Botany 64:375–383.

Malmer, N., Horton, D. G. & Vitt, D. H. 1992. Element concentrations in mosses and surface waters of western Canadian mires relative to precipitation chemistry and hydrology. Ecography 15:114–128.

Masing, V. 1974. Proposal for unified and specified terminology to designate mires meriting conservation. In: Kumari, E. (ed.) Estonian wetlands and their life. Estonian Committee for IBP, No. 7:183–190. Tallinn.

Masing, V. 1984. Estonian bogs: plant cover, succession and classification. In: Moore, P. D. (ed.) European mires, pp. 119–148. London.

Maunu, M. 1984. Simossa tutkitut suot ja niiden turvevarat. Geologian tutkimuskeskus, Maaperäosasto, Rovaniemi, 34 pp.

Moen, A. 1985. Classification of mires for conservation purposes in Norway. Aquilo Ser. Botanica 21:95–100.

Nordhagen, R. 1943. Sikilsdalen og Norges fjellbeiter. Bergen Museums Skrifter 22:1–607.

Okland, R. H. 1989. A phytoecological study of the mire Northern Kisselbergmosen, SE Norway. I. Introduction, flora, vegetation, and ecological conditions. Sommerfeltia 8:1–172.

Okland, R. H. 1990. A phytoecological study of the mire Northern Kisselbergmosen, SE Norway. II. Identification of gradients by detrended (canonical) correspondence analysis. Nordic Journal of Botany 10:79–108.

Okruszko, H. 1979. Wetlands and their classification in Poland. pp. 221–226. Proceedings of the International Symposium on Classification of Peat and Peatlands, Hyytiälä, Finland, September 1979. International Peat Society, Helsinki.

Orme, A. R. 1990. Wetland morphology, hydrodynamics and sedimentation. In: Williams, M. (ed.) Wetlands: a threatened landscape. pp. 42–94. Oxford.

Osvald, H. 1923. Die Vegetation des Hochmoores Komosse. Svenska Växtsociologiska Sällskapets Handlingar 1:1–436. Uppsala.

Påhlsson, L. (ed.) 1984. Vegetationstyper i Norden. Nordiska ministerrådet, Stockholm. 539 pp.

Pajunen, H. 1989. Utajärvellä tutkitut suot ja niiden turvevarat, osa IV. Turveraportti 229. Geologian tutkimuskeskus, Kuopio, 137 pp.

Pakarinen, P. 1976. Agglomerative clustering and factor analysis of south Finnish mire types. Annales Botanici Fennici 13:35–41.

Pakarinen, P. 1979. Ecological indicators and species groups of bryophytes in boreal peatlands. Proceedings of the International Symposium on Classification of Peat and Peatlands, Hyytiälä, Finland, September 1979, pp. 121–134. International Peat Society, Helsinki.

Pakarinen, P. 1985. Numerical approaches to the classification of north Finnish mire vegetation. Aquilo Ser. Botanica 21:111–116.

Pakarinen, P. & Ruuhijärvi, R. 1978. Ordination of northern Finnish peatland vegetation with factor analysis and reciprocal averaging. Annales Botanici Fennici 15:147–157.

Pakarinen, P. & Talbot, S. 1976. Aapa- ja kohosuokasvillisuudesta Suuren Orjajärven lähistöllä (Summary: Observations on the aapa-mire and raised-bog vegetation near Great Slave Lake, Canada). Suo 27:69–76.

38

Rodwell, J. S. 1991. British plant communities. 2. Mires and heaths. Cambridge University Press, Cambridge, 628 pp.

Ruuhijärvi, R. 1960. Über die regionale Einteilung der nordfinnischen Moore. Annales Botanici Societatis Vanamo 31(1):1–360.

Ruuhijärvi, R. 1983. The Finnish mire types and their regional distribution. pp. 47–67. In: Gore, A. J. P. (ed.) Mires: swamp, bog, fen and moor. Ecosystems of the world, Vol. 4B. Amsterdam.

Sjörs, H. 1948. Myrvegetation i Bergslagen. Acta Phytogeographica Suecica 21:1–299.

Sjörs, H. 1950. Regional studies in North Swedish mire vegetation. Botaniska Notiser 1950:173–222.

Sjörs, H. 1963. Bogs and fens on Attawapiskat River, northern Ontario. National Museum of Canada Bulletin 186:45–133.

Sjörs, H. 1983. Mires of Sweden. pp. 69–94. In: Gore, A. J. P. (ed.) Mires: swamp, bog, fen and moor. Ecosystems of the world, Vol. 4B. Amsterdam.

Sonesson, M. & Kvillner, E. 1980. Plant communities of the Stordalen mire – a comparison between numerical and non-numerical classifications. Ecological Bulletins 30:113–125. Stockholm.

Sten, C-G. & Svahnbäck, L. 1989. Parkanon suot ja turvevarojen käyttökelpoisuus. Turveraportti 234. Geologian tutkimuskeskus, Espoo, 178 pp.

Tuhkanen, S. 1984. A circumboreal system of climatic-phytogeographical regions. Acta Botanica Fennica 127:1–50.

Tuittila, H. 1983. Pöytyän turvevarat. Geologinen tutkimuslaitos, Maaperäosasto, Espoo, 37 pp.

Tuomikoski, R. 1942. Untersuchungen über die Untervegetation der Bruchmoore in Ostfinnland. I. Zur methodik der pflanzensoziologischen Systematik. Annales Botanici Societatis Vanamo 17(1):1–203.

Tüxen, R., Miyawaki, A. & Fujiwara, K. 1972. Eine erweiterte Gliederung der Oxycocco-Sphagnetea. pp. 500–520 in: Tüxen, R. (ed.) Grundfragen und Methoden in der Pflanzensoziologie. The Hague.

Tyler, C. 1980. Schoenus vegetation and environmental conditions in south and southeast Sweden. Vegetatio 41:155–170.

Tyler, C. 1981. Geographical variation in Fennoscandian and Estonian Schoenus wetlands. Vegetatio 45:165–182.

Vitt, D. H., Horton, D. G., Slack, N.G. & Malmer, N. 1990. Sphagnum-dominated peatlands of the hyperoceanic British Columbia coast: patterns in surface water chemistry and vegetation. Canadian Journal of Forest Research 20:696–711.

Waren, H. 1926. Untersuchungen über sphagnumreiche Pflanzengesellschaften der Moore Finnlands. Acta Societatis pro Fauna et Flora Fennica 55(8):1–113.

Wells, E. D. 1981. Peatlands of eastern Newfoundland: distribution, morphology, vegetation, and nutrient status. Canadian Journal of Botany 59:1978–1997.

Wells, E. D. & Zoltai, S. 1985. Canadian system of wetland classification and its application to circumboreal wetlands. Aquilo Ser. Botanica 21:42–52.

Wheeler, G. A., Glaser, P. H., Gorham, E., Wetmore, C. M., Bowers, F. D. & Janssens, J. A. 1983. Contributions to the flora of the Red Lake Peatland, northern Minnesota, with special attention to Carex. American Midland Naturalist 110:62–96.

Zoltai, S. C. & Pollett, F. C. 1983. Wetlands in Canada: their classification, distribution, and use. pp. 245–268. In: Gore, A. J. P. (ed.) Mires: swamp, bog, fen and moor. Ecosystems of the World, Vol. 4B. Amsterdam.

Zoltai, S. C., Tarnocai, C., Mills, G. F. & Veldhuis, H. 1988. Wetlands of subarctic Canada. pp. 55–96. In: Rubec, C.D.A. (ed.) Wetlands of Canada. Environment Canada, Ottawa, and Polyscience Publications, Montreal.

Vegetatio **118**: 39–48, 1995.

Inventory and classification of wetlands in India

B. Gopal & M. Sah

School of Environmental Sciences, Jawaharlal Nehru University, New Delhi 110067, India

Key words: Freshwater wetlands, Hydrology, Mangroves, Physiognomy, Saline lakes, Seasonal wetlands

Abstract

The Indian subcontinent has a large variety of freshwater, saline and marine wetlands. Whereas the mangroves are relatively well documented, very little is known about the other wetlands, with few exceptions. Only recently an inventory of these welands has been prepared but no effort has been made to classify them. A vast majority of the inland wetlands are temporary and/or man-made, and they have been traditionally used and managed by the local human populations. In this paper, first, we evaluate the classification schemes of the IUCN, US Fish and Wildlife Services and those of the Australian wetlands, for their applicability to Indian wetlands. Then, we propose a simple hierarchical classification of wetlands based on their location (coastal or inland), salinity (saline or freshwater), physiognomy (herbaceous or woody), duration of flooding (permanent or seasonal) and the growth forms of the dominant vegetation. We stress upon the hydrological factors which determine all the structural and functional characteristics of the wetlands. We consider that the various growth forms of wetland vegetation integrate the totality of hydrological variables and therefore, can be used as the indicators of different hydrological regimes.

Introduction

During the past two decades wetlands have received increasingly greater attention, from the viewpoint of their ecology as well as conservation. Once treated as transitional habitats or seral stages in succession from open water to land, the wetlands are now considered to be distinct ecosystems with specific ecological characteristics, functions and values (Mitsch & Gosselink 1986; Patten *et al.* 1990). Efforts are being made to prepare inventories of wetlands in different parts of the world (Burgis & Symoens 1987; Scott 1989; Scott & Carbonell 1986; Lugo *et al.* 1989; Whigham *et al.* 1993) and identify the more important ones for conservation, particularly under the Ramsar Convention (Anonymous 1990a). The geomorphological, climatic, hydrological and biotic diversity of the Indian subcontinent has ensured a great diversity of its wetlands which include seasonally flooded as well as permanent marshes and swamps in shallow lakes, large river floodplains and littoral zones of large lakes and reservoirs. Along the entire coast there are several lagoons, estuarine backwaters and extensive mangroves in the

deltas and estuaries of numerous rivers. There are also marine wetlands which include coastal beds of marine algae, and coral reefs.

The ecology of both the freshwater wetlands and mangroves of the Indian subcontinent has been studied for many decades (Gopal 1982, 1990; Gopal & Krishnamurthy 1992). Several important wetland sites (e.g., Keoladeo Ghana at Bharatpur, Kaziranga in Assam, and Sunderban in Ganga-Brahmaputra delta) were designated as sanctuaries for the protection of their wildlife long before the need for wetland conservation was voiced by the international organisations. Whereas the distribution, habitat characteristics and biota of the mangroves have been relatively well documented (see Gopal & Krishnamurthy (1992) for a review of literature), the inland wetlands, both freshwater and saline, are, in general, poorly known. Until recently there had been no effort to prepare an inventory of the wetland resources and classify them suitably. An inventory of wetlands is necessary to evaluate the existing resources, their diversity, functions and values as well as the future qualitative and quantitative changes under the impact of various human activities. A classification is essential for the comparison of functions and values of different kinds of wetlands, selection of appropriate sites representative of different wetland types for conservation, and for developing scientifically sound management strategies.

In this paper, we examine some important features of inland wetlands and their recent inventories in India. Then we evaluate some of the wetland classification schemes for their applicability to Indian wetlands, and suggest a tentative scheme of classification.

Important characteristics of inland wetlands

Because the wetlands develop under, and are sustained by their specific hydrologic regimes, it is necessary to understand the peculiarities of the hydrological environment of India. The concentration of precipitation over most of the Indian subcontinent over a short period (June–September) which is preceded by a hot dry spell, and the large variability (20 to 100%) in annual precipitation have a direct bearing on the nature of the wetlands. Large seasonal and year to year variations occur in the water level of rivers, lakes and reservoirs. In terms of wetland habitats, a large majority of wetlands in the region is therefore not only seasonally temporary but many of them often appear and disappear in successive years. A dwarf sedge dominated marsh may turn into a terrestrial meadow or a submerged of floating-leaved macrophyte dominated shallow waterbody in another year.

Second, the vagaries of the monsoon have been responsible for extensive human regulation of water resources throughout the region, primarily for irrigation and domestic supplies, and secondarily for fisheries, hydropower and other uses. The flow of nearly all major and minor rivers has been regulated by large dams and barrages (Rao 1975) which have inevitably resulted in the loss of both upstream and downstream floodplains and their replacement by littoral marshes along the upstream reservoirs. Elsewhere, the surface runoff from small catchments has been impounded into tens of thousands of small reservoirs and village ponds (see also Sharma 1985; Fernando 1984). Still further, the naturally occurring marshes have been converted to fishponds (Jhingran 1982) and paddy fields. Thus, the vast majority of wetlands is man-made or man-modified.

The seasonal variability in precipitation and the regulation of water resources are common also in other regions with a subtropical and monsoonic climate. However, wetlands of the Indian subcontinent are unique due to their interaction with human populations for several millenia. The Indian subcontinent is among the oldest theaters of human history, and the second most populous region of the world. Wetlands and their resources have been an integral part of the social and cultural ethos of human societies in this region. Not only have wetlands influenced ways of human life, the people have traditionally depended on wetlands for millenia. Thus, even the most natural of the wetlands of the region, whether floodplains (e.g., those of the Brahmaputra basin) or marshes (e.g., chaurs of Bihar, lake Kolleru in Andhra Pradesh), swamps (e.g., in the foothills of Himalaya and in south India), coastal lagoons (like Chilika and Pulicat) or mangroves (including Sunderban) have been influenced by human use and management. The continued interaction between wetlands and the humans in various ways keeps their characteristics always changing, often unpredictably.

Wetland inventories

Although a few freshwater sites were listed in the Project Aqua (Luther & Rzoska 1971), it was only after India became a signatory to the Ramsar Convention in 1981, that efforts were initiated to prepare an inventory of wetlands. An earlier report on the estimates of wetland resources of the country (Biswas 1976) included all large lakes and reservoirs as well as small ponds and temple tanks as wetlands. It emphasized upon the nuisance growth of aquatic weeds, especially waterhyacinth, because it apparently relied on the information collected during a survey of the aquatic weeds (Varshney & Singh 1976). In 1984, a questionnaire survey was started by the Wetland Working Group nominated by the Department of Environment, Government of India. It failed to bear any result due to poor response. Later, the Asian Wetland Bureau conducted another survey and relied on the information from a large number of individuals who were involved in the study of different wetlands and waterfowl, and from various publications. This yielded a list of 93 wetlands some of which, in fact, represented a group of different kinds of wetlands spread over a very large area. The inventory included some information on their major biota, especially waterfowl, land use and human impacts as well (Wolstencroft et al. 1989). This inventory has been now updated by the WWF-India to include information on additional 77 wetland sites (DeRoy & Hussain 1993).

Another inventory published recently by the Ministry of Environment and Forests (Government of India) is merely a list of waterbodies larger than 100 ha, without any information about their water depth, biota or any other characteristic (Anonymous 1990b). It includes several hundred temple tanks and large irrigation and multipurpose reservoirs. The mangroves have been treated separately (Anonymous 1987).

These inventories highlight the problem of inadequate understanding of the definition and characteristics of a wetland. Wetlands are indeed difficult to define and none of the definitions has clearly set the limits of what constitutes a wetland. The definition adopted by the IUCN, and followed by the two Indian inventories, considers wetlands simply as bodies of water, albeit of a limited depth, and emphasises their importance as waterfowl habitats (IUCN 1971). The definition is incorrectly interpreted to include all deep water habitats, especially lakes and reservoirs. Neither the fact that wetlands are often associated with deep water, nor the presence of waterfowl alone, confers upon all deepwater bodies the status of a wetland.

The situation is further confounded by the multiplicity of vernacular terms, and more than 20 languages spoken in different parts of the country. Some of the vernacular terms used in northern India are listed in Table 1. Most of these terms do not distinguish between natural and man-made water bodies, between a deep and shallow waterbody or a marsh, or between a large and small waterbody.

It is, however, essential to distinguish between a wetland and a deep water body from the viewpoint of their management. A narrow fringing belt of vegetation characteristic of wetlands may occur along the margins of a large multipurpose reservoir or a temple tank. However, their recognition as wetlands would result in a conflict of interest in their use. Similarly, caution also needs to be exercised in recognising as wetlands the areas which are flooded seasonally for such short periods that the characteristic aquatic vegetation fails to develop there though the area may be visited by a number of waterfowl during the flooding period. Further, from the perspective of conservation, it is doubtful if the intensively managed crop fields (rice paddies and jute fields) as well as fish and shrimp ponds should be included in an inventory although ecologically, they are wetlands and sometimes important habitats of many rare plant and animal species.

Classification of wetlands

The classification of any set of objects depends upon the characteristics considered important to differentiate between units. For example, Euglenoid forms may be classified as plants or animals by emphasising their autotrophic or heterotrophic habit and accordingly placed under green algae or flagellates. In case of wetlands, different classification schemes have variously emphasised upon their geomorphological, hydrological, vegetational, or water chemistry and substrate characteristics, and accordingly, wetlands have been grouped together variously into several types or in a hierarchy of types (Gopal et al. 1990).

One of the simplest classifications is that of the IUCN which has been used in their inventory of Asian wetlands (Scott 1989). It recognises 22 categories of wetlands without a logical basis, and therefore, most of these are artificial and heterogeneous assemblages. For example, a category of 'freshwater ponds, marshes and swamps' is different from the 'shrimp ponds, fish ponds' on one hand and from the 'swamp forest, temporarily flooded forest' on the other. The 'rice paddies' category is distinguished from the 'flooded arable land' whereas the salt pans and salt lakes are placed in different categories.

The most comprehensive system developed todate is that of Cowardin et al. (1979) for the United States Fish and Wildlife Service (USFWS). It defines wetlands as '... lands transitional between terrestrial and aquatic systems where the water table is usually at or near the surface or the land is covered by shallow water'. It further states that 'wetlands must have one or more of the following three attributes: (1) at least periodically the land supports predominantly hydrophytes, (2) the substrate is undrained hydric soil, and (3) the substrate is non-soil and is saturated with water or covered with shallow water at some time during the growing season of the year'. Based on these criteria, an elaborate hierarchical classification has been developed and a series of systems, subsystems and classes has been recognized. The systems and subsystems are based on geomorphological criteria whereas the classes generally emphasise the nature of the substratum and physiognomy of the vegetation. Hydrology, water chemistry and soil characteristics are used as modifiers at class and subclass level.

Besides the fact that it covers also deep water bodies, a major limitation of the USFWS classification (Cowardin et al. 1979) is its complexity for practical purposes, especially due to the difficulty in defining

Table 1. Common vernacular terms used for waterbodies in northern India. (Terms used in Hindi speaking areas are marked with.*)

Bawri*	a large or small village pond
Bheel	a natural lake, generally an oxbow in Assam and West Bengal
(= Beel)	Examples: Deepor bheel, Sareswar bheel, Disama beel, Sanak beel
Bheri	a man-made pond, large or small, in West Bengal
Chaur	a shallow oxbow lake in Bihar
	Examples: Kesaria, Chatia, Bharthua, Bora chaurs
Diara	a periodically flooded sandy to silty island in the river
	(term used particularly in Bihar)
Jheel*	a large lake, generally deep
Khadar*	a periodically flooded, low-lying area, especially the floodplain
	of a river
Phat	a shallow lake in Manipur
	Examples: Loktak phat, Phumlen phat, Kharung phat
Pokhar*	a shallow pond or marsh
Sagar	a large man-made lake (like sea)
	Example: Hussain Sagar
Samand	a large natural or man-made lake
	Examples: Jai Samand and Raj Samand in Rajasthan
Sar	a natural lake in Jammu & Kashmir and Himachal Pradesh
	Examples: Shansar, Waskursar, Hokarsar, Mansar, Surinsar,
	Khushalsar
Sarovar*	a natural or man-made lake
	Examples: Nal Sarovar and Sardar Sarovar in Gujarat
Tal*	a large natural lake or reservoir
	Examples: Nainital, Sattal, Naukuchiyatal, Suraha tal, Gujar tal
	in U.P., Bhopal tal in M.P. and Kabar tal in Bihar
Talab*	a smaller reservoir or pond
Tso	a deep or shallow lake in the Indo-Tibetan region
	Examples: Pongang Tso, Tso Morari

a hydrophyte (see Tiner 1991, 1993). This has been a major problem for the delineation of wetlands in the United States itself where it has been partly overcome by providing a list of plants considered to be hydrophytes (Reed 1988; Sipple 1988). Further, the importance of water regimes, water chemistry, soil properties and human impacts has been relegated to that of a 'modifier' factor below the level of classes. Interestingly, river floodplains are included under the palustrine system whereas the former should be considered a subsystem of the riverine system equivalent to the 'littoral' subsystem of the lacustrine system. It needs to be emphasised that in the tropical regions such as the Amazon basin, and under monsoonic climates with high temporal variability in precipitation, floodplain wetlands may appear to be palustrine systems in different years or even in different seasons. However,

their functions and values are directly influenced by their riverine interactions.

Among several other classification schemes, those proposed for the Australian wetlands deserve some consideration due to some similarities in the monsoonic climate. Stanton (1975) classified wetlands in Queensland into tidal and inland wetlands which were further divided on the basis of either vegetation (mangroves, salt marshes, salt mudflats and saltwater meadows) or the duration and frequency of flooding. Later, Briggs (1981) classified wetlands simply on the basis of their vegetation taking into account their physiognomy (forest, woodland, savanna or herbland) and the dominant taxa. The more recent system developed by Paijmans *et al.* (1985) takes into consideration both vegetation and hydrology. However, they define lakes as 'areas of open water generally more than 1 m deep when full, and with little or no persistent vegetation' whereas swamps are 'less than 1 m deep and have persistent vegetation' which may be woody or herbaceous. The scheme does not distinguish between lacustrine and riverine wetlands. The duration and frequency of flooding have been emphasised.

Although Champion and Seth (1968) recognised a number of kinds of wetland forests in their classification of forest types of India, and the mangroves have been variously classified as well (see Gopal & Krishnamurthy 1992), there has been no effort so far to develop a classification scheme for all Indian wetlands. The USFWS system (Cowardin *et al.* 1979) has also not been evaluated. In the absence of adequate basic data on the hydrology, water chemistry, soils, and vegetation in a number of representative wetlands from all parts of the country, it is indeed difficult to classify these habitats suitably. Under the circumstances, it is considered desirable to adopt a relatively simple classification which can be readily understood and applied in the field.

Proposed classification

We propose here a simple hierarchical classification (Table 2) which is modified from the tentative scheme suggested recently by Gopal and Krishnamurthy (1992). We hope that it would generate some discussion, would be tested in the field and gradually evolve into an acceptable system for the Indian subcontinent. In the proposed scheme, greater importance is attached to the hydrological factors and the vegetation types which develop under specific hydrological regimes.

In the coastal areas, tides constitute a major hydrological variable which distinguishes both the marine and estuarine areas from the remaining inland regions. Therefore, first a distinction is made between the tidal and inland wetlands. Among the water chemistry parameters, salinity is of greater importance in determining the nature of biota than nutrients and other pollutants. Therefore, the inland wetlands are distinguished between the saline and freshwater wetlands. However, the salinity in wetlands under tidal influence does not simply exhibit a gradient but changes with time (due to evaporation) and therefore, a clear distinction between different salinity levels is not possible.

Among other hydrological parameters, the depth, duration, frequency, and season of flooding (including waterlogging) are most important variables that influence all other physical, chemical and biological attributes of different wetlands. The source of water (surface runoff, flooding from rivers, or association with standing water) is of secondary importance as far as the development of specific vegetation is concerned. Wetlands with similar species composition and physiognomy develop in different areas with similar hydrological regimes. The effects of depth and duration of flooding and its seasonal amplitude have been discussed in several publications (Hejny 1971; Gopal & Sharma 1990; Van der Valk 1991; Ekstam & Weisner 1991). However, our knowledge of the responses of different wetland taxa to flooding is too meager and restricted to herbaceous vegetation which includes reeds, cattails, sedges and various submerged and floating leaved species, besides a few mangrove species. It is interesting, however, that both woody and herbaceous species may occur over a wide range of hydrological regimes. Therefore, we recognise Herbaceous and Woody groups among both the saline and freshwater inland wetlands as well as the tidal wetlands. We suggest that as far as possible all herbaceous and woody wetlands should be called 'marshes' and 'swamps' respectively (see Mitsch & Gosselink 1986; Gopal *et al.* 1990). These groups are further divided into permanent and seasonally flooded categories within which a number of wetland types can be recognised.

The permanently flooded tidal wetlands with herbaceous vegetation include coastal beds of kelps and seagrasses, lagoons, estuaries and backwaters. The vegetation is predominantly submerged though some sedges do occur along the margins where water level changes are large. Seasonal tidal flooding may occur in areas influenced by cyclones and high tidal waves during the monsoon season. There is no information on such areas. The halophytic herbaceous growth on dunes behind the sea beaches may as well be considered here. Among woody wetlands, the mangroves of various kinds occur in permanently waterlogged or flooded areas and can be divided along a salinity gradient, besides a zonation within a mangrove type. These types are discussed in detail by Untawale (1987) and Gopal and Krishnamurthy (1992). In seasonally flooded areas behind the mangroves are saline scrubs.

There are relatively few inland saline wetlands which are also limited to certain areas. The seasonally flooded woody type occurs widely in the Rann of Kutch. The littoral zones of several high altitude lakes, particularly those in the Indo-Tibetan region, represent the permanently waterlogged inland saline marshes whereas several salt lakes in Rajasthan (west-

Table 2. Proposed classification of wetlands in the Indian subcontinent.

I. TIDAL WETLANDS
- a. Herbaceous (mostly submerged) vegetation
  - i. Permanently flooded (or waterlogged)
    - Coastal beds of kelps and seagrasses
    - Lagoons (Chilika, Pulicat)
    - Estuaries and Backwaters (many in TamilNadu and Kerala)
  - ii. Seasonally flooded (or waterlogged)
    - (May include areas flooded by very high tides)
- b. Woody vegetation
  - i. Permanently flooded (or waterlogged)
    - Mangroves
    - Mangrove scrub
    - Saltwater mixed forest (*Heritiera*)
    - Brackishwater mixed forest (*Heritiera*)
    - Palm swamp (*Nypa*)
  - ii. Seasonally flooded (or waterlogged)
    - Saline scrubs

II. INLAND WETLANDS
- A. Saline wetlands
  - a. Woody vegetation
    - i. Permanently flooded (or waterlogged)
      - There are none
    - ii. Seasonally flooded (or waterlogged)
      - Saline scrubs (e.g., Rann of Kutch)
  - b. Herbaceous vegetation (submerged or other halophytes)
    - i. Permanently flooded (or waterlogged)
      - Saline high altitudes lakes (littoral zones only)
    - ii. Seasonally flooded (or waterlogged)
      - Saline lakes (e.g., Sambhar, Pachpadra, Deedwana)

ern India) are examples of seasonally flooded inland saline wetlands.

The woody freshwater wetlands, permanent or seasonally flooded, are known only from their brief description provided by Champion and Seth (1968). The updated information on different types of these wetlands (Table 2) is given by Gopal and Krishnamurthy (1992).

The herbaceous freshwater wetlands can be divided into several types on the basis of dominant growth forms of aquatic and semi-aquatic vegetation. Several growth forms have been recognised by Hejny (1971) on the basis of their responses to the duration and depth of flooding. In fact, the growth forms reflect the integrated response of the species to the hydrological conditions over their life span. In other words, the growth forms in general, and certain taxa in particular, can be used as indicators of changing hydrological regimes over time. It should, however, be stressed that most of the

growth forms are adapted to a range of hydrological conditions, and therefore, occur in both the permanent and seasonally flooded wetlands. Factors other than hydrology (such as substratum, nutrients and biotic interactions) also often affect their distribution.

In the proposed classification, we recognise provisionally only eight wetland types related to the dominant growth forms and taxa in the Indian wetlands. These types can be arranged along a combined gradient of depth and duration of flooding as conceptualised in Fig. 1.

The submerged and floating-leaved species occur only in shallow waters and often dominate the relatively deeper parts of the littoral zone which are infrequently exposed and where periodic deep flooding prevents the growth of the emergent plants. They cease to grow vegetatively when the water level drops to a few centimetres above the soil surface. However, most of the submerged plants flower and produce seeds only

Table 2. Continued.

B. Freshwater wetlands
   a. Woody vegetation
      i. Permanently flooded (or waterlogged)
         Myristica swamp (*Myristica* species)
         Submontane hill valley swamp (*Bischofia, Alstonia, Salix*)
         Creeper swamp (incl. cane brakes) (*Magnolia, Eugenia, Calamus*)
      ii. Seasonally flooded
         Eastern seasonal swamp (*Machilus gamblei, Elaeocarpus* sp.)
         Barringtonia swamp (*Barringtonia acutangula*)
         Syzygium cumini swamp (*Syzygium cumini*)
         Seasonal low swamp forest (*Cepahalanthus occidentalis*)
         Eastern *Dillenia* swamp (*Dillenia indica, Bischofia javanica*)
         Riparian fringing forests (*Tamarix dioica, Terminalia* sp.)
         Alder forests (*Alnus nepalensis*)
         Riverine blue pine forests
         Wet Bamboo brakes (*Bambusa, Neohouzeaua*)
   b. Herbaceous vegetation
      i. Permanently flooded (or waterlogged)
         Submerged and/or floating leaved
         Cattails (mainly *Typha angustata*)
         Reeds (*Phragmites karka, P. australis, Arundo donax*)
         Tall Emergent (other than reeds and cattails)
             (e.g., *Ipomoea fistulosa*)
         Tall sedges (*Scirpus, Cyperus, Eleocharis*)
      ii. Seasonally flooded (or waterlogged)
         Submerged and/or floating leaved
         Cattails (*Typha elephantina*)
         Reeds (*Phragmites karka, Arundo donax*)
         Tall Emergent (other than reeds and cattails)
             (e.g., *Ipomoea fistulosa*)
         Tall sedges (*Scirpus, Cyperus, Eleocharis*)
         Short sedges and grasses (*Kyllinga, Eleocharis, Fimbristylis*
               Paspalum, Echinochloa)
         Wet meadows (mostly forbs, *Cynodon*)
         Tall grasses (*Vetiveria, Erianthus, Saccharum*)

in shallow waters. Their vegetative propagules do not survive desiccation but the seeds may give rise to new growth after a prolonged dry period (Chamanlal & Gopal 1993).

Among the emergents, *Typha angustata* occurs in shallow waters as well as on waterlogged soils but does not survive in areas which become completely dry for even a few weeks. *Typha elephantina* thrives well in periodically exposed areas due to its deep seated rhizomes. It occurs in permanent shallow waters (less than 1 m) only in the absence of *T. angustata*. The reed, *Phragmites karka*, requires flooding for several months but continuous flooding adversely affects its growth. Therefore, it occurs on permanently waterlogged and periodically flooded sites but rarely forms large stands in permanent waters. Various populations of *P. karka* and other reeds (*P. australis, Arundo donax, Ochlandra* sp.), however, vary in their responses to different hydrological regimes. Tall emergents, other than reeds and cattails, which grow well in permanent as well as temporary waters include *Ipomoea fistulosa*, a naturalised exotic species.

The tall sedges occur as emergents in shallow water (< 60 cm) habitats which may remain periodically only waterlogged. Their vegetative propagules may survive in small waterlogged pockets in the seasonally

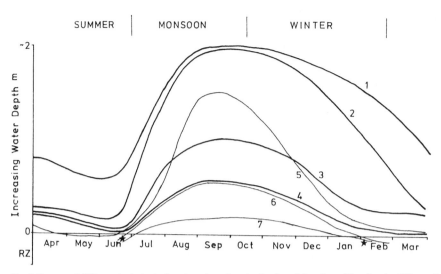

SUMMER | MONSOON | WINTER

Fig. 1. A conceptualised diagram of different hydrological regimes based on the depth and duration of flooding in different freshwater wetlands in the tropical and subtropical climate in the Indian subcontinent. The length of the rainy season varies according to the time of the onset and retreat of the southwest monsoon in different parts of the subcontinent. Habitats with only two metres deep water (including the littoral of large lakes and reservoirs) are considered. Zero (0) marks the level of the substratum. The seven hydrological regimes are: 1. Permanent flooding with small water level changes, 2. Permanent flooding with large water level changes, 3. Permanent waterlogging with periodic moderate flooding, 4. Permanent waterlogging with periodic shallow flooding, 5. Seasonal deep flooding, 6. Seasonal shallow flooding, and 7. Seasonal waterlogging. An * denotes that the water level falls below the surface and the substratum dries upto the rooting zone (RZ). The hydrological regimes (in parentheses) associated with the vegetation types of herbaceous wetlands in the proposed classification are: Submerged and/or floating leaved (1,2,3,5), Cattails (2,3,4,5), Reeds (2,3,4,5), other Tall emergents (1,2,3,4,5), Tall sedges (3,4,5), Short sedges and grasses (6), Wet meadows (6,7), and Tall grasses (7).

dry habitats and give rise to large stands after flooding. Next on the hydrological gradient lie short sedges and grasses. They are confined in permanent wetlands to waterlogged areas which may experience shallow flooding (< 15 cm) for only short periods (few weeks). Most of them propagate by seeds and therefore, occur predominantly in seasonal wetlands with prolonged dry periods. The habitats with prolonged waterlogging but without shallow flooding are dominated by forbs and some grasses like *Cynodon dactylon*. The tail end of the gradient is represented by shallow flooding for such short periods that the substrate does not remain waterlogged for more than a few weeks. These habitats are generally dominated by tall coarse grasses like *Vetiveria zizanioides*, *Erianthus* sp. and *Saccharum spontaneum*. These grasses are also a major constituent of the seasonally flooded savannas which should therefore be included within this type.

Field application on the proposed classification

We believe that the proposed classification is simple enough for quickly recognising a wetland type from its location, physiognomy and vegetation type. However, we recognise the fact that many wetland sites such as floodplains include several physiographically different areas which differ in their hydrological regimes, and therefore, in their vegetation types. Further, depending upon the morphology of a waterbody and the amplitude of seasonal water level changes, zones of two or more different vegetation types may be formed within a wetland. In such situations, all the major vegetation types within a wetland should be recognised together with their distribution or zonation.

We wish to point out that the proposed classification does not emphasise the anthropogenic factors but this does not underestimate their importance. As mentioned earlier, most of the wetlands in the subcontinent have co-evolved with human intervention, and therefore, it is difficult to distinguish between the natural wetlands (like Wular, Kaziranga, Kolleru, Chilika and Sambhar) and those which are either extensively modified by humans (e.g., Keoladeo Ghana at Bharatpur) or are man-made (littorals of reservoirs like Harike) wetlands. However, the role of human beings, particularly the traditional management, needs to be recognised together with other wetland characteristics in all conservation and management efforts (Gopal 1991). Therefore, the distinction between the natural (with traditional use and management) and man-made wetlands should be identified during the preparation of an inventory. This can also be incorporated in the classification as a 'modifier' with most of the types proposed here. Detailed comparative studies are required to bring out clearly the differences in the structure and function of various wetland types.

Acknowledgements

The paper was presented at the INTECOL's IV International Wetlands Conference by the first author (BG) who is grateful to Prof. W. J. Mitsch and Prof. R. E. Turner for providing necessary financial support for his participation in the Conference. We are grateful to Prof. A. van der Valk and the anonymous reviewers for their many valuable suggestions.

References

Anonymous. 1987. Mangroves in India: Status report. Ministry of Environment & Forests, New Delhi. 150 pp.

Anonymous (Ed.). 1990a. Proceedings of the Fourth Conference of the Contracting Parties to the Convention on Wetlands of International Importance Especially as a Waterfowl Habitat, Montreux (Switzerland). Vol. I. Ramsar Convention Bureau, IUCN, Gland, Switzerland. 306 pp.

Anonymous. 1990b. Wetlands of India: a directory. Ministry of Environment & Forests, New Delhi. 150 pp.

Biswas, B. 1976. India – National report. In: Smat, M. (Ed.) Proceedings of the International Conference on Conservation of Wetlands and Waterfowl, Heiligenhafen, West Germany: 108–109. International Waterfowl Research Bureau, Slimbridge, U.K.

Briggs, S. V. 1981. Freshwater wetlands. In: Groves, R. H. (Ed.) Australian Vegetation: 335–360. Cambridge University Press, London.

Burgis, J. J. & Symoens, J. J. (Eds). 1987. African Wetlands and Shallow Water Bodies: A Directory. Editions de l'ORSTROM, Paris. 649 pp.

Chaman Lal & Gopal, B. 1993. Production and germination of seeds in *Hydrilla verticillata*. Aquatic Botany 45: 257–261.

Champion, H. G. & Seth, S. K. 1968. A Revised Survey of the Forest Types of India. Manager of Publications, New Delhi. 404 pp.

Cowardin, L. M., Carter, V., Golet, F. C. & LaRoae, E. T. 1979. Classification of wetlands and deepwater habitats of the United States. US Fish & Wildlife Service, Washington, D.C. 103 pp.

DeRoy, R. & Hussain, S. A. (Eds). 1993. Directory of Indian Wetlands. WWF-India, New Delhi and Asian Wetland Bureau, Kuala Lumpur. 240 pp.

Ekstam, B. & Weisner, S. E. B. 1991. Dynamics of emergent vegetation in relation to open water of shallow lakes. In: Finlayson, C. M. & Larsson, T. (Eds) Wetland Management and Restoration: 56–64. Swedish Environmental Protection Agency, Solna, Sweden.

Fernando, C. H. 1984. Reservoirs and lakes of southeast Asia (Oriental region). In: Taub, F. B. (Ed.) Lakes and Reservoirs. Ecosystems of the World 23: 411–446. Elsevier Science Publishers, Amsterdam.

Gopal, B. 1982. Ecology and management of freshwater wetlands in India. In: Proceedings of the International Scientific Workshop (SCOPE-UNEP) on Ecosystem Dynamics in Freshwater Wetlands and Shallow Waterbodies: 127–162. Center of International Projects, GKNT, Moscow, USSR.

Gopal, B. (Ed.). 1990. Ecology and Management of Aquatic Vegetation in the Indian Subcontinent. Kluwer Academic Publishers, Dordrecht. 250 pp.

Gopal, B. 1991. Wetland (mis)management by keeping people out: Two examples from India. Landscape and Urban Planning 20: 53–59.

Gopal, B. & Krishnamurthy, K. 1992. Wetlands of south Asia. In: Wigham, D. F., Dykjova, D. & Hejny, S. (Eds) Wetlands of the World. I. Inventory, Ecology and Management. Handbook of Vegetation Science 15(2): 345–416. Kluwer Academic Publishers, Dordrecht, The Netherlands.

Gopal, B., Kvet, J., Löffler, H., Masing, V. & Patten, B. C. 1990. Definition and classification. In: Patten, B. C., Jorgenson, S. E., Dumont, H. J., Gopal, B., Koryavov, P., Kvet, J., Löffler, H., Sverizhev, Y. & Tundisi, J. G. (Eds) Wetlands and Continental Shallow Water Bodies: 9–15. SPB Academic Publishing, The Hague.

Gopal, B. & Sharma, K. P. 1990. Ecology of plant populations. I. Growth. In: Gopal, B. (Ed.) Ecology and Management of Aquatic Vegetation of the Indian Subcontinent: 77–105. Kluwer Academic Publishers, Dordrecht, The Netherlands.

Hejny, S. 1971. The dynamic characteristics of littoral vegetation with respect to changes of water level. Hidrobiologia, Bucharest 12: 71–86.

IUCN. 1971. The Ramsar Conference: Final act of the International Conference on the conservation of wetlands and waterfowl. IUCN Bulletin 2, Special supplement: 1–4.

Jhingran, V. G. 1982. Fish and Fisheries of India. 2nd edition. Hindustan Publ. Corp., New Delhi. 660 pp.

Lugo, A. E., Brown, S. & Brinson, M. M. (Eds). 1989. Ecosystems of the World, Forested Wetlands. Elsevier Science Publishers, Amsterdam.

Luther, J. & Rzoska, J. (Eds). 1971. Project Aqua. A source book of inland waters proposed for conservation. IBP Handbook 21. Blackwells, Oxford.

Mitsch, W. J. & Gosselink, J. G. 1986. Wetlands. Van Nostrand Reinhold, New York. 539 pp.

Paijmans, K., Galloway, R. W., Faith, D. P., Fleming, P. M., Haantjens, H. A., Heyligers, P. C., Kalma, J. D. & Loffler, E. 1985. Aspects of Australian Wetlands. CSIRO Division of Water and Land Resources Technical paper no. 44. Canberra, Australia. 77 pp.

Patten, B. C., Jorgensen, S. E., Dumont, H. J., Gopal, B., Koryavov, P., Kvet, J., Löffler, H., Sverizhev, Y. & Tundisi, J. G. (Eds). 1990. Wetlands and Continental Shallow Water Bodies. Vol. I SPB Academic Publishing, The Hague. 759 pp.

Rao, K. L. 1975. India's Water Wealth. Orient Longman, New Delhi. 255 pp.

Reed, P. B. Jr. 1988. National list of plant species that occur in wetlands. National summary. US Fish & Wildlife Service. Washington, D.C. US Biological Report 88(24).

Scott, D. A. (Ed.). 1989. A Directory of Asian Wetlands. IUCN, Gland, Switzerland. 1181 pp.

Scott, D. A. & Carbonell, M. (Eds). 1986. A Directory of Neotropical Wetlands. IWRB, Slimbridge and IUCN, Gland, Switzerland. 714 pp.

Sharma, K. P. 1985. Impact of river basin development projects in India with reference to fisheries development. Proceedings of the International Seminar on Environmental Impact Assessment of Water Resources Projects, vol. 1: 25–32. University of Roorkee, Roorkee, India.

Sipple, W. S. 1988. Wetland Identification and Delineation Manual. 2 vols. US Environmental Protection Agency, Office of Wetlands Protecton, Washington, D.C., USA.

Stanton, J. P. 1975. A preliminary assessment of wetlands in Queensland. CSIRO Division of Land Use Research Technical Memorandum 75/10. Canberra, Australia. 42 pp.

Tiner, R. W. Jr. 1991. The concept of a hydrophyte for wetland identification. BioScience 41: 236–247.

Tiner, R. W. Jr. 1993. Wetlands are ecotones: Reality or myth? In: Gopal, B., Hillbricht-Ilkowska, A. & Wetzel, R. G. (Eds) Wetlands and Ecotones: Studies on Land-Water Interactions: 1–15. National Institute of Ecology, and International Scientific Publications, New Delhi.

Untawale, A. G. 1987. Country reports: India. In: Umali, R. M., Zamora, P. M., Gotera, R. R., Jara, R. S. & Camacho, A. S. (Eds) Mangroves of Asia and the Pacific: Status and Management: 51–88. National Mangrove Committee of Philippines, Quezon City, Manila.

Van der Valk, A. G. 1991. Response of wetland vegetation to a change in water level. In: Finlayson, C. M. & Larsson, T. (Eds) Wetland Management and Restoration: 7–16. Swedish Environmental Protection Agency, Solna, Sweden.

Varshney, C. K. & Singh, K. P. 1976. A survey of aquatic weed problem in India: In: Varshney, C. K. & Rzoska, J. (Eds) Aquatic Weeds in South East Asia: 31–42. Dr. W. Junk Publishers, The Hague.

Whigham, D. F., Dykyjova, D. & Hejny, S. (Eds). 1993. Wetlands of the World. I. Inventory, Ecology and Management. Handbook of Vegetation Science 15(2). Kluwer Academic Publishers, Dordrecht, The Netherlands.

Wolstencroft, J. A., Hussain, S. A. & Varshney, C. K. 1989. India. In: Scott, D. A. (Ed.) A Directory of Asian Wetlands: 367–505. IUCN, Gland, Switzerland.

Vegetatio **118**: 49–56, 1995.

Ecological significance and classification of Chinese wetlands

J. Lu

Institute of Estuarine and Coastal Research, East China Normal University, Shanghai 200062, China

Key words: China, Coastal, Estuarine, Inventory, Lacustrine, Peatlands, Riverine, Wetlands

Abstract

China supports a great variety of wetlands, including some of the most important in the world. However, an appropriate classification system applicable to all wetlands is not available. Based on a preliminary inventory, a new classification system for Chinese wetlands is proposed. This system classifies natural wetlands into three categories on the basis of their natural features and distribution: peatlands, coastal and estuarine wetlands, and riverine and lacustrine wetlands. Each category is divided into several sub-classes. The areal extent of wetlands in each Province has been estimated and their ecological importance assessed.

Introduction

During the past ten years several national studies on the conservation and management and natural functions of wetlands have been made in China (Lu 1988; Lu 1990). Most of these studies require a clear definition of what constitutes a wetland in China. As more and more people become involved in wetland studies the need for an appropriate classification system becomes more essential. However, up to now it has not been possible to agree on a classification system that is acceptable to all Chinese wetland scientists. Traditionally, peatlands have received the most attention (Long *et al.* 1983; Niu & Ma 1985) and these have been classified into three categories (Table 1) according to their peat content and pH value, and 15 sub-types according to floral features. During the 1986–89 wetland inventory the system utilised by Scott (1989) for an Asian wide inventory was adopted. In this inventory 22 categories of wetlands were recognised (Table 2).

Based on a nationwide investigation of the coastline and offshore islands, several comprehensive scientific surveys of remote areas, such as Tibet, Xinjiang and Inner Mongolia, and other general surveys made by the author and colleagues, a preliminary inventory of wetlands in China was completed in 1990 (Lu 1990). During this inventory, the different types of wetlands in 26 of the 30 provinces in the country were surveyed. Individuals and institutions with information on particular sites were contacted or interviewed directly and a new classification of Chinese wetlands proposed (Table 3). This system considers the natural features of the wetlands and their distribution. The system is described in this paper and the ecological significance of wetlands discussed.

Results and discussion

In total, over 62 million ha of wetlands were surveyed; amongst these 25 million ha were natural wetlands and about 36 million ha artificial. The natural wetlands include about 11 million ha of peatland, more than 2 million ha of coastal wetlands, and more than 12 million ha of riverine and lacustrine wetlands. The artificial wetlands include 34 million ha of paddy fields (rice, sugar cane, and areas of aquatic vegetation), more than 2 million ha of aquacultural ponds, and saltpans. The extent of natural wetland in each Province is shown in Table 4.

Peatlands

Peatlands are distributed widely across China. However, three quarters of the total area is found in Eastern China. The Qinghai-Tibetan Plateau has about 20%

50

Table 1. Classification of peatlands in China (long *et al.* 1983).

Type 1:	Eutrophic mire (Peat Content > 7.0%, pH: 5.6–7.7)	
	Subtype 1:	floated grassy mire
		Carex lasiocarpa/Carex pseudocuraica
	Subtype 2:	clumped grassy mire
		Deyeuxia angustifolia, Carex schmidtii/
		Carex meyeriana/Carex muliensis/
		Juncus sp.*, Carex argyi/Carex japonica*
	Subtype 3:	*Cladium* mire
		Cladium chinense
	Subtype 4:	*Scirpus* mire
		Scirpus/Scirpus tabernaemontani
	Subtype 5:	*Phragmites* mire
		Phragmites communis
	Subtype 6:	*Kobresia – Carex* mire
		Kobrisia littledalie, Carex moorcroftii/
		Kobrisia tibetica, Carex muliensis
	Subtype 7:	*Bueckea* mire
		Bueckea tvutesccens
	Subtype 8:	*Alnus* mire
		Alnus sibirica, Alnus hirsuta/
		Alnus trobeculosa
	Subtype 9:	*Betula* mire
		Betula fruticosa, Betula ovalifolia
	Subtype 10:	larch – grassy mire
		Larix olgensis/Larix gmelinii,
		Betula ovalifolia/Betula fruticosa, Carex sp.
Type 2:	Mesotrophic mire (Peat Content 5.1–7.0%, pH: 4.6–5.5)	
	Subtype 11:	clumped grassy – moss mire
		Carex sp. *Sphagnum apiculatum*
	Subtype 12:	floated grassy – moss mire
		Carex sp. *Sphagnum oligoporum*
	Subtype 13:	larch – moss mire
		Larix sp. *Sphagnum* sp. *Polytrichum* sp.
Type 3:	Oligotriphic mire (Peat Content < 5.0%, pH: 3.5–4.5)	
	Subtype 14:	*Sphagnum* mire
		Sphagnum sp.
	Subtype 15:	*Rhynchospora – Sphagnum* mire
		Rhynchospora chinensis, Sphagnum sp.

of the peatlands with another 5% in the remote highlands of the Tian Shan and Altai Shan. Small areas of peatland are scattered all across the country with the largest amounts being found in the mountains and plateaux, with very small amounts on the plains. For example, peatlands are well developed in Da Hinggan Ling, Xiao Hinggan Ling and Changbai Mountains in north-east China, Tian Shan, Altai Shan, Qilian Mountains in north-west China, Yan Taihang Mountains in north China, Xi Shan, Lu Shan, Huang Shan, Jinggang Shan, Wugong Shan, Qin Ling and Shennongjia Mountains in central China, and Daliang Shan and Xiaoliang Shan in south-west China. Mountain peatlands account for about 60% of the total area compared to 23% in the plateaux. Two thirds of the peatlands on

Table 2. Classification of wetlands in China used in 1986–1989 inventory (Scott 1989).

1.	shallow sea bays and straits (under six metres at low tide)
2.	estuaries, deltas
3.	small offshore islands, islets
4.	rocky sea coasts, sea cliffs
5.	sea beaches (sand, pebbles)
6.	intertidal mudflats, sand flats
7.	mangrove swamps, mangrove forest
8.	coastal brackish and saline lagoons and marshes
9.	salt pans (artificial)
10.	shrimp ponds, fish ponds
11.	river, streams – slow-flowing (lower perennial)
12.	river, streams – fast-flowing (upper perennial)
13.	oxbow lakes, rivering marshes
14.	freshwater lakes and associated marshed (lacustrine)
15.	freshwater ponds (under 8 hectares), marshes, swamps (palustrine)
16.	salt lakes, saline marshes (inland drainage systems)
17.	water storage reservoirs, dams
18.	seasonally flooded grassland, savanna, palm savanna
19.	rice paddies
20.	flooded arable land, irrigated land
21.	swamp forest, temporarily flooded forest
22.	peat bogs

the plains are on the Shan Jiang Plain (Three Rivers Plain) in north-east China.

In eastern China there is a greater variety of peatlands than occurred in western China. Forested, grassy or moss dominated mires with eutrophic, mesotrophic or ologotrophic conditions are common in the eastern mountains. Only grassy or reed dominated mires with eutrophic conditions are found in the western regions.

Peatlands not only provide rich natural materials, such as peat, reed, medicinal herbs etc., but also improve the local climate. Recently, the ability of peatlands to ameliorate waste waters has received some attention.

Coastal and estuarine wetlands

The coastline of China extends for about 18 000 km along the shores of the Yellow Sea, East China Sea and South China Sea, and includes some 5000 offshore islands (Chen *et al.* 1989). Seven major types of wetlands are included in this broad category: deltas and bays, tidal mudflats, grassy and reed-bed salt marsh-es, mangrove swamps, sand beaches, rocky sea coasts, and offshore islets. It is difficult to estimate the total area of wetland under the criteria used by Scott (1989) which includes all areas down to 6 m depth at low tide. An estimated 2 million ha of wetland occur down to the low tide mark. These wetlands vary with those north of hangzhou Bay being sandy or muddy, except for the Liaodong and Shandong Peninsulas which are rocky. About 1.6 million ha of coastal marshes and mudflats occur in this region: in the estuaries of the rivers along the coastline of Liaoning Province, around the estuary of the Yellow River in Bohai Gulf (Tianjin Municipality, Hebei Province and Shandong Province), and in the estuary of the Yangzte River and along the adjacent coast of Jiangsu Province.

Most of the rivers flowing into the Yellow Sea carry large amounts of sediment, resulting in rapid accretion of deltas and the continuous creation of new wetlands. In some areas (e.g. the coast at Dongtai County, Jiangsu Province) the rate of coastal accretion exceeds 400 mm y^{-1}. Mudflats in this area support a variety of molluscs and are highly productive. Besides providing wintering areas for large numbers of waterfowl (swans, geese and ducks), the coastal marshes and mudflats are extremely important as staging areas for shorebirds during the migration season, and as breeding habitat for a variety of species, including the rare Saunder's Gull *Larus saundersi*. The coastal marshes also produce a large amount of reed and grass that is sued as stock feed and the raw material for paper-pulp manufacture. Recent studies have indicated that *Spartina* can be planted in this biome and provide protection for the coastline and promote siltation (Zhong 1985).

The coast to the south of Hangzhou Bay is mainly rocky, with extensive wetlands confined to the mouths of the larger rivers, e.g. in the estuaries of the Pearl River in Guangdong Province, and the Min River in Fujian Province. Mangroves (with 30 species, Table 5) occur patchily along the coast as far north as central Fujian Province and around islands in the South China Sea. The total area of mangroves has declined from 67 000 ha in the 1970s to less than 15 000 ha in 1990, and much of that in a very degraded condition (Lin 1984, 1990). Some of the best remaining stands are found on Hainan Island, e.g. in Dongzaigang Natural Reserve which has now been designated as internationally important under the Ramsar Convention. The whole region is very important for fisheries production. Locally, the mangroves, fish ponds and rice paddies support large numbers of herons and egrets (e.g. in the Pearl River delta), while coastal mudflats are

Table 3. A proposed classification system of wetlands in China.

1. Natural wetlands		
System 1:	coastal and estuarine wetlands	
	Type 1:	deltas and bays
	Type 2:	tidal mud and sand flats
	Type 3:	grassy and reed-bed salt marshes
	Type 4:	mangrove swamps
	Type 5:	sea beashes (sand, pebbles)
	Type 6:	rocky sea coasts, sea cliffs
	Type 7:	small offshore islands, islets
System 2:	riverine and lacustrine wetlands	
	Type 8:	lake and river islets
	Type 9:	shollow freshwater lakes
	Type 10:	shollow brackish and saltwater lakes
	Type 11:	flood plain
	Type 12:	seasonal lakes and riverbeds
	Type 13:	wet meadows
	Type 14:	temporaily flooded forest
	Type 15:	silted reservoirs
System 3:	peat bogs	
	Type 16:	floated grassy bogs
	Type 17:	clumped grassy bogs
	Type 18:	stretched grassy bogs
	Type 19:	reed bogs
	Type 20:	shrub bogs
	Type 21:	wood bogs
	Type 22:	moss bogs
2. Artificial wetlands		
	Type 1:	paddy fields (rice, surgecane, aquatic vagetable)
	Type 2:	aquatic culture ponds (shrimp, fish and clam ponds)
	Type 3:	water storage reservoirs
	Type 4:	salt pans

important as staging and wintering areas for migratory shorebirds, gulls and terns.

Riverine and lacustrine wetlands

China has more than 50 000 rivers with a catchment basins in excess of 100 km², and over 1500 with a basin area of more than 1000 km². Wetlands along these rivers can be subdivided into two types. First are those in peripheral drainage regions, and second, those in endorheic regions (Fig. 1). Both include river and lake islets, shallow lakes (freshwater, brackish or saltwater), flood plains, seasonal rivers and lakes, wet meadows, temporarily flooded forests, and silted reservoirs.

Wetlands in peripheral drainage basins

Peripheral drainage basins cover 6.12 million km², about 64% of the total territory of China. This encompasses 23 major rivers; 18 of these are part of the hydrographical network of the Pacific Ocean, 4 of the Indian Ocean and 1 of the Arctic Ocean. The hydrographical net of the Pacific Ocean includes the Nenjiang-Songhuajiang-Heilongjiang Basin in northwestern China and the Yangtze-Yellow-Huaihe Basin. The largest freshwater marshes in China are found in the former basin, in the Provinces of Heilongjiang,

Table 4. Estimate of wetland area in each province of China.

	Province	TOTAL AREA (1,000 ha.)	WETLAND AREA (1,000 ha.)	%
1.	Anhui	13,900	400 – 500	3.0
2.	Beijing	1,700	30 – 40	2.1
3.	Fujian	12,100	100 – 120	0.9
4.	Gansu	45,100	20 – 25	0.05
5.	Guangdong	19,00	500 – 550	2.7
6.	Guangxi	23,600	150 – 200	0.7
7.	Guizhou	17,000	15 – 20	0.1
8.	Hainan	3,200	10 – 15	0.4
9.	Hebei	19,000	200 – 250	1.2
10.	Heilongjian	46,900	5,800 – 6,100	12.8
11.	Henan	16,700	250 – 300	1.6
12.	Hubei	18,000	300 – 350	1.9
13.	Hunan	21.000	600 – 650	3.0
14.	Inner Mongolia	120,000	1,100 – 1,250	1.0
15.	Jiangsu	10,300	1,500 – 1,700	14.6
16.	Jiangxi	16,700	550 – 600	3.4
17.	Jilin	18,700	600 – 650	3.2
18.	Liaoning	14,600	350 – 400	2.4
19.	Ningxia	6,600	40 – 50	0.6
20.	Qinghai	72,400	1,000 – 1,150	1.4
21.	Shaanxi	19,500	40 – 50	0.2
22.	Shandong	15,300	800 – 900	5.2
23.	Shanghai	600	25 – 30	4.5
24.	Shanxi	15,600	50 – 60	0.3
25.	Sichuan	56,700	400 – 450	0.7
26.	Taiwan	3,600	10 – 10	0.3
27.	Tianjin	1,100	70 – 80	6.8
28.	Xinjiang	160,000	3,800 – 4,000	2.4
29.	Xizang	120,000	3,800 – 4,200	3.3
30.	Yunnan	39,400	150 – 200	0.4
31.	Zhejiang	10,200	150 – 200	1.6
	Total	958,500	22,810 – 25,100	2.5

Jilin, and north-eastern Inner Mongolia. There are about 2 million ha of wetlands on the Sanjiang (Three Rivers) Plain alone. This region of north-eastern Heilongjiang consists of a vast complex of shallow freshwater lakes, reed marshes and peat bogs in the lowlands near the confluence of the Heilong (Amur), Sungari and Wusuli (Ussuri) Rivers. Other extensive systems of freshwater lakes and marshes occur near Qiqihar in south-western Heilongjiang, around Tongyu and Baicheng in western Jilin. There are two very large lakes in this region: Hulun Nur in north-eastern Inner Mongolia and Xingkai Hu (Lake Khaka) straddling the border between Heilongjiang and Russia. Other important wetlands include the mountain bogs of the Changbai Shan on the Korean border. The wetlands of the north-east are extremely important as breeding habitat for waterfowl, notably ducks, geese and cranes,

Table 5. Species of mangrove in China (Lin 1990). Thirty species of mangrove belonging to 20 genera in 16 families occur in Guangxi (GX), Hainan (HN), Guangdong (GD), Fujian (FJ), and Taiwan (TW) provinces.

	GX	HN	GD	FJ	TW
Acanthus ebracteatus	-	+	-	-	-
Acanthus ilicifolius	+	+	+	+	-
Acrostichum aureum	+	+	+	-	-
Acrostichum speciosum	-	+	-	-	-
Aegicera corniculatum	+	+	+	+	-
Avicennia marina	+	+	+	+	+
Barringtonia racemosa	-	+	+	-	+
Bruguiera cylindrica	-	+	-	-	-
Bruguiera gymnorhiza	+	+	+	+	+
Bruguiera sexangula	-	+	-	-	-
Cerbera manghas	+	+	+	-	+
Ceriops tagal	+	+	+	-	+
Excoecaria agallocha	+	+	+	+	+
Heritiera littoralis	-	+	-	-	+
Hibiscus tiliaceus	+	+	+	+	+
Kandelia candel	+	+	+	+	+
Lumnitzera littorea	-	+	-	-	-
Lumnitzera racemosa	+	+	+	-	+
Nypa fruticans	-	+	-	-	-
Pemphis acidula	-	-	+	-	+
Rhizophora apiculata	-	+	-	-	-
Rhizophora mucronata	-	-	-	-	+
Rhizophora stylosa	+	+	+	-	+
Scyphiphora hydrophyllacea	-	+	-	-	-
Sonneratia alba	-	+	-	-	-
Sonneratia caseolaris	-	+	-	-	-
Sonneratia hainanensis	-	+	-	-	-
Sonneratia ovata	-	+	-	-	-
Thespesia populnea	-	+	+	-	+
Xylocarpus granatum	-	+	-	-	-

and are also very important for their fisheries and reed production.

The great plains of the Yangtze-Yellow-Huaihe Basin in eastern China contain some of the largest wetlands in China, and the greatest concentration of freshwater lakes. The total area of lacustrine wetlands in this region is estimated at over 4 million ha. The Yangtze Basin, in particular, is fames for its lakes; these include the Dongting Lakes in Hunan Province, the Wuhan Lakes in Hubei Province, Poyang Lake in Jiangxi Province, and a chain of large lakes, including Shengjin, in south-western Anhui Province. Cao Lake in central Anhui Province, and Tai Lake, Hongze Lake and Gaoyou Lake in Jiangsu Province. Many of the lakes are fringed with marshes, and there are also extensive marshes in the dried out beds of ancient lakes, in old river channels (particularly along the Yellow River), and in seasonally flooded areas.

The wetlands in the Yangtze-Yellow-Huaihe Basin are of great importance for wintering waterfowl, particularly herons and egrets, storks, ducks, geese and coots. Several species of cranes winter in large numbers, including the rare Siberian Crane *Grus leucogeranus*, White-naped Crane *Grus vipio* and the Hooded Crane *Grus monacha* (Ma 1986). A small population of the endangered Baiji or Chinese River Polphin *Lipotes*

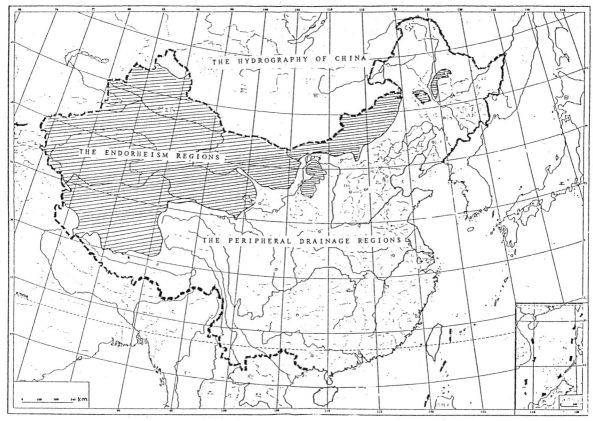

Fig. 1. Regions in China with endorheic and peripheral drainage.

vexillifer still survives along the lower-middle reaches of the Yangtze River. Recent estimates suggest that as few as 100 individuals remain (Zhou 1987). The lower Yangtze valley in Anhui Province is also the last stronghold of the endangered Chinese Alligator *Alligator sinensis*. The region includes some of the country's major freshwater fish producing areas; the lower Yangtze Basin, with its many large and small lakes, is the most important and includes the provinces with the highest fish production, Jiangsu and Hubei (Wang *et al.* 1989; Wang 1987).

A chain of freshwater lakes and marshes in the headwaters of the Brahmaputra (Yarlung Zangbo Jiang Basin), Ganges and Indus Rivers in the main Himalayan ranges in south-eastern Tibet, is part of the Indian Ocean hydrographic net (Gua *et al.* 1984). These wetlands are known to be of considerable importance for breeding waterfowl, notably Bar-headed Geese *Anser indicus* and Black-necked Cranes *Grus nigricollis* (Zhen 1983). A recent survey showed that there are also a considerable number of wintering waterfowl in these areas (Lu 1990). The wet meadows

also support the main pastoral area in Tibet (Zheng 1985).

There are a few large lakes and extensive marshes in the highlands of the Altai Shan. Chianassu Hu in the central Altai and a group of small lakes at the southern end of the Altai lie in the upper drainage of the Ertix He (Irtysh River), and are thus part of the Arctic Ocean hydrographic net. A nature reserve has been established around some of the Altai lakes in order to protect the beaver *Castor fiber* (Liang 1986).

Wetlands in endorheic drainage regions
The endorheic regions in China cover 3.5 million km², about 36% of the country. They are spread over 12 regions with all except the Wuyur He and Baicheng in north-eastern China being located in the western part of the country. Wetland information has been received from two of these regions: the Qinghai-Tibetan Plateau and Tian Shan Mountains and desert areas in Xinjiang. The Qinghai-Tibetan Plateau is the largest high altitude plateau in the world, most of it being between 4000–5500 m above sea level and bounded to the north by the

Kun Lun Shan and Nan Shan, and to the south by the great Himalayan Range. The entire plateau is dotted with innumerable lakes, ponds and bogs, and includes the sources of a number of great rivers: the Yellow, Yangtze, Mekong and Salween in the east; and the Indus, Ganges and Brahmaputra in the south (Zhang 1982). However, much of the plateau consists of inland drainage systems and most of lakes are saline. Qinghai Lake covers 495 200 ha and is the largest saline lake in China. The total area of wetlands exceeds 5 million ha. The wetlands fall into two major groups: i) the hundreds of large, mostly saline lakes in the north plain of Tibet (Qiangtang area); and ii) a chain of large saline lakes in the Zaidan Basin, north of Kun Lun Shan. Some of the wetlands are very important for breeding waterfowl, notably bar-headed Geese and Black-necked Cranes (Lu 1990).

The great inland drainage systems of the Xinjiang deserts in north-western China include several large freshwater and saline lakes with associated brackish marshes surrounded by sandy desert. The largest of the freshwater lakes is the 100 000 ha Bosten Hu near the northern edge of the Tarim Basin. The Bo Hu marshes to the south-west of Bosten Hu cover some 30 000 ha, and include many small lakes and ponds, and the extensive reed beds support a major paper industry. Other major wetlands include the Tarim Liuchang Lakes in the Tarim Basin, Aiding Hu in the Turpan Depression, Manasi Hu in the Junggar basin, and numerous intermittent rivers and streams rising in the surrounding mountains along the edges of the Tarim and Junggar Basins. Recent surveys show that the marshes along the Tarim River support one of the largest breeding populations of Black Stork *Ciconia nigra*. There are also many small lakes, marshes and streams in the Tian Shan. The Bayinbuluke complex of small lakes and marshes along the Kaidu River at more than 2400 m in the Tian Shan is known to be an important breeding area for waterfowl, notably Whooper Swans *Cygnus cygnus*. The areal extent of wetlands in Xinjiang Region exceed 5 million ha.

References

Chen, J., Wang, B. & Yu, Z. 1989. Developments and Evolution of China's Coast (in Chinese). Shanghai Scientific & Technical Publishers, Shanghai, China. 555 pp.

Liang, C., 1986. *Castor fiber* in Buergen, Xinjiang (in Chinese). Wildlife. 86:1–4.

Lin, P. 1984. The Mangrove (in Chinese). Ocean Press, Beijing. 104 pp.

Lin, P. 1990. Mangrove research papers (1980–1990) (in Chinese). Xiamen University Press, Xiamen, China. 230 pp.

Long, H., Zhu, W. & Jin, S. 1983. Chinese mires (in Chinese). Shandong Science and Technology Press, Jinan, China. 269 pp.

Lu, J. 1988. Wetlands and the strategies for wetland ecosystem managements (in Chinese). Rueal Eco-environment 2:39–42.

Lu, J. 1990. Wetlands in China (in Chinese). East China Normal University Press, Shanghai, China. 177 pp.

Ma, Y. 1986. Crane research and conservation in China (in Chinese). Heilongjiang Education Press, Harbin, China. 253 pp.

Niu, H. & Ma, X. 1985. Peatlands in our country (in Chinese). Commercial Press, Beijing, China. 180 pp.

Scott, D. A. 1989. A directory of Asian wetlands. IUCN, ICBP & IWRB, Cambridge University Press, Cambridge UK. pp. 129–294.

Wang, H. 1987. Water resource of Chinese lakes (in Chinese). Agricultural Press, Beijing, China. 149 pp.

Wang, H., Dou, H., Yan, J., Shuo, J. & Zhang, Y. 1989. Lakes in China (in Chinese) Science Press, Beijing, China. 254 pp.

Zhang, R. 1982. Natural Geography of Xizang (in Chinese). Science Press Beijing, China. 200 pp.

Zheng, Z. 1983. The Avifauna of Xizang (in Chinese). Science Press, Beijing, China. 353 pp.

Zheng, Z. 1985. Qingzang, Plateau in China (in Chinese). Science Press, Beijing, China. 350 pp.

Zhong, C. 1985. Advances of *Spartina* Research-Achievement of Past 22 years (in Chinese). Nanjing University Press, Nanjing, China. 358 pp.

Zhou, K. 1987. A study on the distribution of *Lipotes vexillifer* (in Chinese). Acata Zoologica Sinica Vol 23:72–79.

Vegetatio **118**: 57–79, 1995.

Developing wetland inventories in Southern Africa: A review

A.R.D. Taylor[1], G.W. Howard[2] & G.W. Begg[3]

[1]*Environment Department, Somerset County Council, Taunton TA1 4DY, UK*
[2]*IUCN — World Conservation Union, Regional Office for Eastern Africa, P.O. Box 68200, Nairobi, Kenya*
[3]*Environmental Advisory Services, P.O. Box 37363, Overport 4067, South Africa*

Key words: Inventory, Southern Africa, Wetlands

Abstract

The status of wetland inventory effort and availability of maps and other data sources is reviewed for the ten countries of southern Africa: Angola, Botswana, Lesotho, Malawi, Mozambique, Namibia, South Africa, Swaziland, Zambia and Zimbabwe. The aims and strategies for inventory are discussed and the main survey methods compared. Prior to commissioning new inventory work, careful collation of existing maps and imagery is recommended together with targeting of strategic inventory at Province level, reserving high resolution effort only for certain important sites.

Introduction

The value of wetlands in Africa is beginning to be realised as countries struggle with natural resource management decisions and new pressures imposed by structural adjustment programmes and National Environment Action Plans. This realisation has led to the development of programmes to manage wetlands, all of which are held back by the need to know the extent and nature of national wetland resources — both to evaluate their worth and to plan their management priorities.

The purpose of this review is to assess the status of development of wetland inventories in southern Africa and to compare progress in different states with a view to assisting the exchange of experience on wetland issues within the Region.

Why perform inventories?

We would argue that inventories of wetland, provide an indication of the location of land with the highest biological productivity, greatest potential of multiple use, the ecotone of which possesses the most biodiversity of any land. Southern Africa's wetlands are among the most diverse, both physically and biologically of any in the world, and possess multiple values, yet their rate

of loss is suspected to be very high. For example, Begg (1988) showed that in the Mfolozi catchment of Natal, South Africa, 58% of the original wetland area had been lost by the mid 1980's. Inventories are required to establish baseline data on wetland area, distribution, seasonality, characteristics and values, before rational management plans can be designed.

Wetlands are distributed from the top of the catchment to the deltas at the bottom, indicating clearly the networked interdependence of water resources. Wetlands are often indicators of substantive groundwater resources and fluvial buffering, in particular, important in sediment and nutrient trapping. Knowledge of wetland location has particular relevance for water supply and runoff interception in many African countries, because whole communities are directly dependent upon the biological and physical filtering and water supply buffering capacity of larger wetlands. Watersheds are basic units for wetland management, often crossing national boundaries in common with many wetlands and major rivers of the region, thus able to promote cooperation between countries needing to achieve effective wetland management.

Information concerning the distribution and status of wetlands in a country, greatly assists those concerned with wise use of natural resources to reach balanced decisions about wetland exploitation and conservation. The inventory process itself can promote greater cooperation between the often narrow interests of managers of agriculture, forestry, and water resources and may provide influential input to national conservation strategies, now under development in many southern African countries. An important lead has been provided by the recent SADCC Wetlands Conservation Conference for Southern Africa (Matiza & Chabwela 1992) and by the SADCC Wetlands Programme. This regional focus is establishing a framework for greater awareness about wetland conservation and inventory.

Most of the larger wetlands of southern Africa have now been described, Hughes and Hughes (1992) in particular, list the major characteristics of them, including their fauna, flora, human impact and utilisation (but see also sections in Denny 1985; Burgis & Symoens 1987; Whigham *et al.* 1993). However, while such wetland directories are invaluable, they do not alone provide a satisfactory substitute for systematically conducted national wetland inventories, which must also produce mapped wetland boundaries and at least sufficient detail of current status to allow baseline assessment. These inventories must also take account of smaller wetlands and their combinations, which are often more important to national development than the large ones that are already described.

We understand the purposes of wetland inventory to include:

- quantifying the amount of wetland with indications of categories (or types), constituents and current uses
- quantifying the extent and rate of alteration of wetlands
- disseminating this information to persons who have the responsibility of managing and conserving southern Africa's wetlands.

Wetland definitions and scope of paper
For this review, the Ramsar Convention definitions will be used, which include estuaries, open coasts (including mangroves), riparian zones and floodplains, lakes, swamps and swamp forests, marshes, endorheic basins and pans, headwater wetlands (dambos), upland and lowland peat bogs. Many competing classification systems have been published; we use the hierarchical one proposed by Dugan (1990), based on Scott's (1989) Ramsar database form, which in turn is derived from the scheme of Cowardin *et al.* (1979). It is important

to note that published literature often artificially differentiates wetland as a category distinct from coastal and riparian zones, we do not subscribe to this narrow definition.

We use the terms Region, and Province: 'Region' refers to the southern African region comprising ten countries, while 'Province' is used to refer to a large sub-division of a country, (alternatively known as a 'District' in some southern African countries). We use the heading 'International designations' below to refer to the Ramsar Convention on Wetlands (site, Jones 1993), or to UNESCO World Heritage recognition of a site.

This review does not attempt to detail the characteristics of southern African wetlands, it merely reports the progress made in inventorying the wetlands of the region and contrasts the methods used and examines possible ways forward. The maps in this review (Figs 1 to 11) therefore are merely indicative of the main wetland areas and also outline the principal rivers of the drainage system. The diversity of southern African wetlands and the limited resources available for the inventory effort demands a systematic approach to inventory that leads to a strategic view of these resources, however, the inventory process is vulnerable to arguments concerning its spatial (mapping) resolution, seasonality, and targeting.

Taking these issues in order, except for the smallest countries (Lesotho, Swaziland) attempting to map a whole country's wetlands on to maps with scale more detailed than 1:50 000 is self-defeating, particularly in countries possessing complex interlocking endorheic wetlands, for the closer one looks, the greater the complexity revealed. At the other extreme, maps at a scale less detailed than 1:1 000 000 cannot show human settlement with sufficient resolution to examine wetland utilisation and impacts. We suggest that for southern Africa, national inventories should aim initially at producing maps at a scale of 1:500 000 (or 1:250 000 if necessary), leaving higher resolution mapping until later. While it is rare for a wetland aerial survey set to exist, our experience suggests that within each of the southern African countries there are already likely to be aerial survey photograph sets concerned with forestry resource mapping, agricultural land use, etc, which may be suitable for wetland assessment.

Wetland recognition is the second main problem, discussed below and it is vital that inventory takes the long term view in southern Africa. Huge numbers of wetlands occur in otherwise semi-arid areas on a shifting mosaic basis, dependent upon uncertain and uneven rainfall, with return periods often greater than five years, yet Breen (1991) argues that these seasonal pans may be important conservation sites. Similarly Scoones (1991) argues that dambos are important sustainable agricultural and pastoral resources within dry lands.

The targeting of inventory effort relies upon knowledge of past effort and it is hoped that this review will assist in future targeting, because when planning wetland inventories, it is worth first considering whether the effort should be aimed at obtaining a strategic overview, or more detailed local, perhaps subcatchment information. Current experience suggests that in southern Africa, provincial or national inventories should take precedence, with sample inventories subsequently conducted within each major identified wetland type, to confirm the reliability of the strategic view.

Regional introduction and description including drainage systems, climate, geography and main vegetation types
Drainage. Southern Africa, shown in Fig. 1, consists mainly of a broad swath of high land, the Central African Plateau, mostly above 1,000 metres, draining to the north west into the Zaire basin and east to the Zambezi, which passes through the comparatively low land of Mozambique to join with the drainage from the southern part of the Rift Valley, before passing to the sea. In the centre of southern Africa, Botswana's Okavango Delta wetland lies in the Cubango and Kalahari internal drainage basin, overflowing to the Zambezi when Angola's rainfall is very high. Further south the Orange river drains a large basin falling away westwards from the high mountains of Lesotho and the Drakensberg. All around the Central African Plateau, hundreds of rivers drain the catchment, which often originate in dambos, their overall distribution (see Acres *et al.* 1985) is also shown in Figure 1. Notably in Mozambique, mangrove forest forms along the coast, towards the equatorial regions. Even the most arid of the southern African countries, Namibia, has a significant coastal wetland system. To a great extent the ten countries of the region either share their borders with great rivers or have them passing across international boundaries. The Zambezi in particular is, to a varying extent, shared by Angola, Zambia, Zimbabwe, Botswana, Namibia, and reaching the sea in Mozambique.

Climate. The region extends from 6° S to almost 35° S, and therefore experiences a range of tropical to temperate conditions, determined by the prevailing wind temperature and moisture content. Rainfall in the region is mainly driven by the *Inter Tropical Convergence Zone* (ITCZ). The ITCZ effect diminishes southwards towards South Africa, where rainfall is dominated by southern hemisphere (antarctic) fronts. The west coast experiences the northward cold Benguela current, while the warm Agulhas Current flows southwards along the eastern coast. The winds along the west coast are cold and therefore low in moisture, resulting in arid conditions in Namibia and further north into Angola, while the south-eastern and eastern coast experiences warm moist south-moving air, precipitating on the coastal plains and onto the high mountains of the Drakensberg and Maloti, where peak rainfall is greater than 2,000 mm annually. Temperatures vary with season and altitude, ranging from minus 18 °C in winter in the Drakensberg mountains of Lesotho, to over 40 °C in summer in Botswana.

Wetland vegetation and associated communities. Very large areas of the southern African highland plateau have at least some *miombo* woodland, which is dominated by *Brachystegia* and *Julbernadia* species, that often coincides with granitic basement rock. Dambos are also commonly associated with this highland and possess a specialised range of freshwater flood tolerant woody species including *Syzygium* species and palms. Coastal mangrove forests are especially widespread in Mozambique, while herbaceous swamps in the region are widespread and common, with papyrus and reeds. There are relatively small areas of high altitude swamps, bogs and mires, which are characterised by Afro-alpine vegetation which is specialised and often endemic. Open water may have many submerged potamogetons and is often colonised by floating *Pistia stratiotes*, with fringing *Vossia cuspidata* and *Cyperus papyrus* entering stands of *Typha domingensis*, or in shallow water, *Echinochloa scabra*. Clear water rivers downstream of wetlands such as the Okavango Delta, contain much submerged vegetation, including water lilies (*Nymphaea* sp.), the tubers of which are often harvested for food. Most wetland plant species have some economic value, for example as food or more usually as thatching, screen or carpet making material.

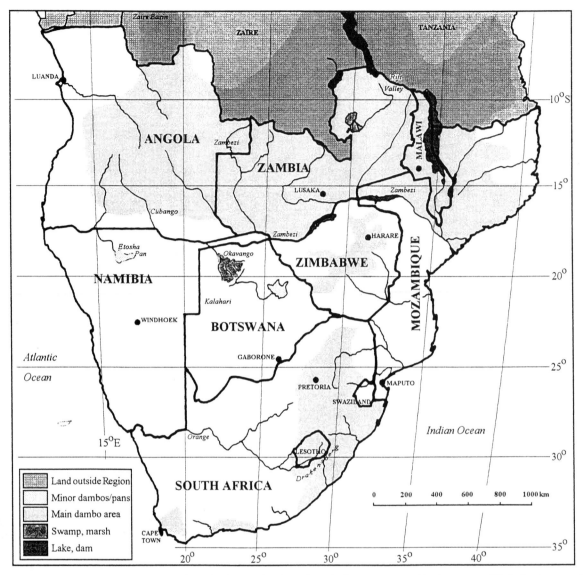

Fig. 1. Map of southern Africa illustrating the major rivers and regions where dambos are common (dambo areas modified after Acres *et al.* 1985).

Inventory methodologies and sampling techniques

Inventory methods are sampling exercises informed by statistical tools to establish, with varying degrees of confidence, the area, distribution and characteristics of wetland in a geographical area. A broad area surveyed at low resolution may yield data about large wetland locations and catchment relationships, but little about their composition. Conversely, a small area surveyed to reveal the distribution of wetland plant species will provide higher confidence about (plant) composition, but little about the relationship of that wetland to its catchment. This review will not attempt a rigorous analysis of sampling statistics, but those planning an inventory should carefully define what is to be achieved through it, by all means maximizing the information return for the resource outlay but remaining mindful of the uncertainties.

Inventory is a mix of objective and subjective methods, often relying on the experience of photo-interpreters. In the USA, Ducks Unlimited using a computerised spectral class recognition system to attempt to eliminate subjectivity, were able to quantify wetland recognition reliability depending upon wetland size (Ducks Unlimited 1992). Using Landsat TM satellite data, the following results were obtained:

Wetland size class (hectares)	Percent of wetlands recognized
0.0–0.75	22
0.76–1.9	70
2.0–3.9	91
4.0–9.9	96
>10	100

These results emphasise that even with objective analytical methods, recognition of small wetlands can be unreliable, thus underlining the need for key decisions to be made by planners of inventories concerning imagery type, its scale, who will interpret it and ground truth (verify) data inferred from imagery.

Remote sensing: satellite, high and low altitude aerial photography
Remote sensing is conventionally defined as the use of satellite-borne sensors to image the earth's surface, however, making a distinction between radiometer sensors carried by satellite and aeroplane is increasingly difficult, with the advent of portable multi-spectral units that may be used from low altitude flights. Aerial photography usually employs large format film exposed in vertically fixed cameras looking down on the flight path, while satellite photography uses digitising scanners converting a focused radiometer image to a datastream, transmitted to a ground receiving station. There are radiometer sensors for each of several narrow wavelength bands, depending upon the manufacturer.

Three satellite systems have been widely used for commercial remote sensing. The US Landsat satellite series 1 to 3, carried the Multi Spectral Scanner (MSS); Landsat 4 to 6 were equipped with the Thematic Mapper (TM), while the French Satellite Pour l'Observation de la Terre (SPOT) has a panchromatic and multispectral mode and a new radar image product. Landsat scenes cover approximately 185 km × 185 km, while SPOT full scenes cover 60 km × 60 km. Details of spectral bands, resolution and other data are given in Table 1.

The operators of these satellites have over several years built up a large library of image data for most of the world's surface, including southern Africa, therefore it is possible for those conducting inventories to obtain useful images from the past in a variety of scales and formats including panchromatic (black and white) or multispectral false colour film or prints, computer tapes or discs. Recently, EOSAT began offering TM data in Geographic Information System (GIS) format suitable for ARC/INFO, and in addition offers three band 56 km × 56 km map-oriented images suitable for GIS work. In addition to the above widely marketed products obtainable through EOSAT or SPOT Image, Russian satellite imagery is now available. Contact addresses for obtaining further details and advice for southern Africa are given below.

In a recent review of the application of satellite data to wetlands (Federal Geographic Data Committee 1992), it was concluded that while satellite remote sensing was easy to integrate into a GIS, manual interpretation of aerial photographs was still more accurate in wetland delineation than equivalent satellite spectral imagery. The main controversies over satellite imagery as compared to conventional aerial photography relate to its relatively poor spatial and wetland class (spectral) resolution, and the relative usefulness of the re-visit capability of satellites.

Theoretically, the maximum spatial resolution, ie discrimination between ground features based on spectral radiance, is obtained by comparing two adjacent image pixels. For Landsat TM data, 1 pixel equates to 30 m × 30 m ground area, so at least two pixels are required. In practice, an object requires a 3 × 3 pixel box for reliable resolution, equivalent to about a hectare in area. Even for SPOT Panchromatic data, features less than 30 m across may not be reliably detected. It is considered that riparian wetland strips less than 60 to 90 m wide cannot be reliably detected with either current Landsat TM or SPOT data (Federal Geographic Data Committee 1992).

Wetlands are best detected through their infra-red spectral reflectance, which detects vegetation and soil moisture related features. Landsat TM Bands 4,5 and 7 and the SPOT Multispectral 0.79–0.90 μm band are the infra-red bands available. Most wetland discrimination has used TM Band 5 (1.55–1.75 μm), however, TM Band 1 (0.45–0.52 μm) can be used to detect open water, therefore use of TM Bands 1 and 5 can discriminate wetland vegetation from open water. There is difficulty in separating wetland vegetation from other vegetation in the wet season, so differential use of TM

Table 1. Characteristics of the main commercial remote sensing satellites.

Satellite	Landsat 1,2,3	Landsat 4,5,6		SPOT 1,2
Launch date	1972,75,78	1982,84,93		1984,91
Data supplier	EOSAT	EOSAT		SPOT Image
Sensors	Multi Spectral Scanner (MSS)	Thematic Mapper (TM)		Multispectral, Panchromatic Almaz Radar
Path width of image	185 km	185 km		60 km
Spectral bands of sensors in μm, (with TM band numbers)	Landsat 3 only 0.50–0.60 0.60–0.70 0.70–0.80 0.80–1.10 10.40–12.60	TM band 0.45–0.52 0.52–0.60 0.63–0.69 0.76–0.90 1.55–1.75 10.40–12.50 2.08–2.35	Band 1 2 3 4 5 6 7	Multispectral 0.50–0.59 0.61–0.69 0.79–0.90 Panchromatic 0.50–0.90
Ground resolution of sensor (= pixel)	80 m	30 m		20 m Multispectral 10 m Panchromatic

Bands 4 and 5 can be useful, however, it is much better to use imagery from seasons when wetland vegetation is still moist, while surrounding land is dry. For riparian strips, there is a persistent problem distinguishing between wetland and overhanging tree canopy spectral reflectance, which can only be resolved if the trees are seasonally deciduous, which is the case in temperate regions of southern Africa.

Manual interpretation of conventional aerial photographs is currently unbeatable for defining wetland boundaries and associated topography. Where resources permit, it will be preferable to use any reasonably recent photography to delineate wetlands in this way, particularly for riparian wetland strips, however the date or season of this photography will affect recognition of seasonal wetlands and their moving boundaries. Satellite imagery, either from Landsat or SPOT, is then most useful in detecting change over time, and in particular for seasonal re-visits to the area, having first delineated the wetland. Aerial photography and its subsequent manual interpretation is, however, a time-consuming and therefore expensive task, which might be justified for relatively small areas, sensitive wetlands or for riparian survey, but is unlikely to be cost-effective for exploratory national wetland inventory, unless the cost is shared with agencies intending to use it for other strategic purposes.

Finally, comparing the cost of satellite imagery with that of conventional low or high altitude aerial photography is not at all straightforward. SPOT or Landsat imagery can be obtained as photographic prints, which may seem to compare directly with aerial photographs, however, the effective use of satellite data depends upon the use of computers to manipulate spectral bands, in order to resolve wetland features reliably. Purchase of Band 5 TM data or SPOT Panchromatic data as photographs *may* be cheaper per unit area of ground surveyed than aerial photography, but the investment made would, in the long term, be better directed towards purchase of powerful PC computers and the original Computer Compatible Tapes (CCT) for the imagery. A recent study by Nakayama (1992) used PC-based IDRISI software to process SPOT, Landsat MSS and TM data and reviewed its use in developing countries.

Agencies contemplating purchase of PCs and software for processing of satellite imagery leading to map products, must also take into account long term service and training costs. Capital costs for a PC and all necessary software and peripheral equipment may be £20 000 (1993 prices), to which training and service costs must be added. The United Nations Environment Programme (UNEP) Global Environment Monitoring

Table 2. Available map resources for the countries of southern Africa.

Country	Topography	Geology	Soils/Agriculture	Vegetation	Other
Angola	1:250,000 (date?) 1:100,000 (date?)	1:250,000 (1961) 1:100,000 (1971)	1:1,000,000 & 1:500,000 (1959-72)	1:1,000,000 (1982)	N/A
Botswana	1:500,000(1965-69) 1:250,000 (1986) 1:125,000 part cover (1958-66) 1:100,000 part cover (1981-83) 1:50,000 (1967-82)	1:1,000,000 (1984) 1:1,000,000 photomap (1979) 1:250,000 & 1:125,000 (1963-86)	1:500,000 Eastern only (1963) 1:250,000 (date?) 1:50,000 Okavango (1978 & 86)	1:500,000 North only (1968)	1:250,000 & 1:100,000 Okavango ecological zoning (1978)
Lesotho	1:250,000 (1978) 1:50,000 (1979-82) 1:25,000 (1977) Thaba-Tseka only	1:250,000 (1982) 1:100,000 & 1:50,000 (1980-82)	1:250,000 (1968) 1:250,000 Agricultural potential map (1967)	N/A	N/A
Malawi	1:250,000 (1974-88) 1:50,000 (1986)	1:1,000,000 (1966) 1:250,000 (1970) 1:100,000 (1958-84)	1:500,000 (1965-71)	N/A	N/A
Mozambique	1:500,000 (date?) 1:250,000 (date?) 1:50,000 (date?)	1:1,000,000 (1987) 1:250,000 (1959-75)	1:4,000,000 (1972)	N/A	N/A
Namibia	1:250,000 (1975) 1:50,000 (1975)	1:1,000,000 (1980) 1:250,000 (1977) 1:125,000 (1938-42)	N/A	N/A	N/A
South Africa	1:250,000 (1950-72) 1:50,000 (1950-81)	1:1,000,000 (1984) 1:250,000 (1957-81) 1:125,000 (1910-72) 1:50,000 very few (1948 onwards)	1:250,000 Agricultural land types (1978 on)	1:1,500,000 Veld types (1988)	1:750,000 Priority Wetlands of Natal (1989)
Swaziland	1:50,000 (1954-71)	1:250,000 (1982) 1:50,000 (1961-80)	1:125,000 (1968) 1:50,000 Soils and Irrigability for Lower Usutu only (1961)	N/A	1:250,000 Satellite map (1984)
Zambia	1:250,000 (1971-76 + some late 1980's) 1:100,000 (part only 1979) 1:50,000 (1949 on)	1:250,000 (1984) 1:100,000 (1959 on)	1:250,000 Bangweulu and Kalasa Mukosa Flats (1982)	+ Climate 1:1,250,000 (1988) 1:500,000 (1976)	
Zimbabwe	1:250,000 (1973-75) 1:50,000 (1954-68)	1:100,000 (1920-82)	1:1,000,000 (1979) 1:1,000,000 (1980)- Agriculture only 1:50,000 (1980)	N/A	N/A

System (GEMS), based in Nairobi will be useful source of practical advice.

Reconnaissance surveys and ground truthing
There is a great deal of merit in conducting a preliminary limited wetland inventory targeted only at a province, in order to get good wetland management started, rather than waiting for a full national inventory, which may take several years and lose time for vital early strategic decisisons to be taken. This is because it is usually at Province (or District) level that decisions on wetland use will be made in the long term.

Secondly, to get a whole country inventoried even at a scale of 1:500 000 will consume more resources than are often available, while investing much effort in basing inventory upon the often very dated photo-interpretation information contained in maps produced in colonial times 30 or more years ago, is a waste of resources.

Even comparatively recent detailed maps are sometimes unreliable. In a study by Begg (1988) the 'marshes, swamps and vleis' shown on standard 1:50 000 topographic maps of South Africa were compared with the ground truthed result of a proper wetland inventory of an area in the Mfolozi catchment. Of the 996 extant wetlands found (or fragments thereof), only 33 (3%) of these sites appeared on the maps, (made 5–10 years ago). Despite this, Silberbauer and King (1991) have recently drawn up a preliminary distribution and classification of wetlands in the south-western Cape (an area approximately 100000 km^2 in size) by summarising the information contained on 1:50 000 topographical maps.

In contrast to topographic maps, soil maps showing hydromorphic soils are a comparatively useful source of information, for example, the soil maps of the Natal Tugela Basin, shown as an inset to Figure 8 (van der Eyk *et al.* 1969), of the Howick area (Scotney 1970) and of Botswana's Okavango (Staring 1978) are very useful, while alluvium shown on some geological maps generally coincides with floodplains.

It would, however, be wrong to dismiss existing topographical maps; they show drainage patterns, contours and altitudes, and are useful in providing a framework for inventory information, but as with any inventory based upon existing maps, it is vital to ground truth random sample areas to determine the potential usefulness of the maps. Reconnaissance surveys should therefore be undertaken, which use a pragmatic combination of soil and geological maps, broad scale aerial and satellite imagery, selective ground truthing and by all means the use of existing topographical maps to place the results on. An example of such a reconnaissance survey, using 1:80000 panchromatic aerial photographs and existing 1:50000 map sheets, was conducted by Whitlow (1984) for Zimbabwe. He used 180 km^2 grid squares in combination with the aerial photographs. Estimates of wetland area were obtained by means of 25 randomly selected sample points on each photograph, each of which represented an estimated 4% of the area. Percentage and absolute wetland areas were computed from these measurements and on the basis of frequency distributions within the data,

three wetland area classes were defined and ground truthed in a sample of the area inventoried.

Ground truthing of aerial/satellite photography and existing maps (if used) is essential. Its aim is to provide assurance that wetlands and their pre-defined subclasses have been identified with a high degree of confidence and consists of visits made to representative areas to determine physical and biological characteristics. The location of visits to presumed wetland should be selected randomly, but often have to be restricted to those accessible by road, track or boat. All expected wetland classes/communities must be ground truthed, but so should features that may be mistaken for wetland plants, such as some tree canopies, therefore locations should be randomly selected using a grid overlay on a photograph (or use a map grid).

A summary of the main types of maps known to exist for each southern African country is given in Table 2. In addition to the maps referred to, there will be many maps at province or district level, dealing with specialist information that are not mentioned in the Table, e.g. the soils map of the Tugela basin, South Africa (van der Eyck *et al.* 1969).

Seasonality and sampling problems
In many southern African countries, seasonal wetlands may remain without water for several years or even decades, yet when water arrives, fauna and flora characteristic of wetlands soon reappears, including fish, frogs, and hydrophytes. One of the biggest such wetlands is Makgadikgadi Pans of Botswana, easily visible on Landsat TM images and at least 1 million ha in area (Hughes & Hughes 1992). For such large, coherent wetlands, inventory is easy compared to the problem of inventorying the millions of small pans or dambos scattered over southern Africa. For practical purposes, seasonal wetlands smaller than about 10 ha are unlikely to be recorded; if they occur scattered across a large area, an inventory can only report these wetlands as a kind of probability estimate for the larger area.

A similar problem is met with small semipermanent wetlands that intercalate with dry land. In Zambia, Malawi and Zimbabwe especially, small dambos account for 10 to 15% of land in some regions, and are probably among the most significant in terms of human utilisation, but due to their small individual size are mostly not recorded. In such places, an inventory should attempt an assessment of the wetland area density and define the whole area as wetland-related, see for example Agnew's map (unpublished, in Meadows 1985) for Malawi and Whitlow's (1985) dambo distribution map for Zimbabwe.

Overview of the known distribution and characteristics of wetlands in the region

Angola

Wetland distribution. Total area of Angola: 1 246 700 km^2, approximate wetland area (Hughes & Hughes 1992): 4,750 km^2, or 0.3%. Note that most upland sponges, most dambos, river floodplains and lakes are not included in the estimate. The central plateau dominates the drainage system, from which many perennial rivers flow. Most wetland types exist in Angola, including up to 70 000 ha of marine intertidal wetlands. Figure 2 illustrates the large areas where dambos occur, note the extensive drainage system draining the highland.

Inventory status. Owing to many years of civil and military strife, systematic inventory has not been possible, however, the wetland resources of Angola are so extensive and diverse that a national inventory must be fully encouraged.

Wetland characteristics. Angola has come of the most complex wetland systems of any country, with a high density of upland palustrine emergent types, riverine and lacustrine wetlands, with marine and estuarine intertidal mangroves and salt marshes.

International designations. None.

Botswana

Wetland distribution. Total area of Botswana: 569 582 km^2, approximate wetland area (Hughes & Hughes 1992): 28 310 km^2, or 4.9%. Most of the country's wetlands are associated with the Okavango and Zambezi rivers in the north, while the large pan complex of Makgadikgadi derives its water from the Boteti, direct rainfall and in the case of Sowa Pan, Zimbabwe's Nata River. As Figure 3 illustrates, the country possesses few permanent wetlands, however, there are widely scattered seasonal pan wetlands across the arid interior.

Inventory status. Controlled photomosaic maps (at scales of 1:100 000 and 1:50 000) are available for the whole country, while there are ecological zoning maps for the Okavango Delta at scales of 1:250 000 and 1:100 000 (SMEC/Kalahari Cons. Soc. 1989) and on soils (Staring 1978; FAO/UNDP/GOB 1986). These maps and others on soils and vegetation, indi-

cate wetland location. Thus, although the wetlands of Botswana have not been formally inventoried, their distribution has been established.

Wetland characteristics. The best researched wetland is the Okavango Delta, the subject of a major symposium (Botswana Society 1976), numerous hydrological and geophysical studies (e.g. McCarthy 1992) and the subject of a recent major review by the IUCN (Scudder *et al.* 1993). The Okavango Delta and Linyanti/Chobe River wetlands are the only permanent natural wetlands. Other wetlands are much less well known, however, and systematic inventory is required, particularly for the numerous small pans.

International designations. None.

Lesotho

Wetland distribution.
Total area of Lesotho: 30 344 km^2, approximate wetland area: $\sim$200 km^2, or 0.6%. Wetlands occur in highest frequency in the mountains (>2,000 m) and decline in density towards the southern and western lower areas. The impermeable basalt rock above 2,300 m develops the greatest mire density. In general, mires occur above 2,750 m, while marshes are more often found below this level. Figure 4 illustrates that the mires mostly run along the north east boundary of the country, while the Orange River originates in the Lesotho highlands.

Inventory status. Ground surveys of mires on the central Thaba-Putsoa mountain range were conducted by Backeus (1988), while Schwabe (1989) and Schwabe and Nthabane (1989) report an inventory using SPOT satellite imagery for the Maluti/Drakensberg mountains. There have also been studies of the sub-alpine zone wetlands done by Loxton, Venn and Associates (Scudder 1993, pers. comm.) in support of the Lesotho Highlands Water Project pump storage scheme. No national wetland inventory has been conducted.

Wetland characteristics. All of Lesotho is above 1,500 m and 80% of its area is above 1,850 m (Hughes & Hughes 1992), so its wetlands are primarily montane mires, with riverine reed or sedge swamps in the lower areas. Schwabe's (1989) work included a functional assessment of the montane wetland using the Wetland Evaluation Technique (Adamus *et al.* 1987).

International designations. None.

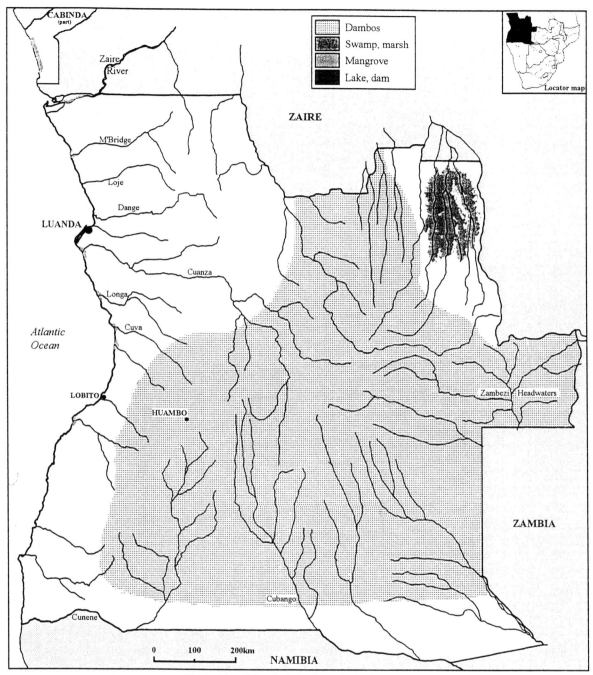

Fig. 2. Map of Angola, indicating main dambo areas.

Malawi

Wetland distribution. Total area of Malawi: 118 484 km^2, approximate major wetland area (Hughes & Hughes 1992): 2,730 km^2, plus Lake Malawi: 24 208 km^2, Lake Chilwa: 1,850 km^2, Lake Chiuta: 130 km^2, total: 28 918 km^2, equivalent to 2.9% (non-lake wetlands) or 24.4% in total. Alternatively, from Agnew's map (in Meadows 1985), it can be calculated that seasonally waterlogged areas total about 15 000 km^2, thus total wetlands amount to 15.9% of the national area, excluding Lake Malawi. Lake Malawi and its associated wetlands make up the majority of

67

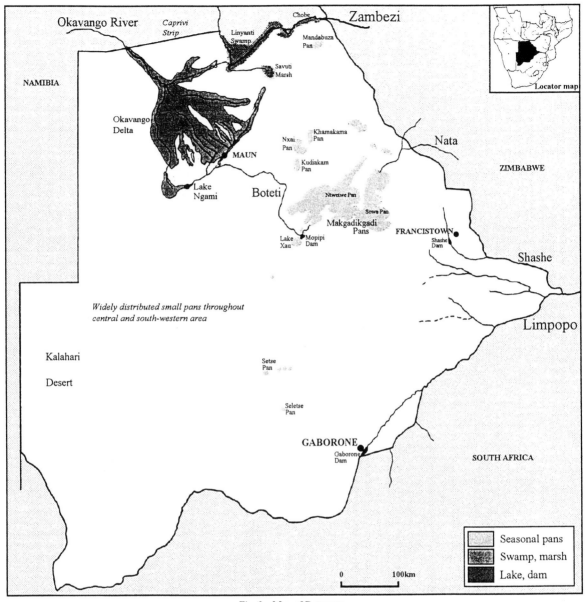

Fig. 3. Map of Botswana.

Malawi's wetlands, the rest are other lakes, dambos and floodplains. Apart from the lakes and their associated floodplains, dambos are the dominant upland (over 1,000 m) wetland type in Malawi and are widely distributed as headwater wetlands, extensively used for agriculture. Lake Malawi and its wetlands are by contrast lowland (from 475 m) areas. The general location of the dambos is shown in Figure 5, which also indicates the location of the main swamps.

Inventory status. Mzembe (1990) describes the dambos of Malawi as covering 2,590 km², mostly under traditional use and cover about 12% of the total land available for agriculture, including grazing, cultivation and rice irrigation. Mzembe's figure is, however, much less than the estimate obtained from Agnew's map (in Meadows 1985) showing seasonally waterlogged areas in quarter degree squares for the country. The differences between these sources may be explained by the wetland definitions used in the pub-

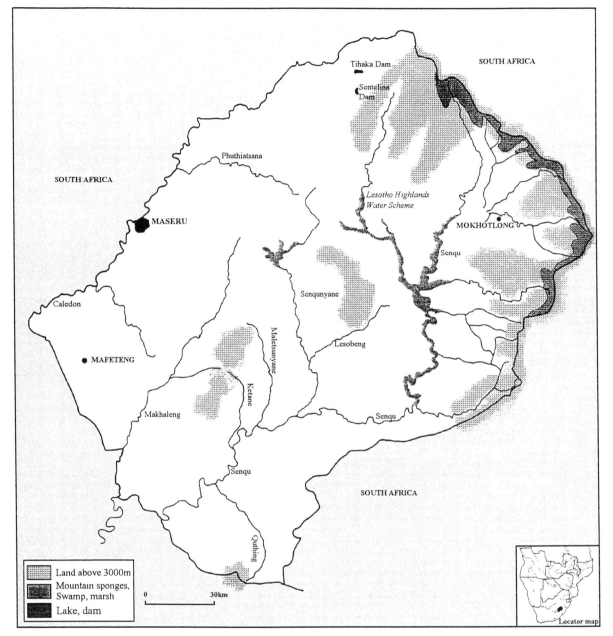

Fig. 4. Map of Lesotho.

lications concerned and it should be possible to arrive at a fairly reliable estimate of wetland type and distribution from published sources and available maps.

Wetland characteristics. Dambos have been studied in respect of drainage (Agnew 1973), and as indicators of environmental change (Meadows 1985), while Mzembe (1990) has reviewed dambo use for agriculture. There is extensive published work on limnolog-

ical and fisheries related research (e.g. Tweddle *et al.* 1978; Tweddle & Mkoto 1986).

International designations. None.

Mozambique
Wetland distribution. Total area of Mozambique is 784 755 km^2, approximate wetland area: 12 529 km^2, plus Cahora Bassa: 2,665 km^2, Lake Malawi:

7,000 km², other lakes total: 1,928 km², total: 24 122 km², equivalent to 1.6% (non-lake wetlands) or 3.1% in total. These totals are a considerable under-estimate, for Mozambique has extensive dambos, endorheic pans and swamps scattered across very large areas. There are extensive mangrove swamps along the coast and the coastal plain also has the most important wetlands, which are mostly riverine floodplains. The Zambezi is associated with extensive floodplain wetlands and its delta used to be regularly flooded before impoundments upstream such as Cahora Bassa reduced flows. There are numerous other impoundments. Figure 6 indicates the main swamps, lakes and mangrove locations, but not the dambos, which are distributed sporadically.

Inventory status. No national wetland inventories have been performed and while a good range of topographical maps exist, their age and quality is variable. Hughes & Hughes (1992) cite some work on coastal systems performed in 1969 and 1974, but little other work is available from the literature, except for a wildlife study in part of the Zambezi Delta by Anderson *et al.* (1990).

Wetland characteristics. All wetland types occur in Mozambique and there is every reason to suppose that their biological and physical diversity is extensive, though poorly studied or reported.

International designations. None.

Namibia

Wetland distribution. Total area of Namibia: 824 295 km², approximate wetland area: 11 807 km², or 1.43% The country is predominantly arid to semi-arid and has no perennial rivers, except those that run along the north and south borders. The majority of the wetland area is made up of pans, fed by seasonal streams. The notable wetlands are along the 1,400 km long coastal strip, and along the Linyanti-Chobe-Zambezi river system to the north (including 6,480 km² of wetland), the Oshakati pan system (about 600 km², the Etosha Pan (4,600 km²) and many other pans and impoundments throughout the country. Figure 7 shows the main rivers and pans, but does not indicate the large number of small pans scattered across the interior of the country.

Inventory status. A complete set of 1:50 000 topographical maps are available, with useful wetland-related information. The wetland inventory process was stimulated by a wetlands workshop held in 1988, which drew together the available knowledge in one

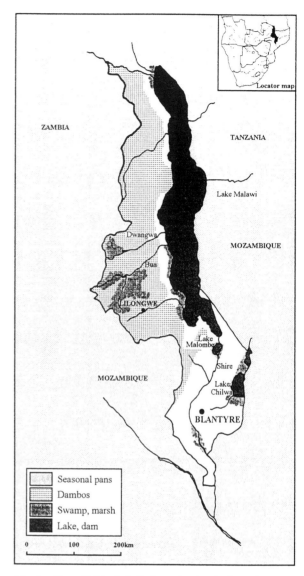

Fig. 5. Map of Malawi, indicating extensive dambo and swamp distribution.

volume (Simmons *et al.* 1991). A national wetland resource review has now been carried out by the Ministry of Wildlife, Conservation and Tourism (see IUCN 1992), however, maps which specifically delineate wetlands are currently lacking.

Wetland characteristics. Namibia's wetlands are predominantly extensive seasonal pans inland, saline lagoons at the coast and several artificial impoundments. The relative importance of Namibia's wetlands in such an otherwise arid area is considerable, especially for migrating waterbirds.

70

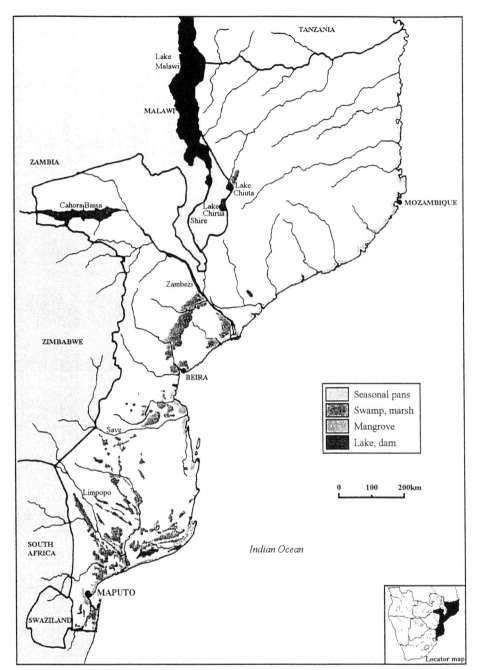

Fig. 6. Map of Mozambique, note the widely distributed mangrove swamps along the coast.

International designations. Namibia has applied to become a Contracting Party to the Ramsar Convention.

South Africa
Wetland distribution. Total area of South Africa: 1 184 825 km², approximate wetland areas natural:

2,500 km², man-made: 2,100 km², total: 0.39%. Throughout the country there are man-made dams and vleis (=dambos; seasonal emergent palustrine wetlands), with dambos in the central highlands, while most of the riverine floodplain is along the south and east coast, associated with many rivers draining the central highland plateau. In addition, remnants of once

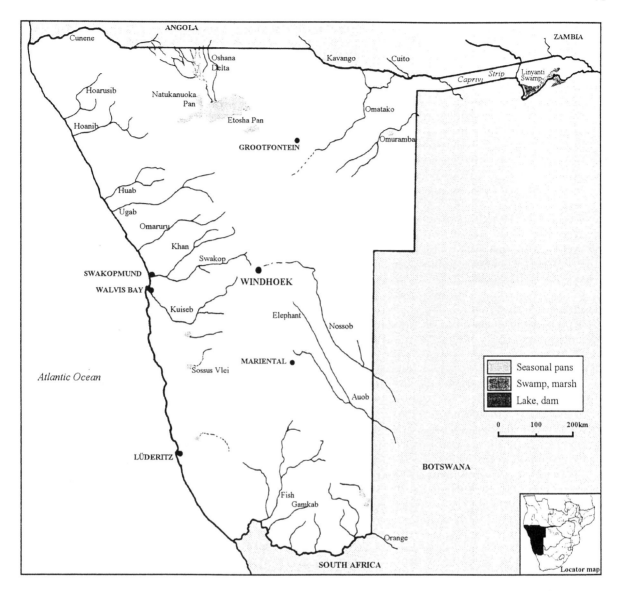

Fig. 7. Map of Namibia

extensive mangroves are found in estuaries of Natal. South Africa's wetlands are relatively small compared to those of neighbouring countries such as Mozambique, but their distribution as shown in Figure 8, is widespread. However, it is difficult to illustrate the many small wetlands that together are significant, for example the Natal wetlands (Begg 1989).

Inventory status. South Africa is well supplied with maps of all types and the information base would allow wetland inventory to proceed with few technical problems. Noble and Hemens (1978) compiled the first inventory of about 180 natural endorheic pan wetlands

and floodplain across the country, while the Orange Free State pans, which are important for wildfowl, were classified by Geldenhuys (1981). Overall, Natal has received the most attention. Although generated for a different purpose, a good impression of wetland distribution may be gained from soil maps, such as for the Tugela basin (van der Eyk *et al.* 1969). These suggest that before human influence became significant, wetlands occupied 16.5% of the basin, an area of approximately 30 000 km^2, and shown as an inset to Figure 8. Similarly, Scotney and Wilby (1983) estimated that on a soils basis, up to 15% of the landscape in Natal was

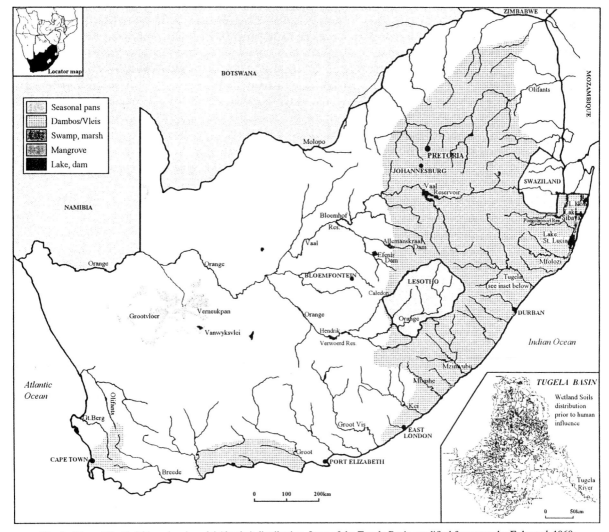

Fig. 8. Map of South Africa, showing vlei (dambo) distribution. Inset of the Tugela Basin modified from van der Eyk *et al.* 1969.

classified as wetland. Today it is estimated that over 90% of the original wetland resources have been lost in parts of the basin (Scotney 1978). In a systematic and detailed study of one 10 000 km², catchment — the Mfolozi, Begg (1988) demonstrated that 58% of the original wetland area (502 km²) had been lost. Begg (1978, 1983) earlier surveyed the estuaries of Natal. Partial inventories have been performed for the larger wetlands, for example, the Pongolo floodplain (Heeg & Breen 1982), and the Greater Mkuze Swamp (Stormanns & Breen 1986), while Begg (1989) followed up his earlier work in a study of the priority wetlands of Natal. Recently V. and C. Semeniuk (1993), have conducted a further inventory of the wetlands of the eastern shores of Lake St Lucia. Natal. In an ongo-

ing programme, the estuaries of the Cape are identified and partly assessed by the Coastal Development Programme of the CSIR (see for example Heydorn & Tinley 1980; Heydorn & Grindley 1981–85, etc.).

A workshop on inventory and classification of wetlands in South Africa was held in 1988 (Walmsley and Boomker, 1988), which concluded that a national inventory was overdue and that such an inventory should drive a national strategy to improve wetland conservation, this aspect was further discussed by Breen and Begg (1989) in a second workshop held on biotic diversity held the same year. The Department of Environment Affairs has announced an interest in undertaking a national wetland inventory, but methodology has not yet been agreed.

73

Wetland characteristics. All the types of wetlands occur in South Africa and many of their biotic characteristics have been well studied, for example waterfowl, invertebrates and fishes. For rivers, O'Keeffe *et al.* (1989) cites Harrison's (1959) hydrobiological regions and proposes an updated hierarchical system for classifying them, which could also apply to riverine wetlands and their associated lake systems. It appears that primary data for most wetlands in the country exist, but that owing to a lack of an agreed national inventory process, the collation and review of strategically important data has not been done. Begg (1992) suggests that wetland conservation is a political process that requires a well informed public to succeed. Sadly as in many countries, South Africa has not yet developed adequate awareness.

International designations. Ramsar Convention (acceded 1975): 12 wetlands with total designated area 228 344 km².

Swaziland

Wetland distribution. Total area of Swaziland: 17 365 km², approximate wetland area: no reliable estimate is available, probably >100 km². The wetlands are small sponges above 1,400 m (Hughes & Hughes 1992) and occur as headwater wetlands and as bogs along the streams, the main wetland areas are shown in Figure 9.

Inventory status. No systematic inventory has been conducted, Hughes & Hughes (1992) have given a general account of the wetlands.

Wetland characteristics. Mostly upland peat sponges and bogs forming the headwaters of streams.

International designations. None.

Zambia

Wetland distribution. Total area of Zambia: 752 972 km², approximate area of large wetlands including shallow open waters: 38 000 km², or 5%, with at least a further 10% area composed of thousands of dambos (Chidumayo 1992). The larger wetlands (riverine floodplains, perennial swamps and shallow lakes) are mostly associated with the major tributaries of the Zambezi River (Upper Zambezi, Kafue, Luangwa) or the Luapula River which drains to the Zaire basin. Man-made Lake Kariba is deep and has a significant drawdown which prevents edge wetland formation, while Lake Tanganyika has very steep shores that cannot accommodate significant wetlands. The whole

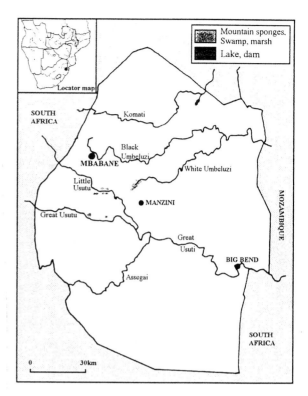

Fig. 9. Map of Swaziland

country has dambos of varying density, their distribution is not shown in Figure 10, however, the major swamps and floodplains are included.

Inventory status. Zambia has an incomplete but extensive coverage of 1:50 000 topographical maps derived mainly from aerial photography of the 1960's. There is a complete set of 1:250 000 maps, parts of which are based on later photography and some sheets of which have been updated and reprinted in the late 1980's. Various areas have received more detailed attention from aerial reconnaissance and photography in the last two decades.

The larger wetlands were described for the Ramsar meeting of 1971 by Clarke (1973) who drew attention to their importance for wildlife as well as for wildfowl and fisheries. The WWF-Zambia Wetlands Project (developed in 1986, Jeffery *et al.* 1992) drew national and regional attention to the wildlife and conservation values of two of Zambia's large wetlands — the Kafue Flats and Bangweulu basin — and in doing so established wetlands as *bona fide* ecosystems within government. Wetlands featured in Zambia's National Conservation Strategy, which resulted in the development of a National Environment Council that is now promoting wetlands conservation and management while contemplating a national wetlands policy. A recent (June, 1992) three-day workshop on the wetlands of Zambia gathered contemporary information on status, distribution, uses and development prospects for these ecosystems and resolved that a national inventory was required.

Wetland characteristics. Zambia's wetlands are diverse and widespread despite the dominance of relatively flat topography and extensive *miombo* woodland. Most wetlands are based on freshwater derived from a rainfall which varies from 500 to 2,000 mm per annum, and from the drainage basins of the Zambezi and Luapula Rivers (with a very small portion which drains to the Zaire system *via* Lake Tanganyika). Hughes & Hughes (1992) have reviewed the major water systems and wetland types, which include: permanent and seasonal herbaceous swamps and swamp forests, marshes, riverine wetlands, floodplains, pans shallow lakes, lake edges and man-made wetlands. The larger floodplains were researched and described in the 1960's because of their potential value as grazing areas for livestock (e.g. Vesey- Fitzgerald 1965; van Rensburg 1968) and there have been significant studies of the wetland fisheries of Zambia (e.g. Muyanga & Chipungu 1982; Bernacsek 1985; Subramaniam 1992).

The wetland that has received the most attention is the Kafue Flats, a floodplain of around 6,000 km^2 containing a complex of swamps, oxbows, small lakes and riverine wetlands as well as extensive seasonally-flooded grasslands (e.g. Ellenbroek 1987; Howard 1985; Howard & Williams 1982). The dambos dominate most of the plateau water courses and are, together, the most significant type of wetland in Zambia (e.g. Perera 1982) as they are not only widespread but are varied in hydrological vegetative characteristics and the uses to which they are put by Zambians.

International designations. Ramsar Convention (acceded 1991): 2 sites, part of Kafue Flats and the Bangweulu Swamps, total designated area: 333 000 km^2. World Heritage site list includes Victoria Falls/Mosi oa Tunya, which includes some riverine wetlands upstream of the cataracts; Zambia (in partnership with Zimbabwe) plans to designate parts of the middle Zambezi valley.

Zimbabwe

Wetland distribution. Total area of Zimbabwe: 390 310 km^2, approximate wetland area (Whitlow 1985): 12 800 km^2 or 3.3%. The lowland of Zimbabwe is dominated by the drainage systems of the Zambezi in the north, Save (east), and the Limpopo (south), while the Nata River drains to the Makgadikgadi Pans in Botswana. The central highland spine of the country has many man-made dams and dambos (seasonal wetlands located in the headwaters of catchments). Figure 11 illustrates the main dambo areas (after Whitlow 1985), however, other swampy areas are too small to include.

Inventory status. A complete set of 1:50 000 topographical maps exists for Zimbabwe, together with aerial photography dating from 1962–1973 at 1:80 000 (18 zones) and 1976 at 1:50 000 (3 zones), (Bullock 1993, pers. comm.). Bullock (1992b) used Landsat TM imagery relating to a basement aquifer study dating to 1989, to assess dambo evaporation rate. A country-wide survey of dambos was conducted by Whitlow (1984, 1985), who concluded that there are about 1.3 million ha of wetlands in Zimbabwe (3.3% of its total area), mostly associated with the headwater valleys of river systems draining the central watershed system. As Figure 11 illustrates, dambo distribution is widespread, but within each area, wetland and dryland are interspersed, making inventory difficult.

Hughes & Hughes (1992) record a different spectrum of wetlands, mainly associated with the lowland rivers and their floodplains, amounting to 55 000 ha of wetland. These partially overlapping, yet different views of Zimbabwe's wetlands emphasise the importance of conducting a national wetland inventory, for no national wetland map yet exists, although one could be constructed at a scale of 1:250 000 based primarily upon the data of Whitlow (1984) and the existing 1:50 000 topographical series.

Wetland characteristics. No systematic review has been attempted, Fish and benthic fauna have been studied for some lakes (Marshall 1978, 1979). For dambos, Bullock (1992a) has examined critically their role in buffering river flows in 110 gauged catchments of six major hydrological zones. He concluded there was no evidence that dambos maintained dry season flows or moderated most flood events. Thus the principal value of headwater dambos appears to be as a source of wetland vegetation and grazing. In a related review of dambo function in southern Africa comparing data from Zimbabwe to other published data (Bullock 1992b), it was concluded that dambos often act to reduce dry season flows, as compared to similar catchments without headwater wetland. However, these views are controversial, for dambos at high alti-

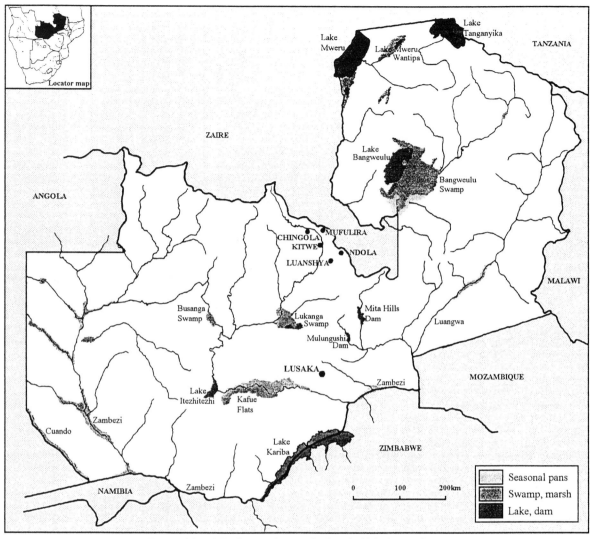

Fig. 10. Map of Zambia. Dambos are distributed widely, but not shown.

tudes (e.g. Lesotho, see Schulze 1979) can exhibit buffering and low flow maintenance.

International designations. World Heritage Area designation sought for Mid-Zambezi Valley and Mana Pools.

Recommendations and conclusions

The wetland inventory process can be a useful focus for developing national conservation strategies and for catalysing the mapping and assessment of other categories of natural resources. We emphasise that establishment through inventory of a baseline for quantify-

ing the extent and rate of alteration of wetlands and other resources, is most valuable in providing a sensitive indicator of the environmental health of nations. We discussed the relative advantages and disadvantages of inventory methodologies, however, before inventory projects embark upon purchase of expensive new aerial or satellite photography, we strongly recommend persistent enquiries to all agencies in a country to establish what imagery already exists. For example in East Africa, Kampala was found to have been photographed within the space of two years by three separate aerial photography projects. It may be that no recent imagery is available, however, existing photography no matter how old, also has value in

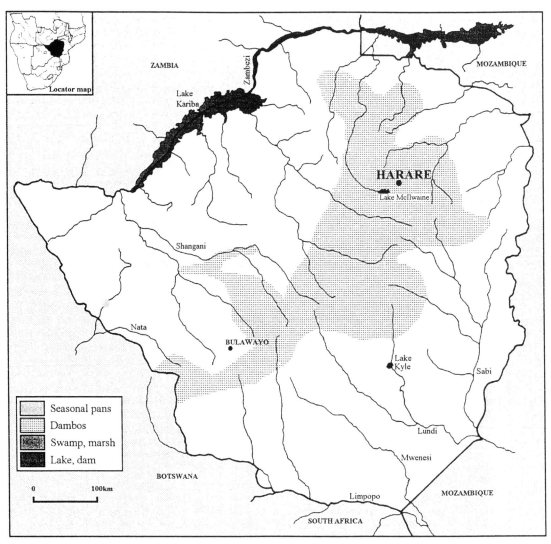

Fig. 11. Map of Zimbabwe. Dambo distribution after Whitlow (1985).

determining the rate of change of the area of wetland and its use.

Our principal recommendation therefore, is that more coordinated effort be put into collation of information that is already available, for there is a wealth of site-specific reports on wetland sites across southern Africa, but, with some notable exceptions, very little attempt to integrate this in order to provide countries with a baseline product suitable for strategic planning. This lack is especially evident in Lesotho and South Africa, which is surprising given their relatively well-developed scientific base. By contrast, Botswana, Malawi, Namibia and Zambia have achieved relatively good results in collation of information. Coun-

tries which have suffered prolonged conflicts, however, such as Angola and Mozambique, urgently require intensive efforts to recover such information, for their wetlands are extensive, of major natural resource value and hence important for the rehabilitation of such nations.

Designers of inventory work must clearly set out their aims at the outset, determine the desired sampling reliability and be aware of a range of factors, the most important being the sampling season and availability of trained staff. Finally, it is no use pouring resources into inventory effort unless the product will be useful, we again emphasise that broad scale assessment has far greater value than painstaking high resolution assessment, hence reconnaissance surveys should be targeted at whole provinces, with more detailed assessment directed initially only at 'hot spots', where wetland use immediately requires better management.

Acknowledgments

Thanks are due to the International Waterfowl and Wetlands Research Bureau, Slimbridge for providing library facilities, also, to the World Conservation Monitoring Centre, Cambridge for providing wetland distribution information for some of the ten countries covered and to the many correspondents contacted in southern Africa and elsewhere.

Addresses of the main remote sensing information providers in southern Africa: General advice on remote sensing: United Nations Environment Programme (UNEP): Global Environment Monitoring System — P.O. Box 30522, Nairobi, Kenya; CSIR, Mikontek, P.O. Box 395, Pretoria 0001, South Africa (SPOT and EOSAT imagery); Satellite Remote Sensing Centre, P.O. Box 3718, Johannesburg 2000, South Africa.

References

Acres, B. D., Blair Rains, A., King, R. B., Lawton, R. M., Mitchell, A. J. B. & Rackham, L. J. 1985. African Dambos: their distribution, characteristics and use. Z. Geomorph. N.F. Supp-Bd. 52: 63–86.

Adamus, P. R., Clairain, E. J., Smith, D. R. & Young, R. E. 1987. Wetland Evaluation Technique (WET); Vol. 2, Operational Draft Report Y-87, US Army Waterways Experiment Station, Vicksburg, Mississippi.

Agnew, S. 1973. Dambo drainage and its effect on land use. East Africa Geographical Review, 11: 43–51.

Anderson, J., Dutton, P., Goodman, P. & Souto, B. 1990. Evaluation of the wildlife resource in the Marromeu complex with recommendations for its future use. Lomaco, Maputo.

Backeus, I. 1988. Mires in the Thaba-Putsoa range of the Maluti, Lesotho. Studies in Plant Ecology, 17. Almqvist and Wiksell International, Stockholm.

Begg, G. W. 1978. The Estuaries of Natal. Natal Town and Regional Planning Report, 41: 657 pp. Pietermaritzburg.

Begg, G. W. 1983. The Estuaries of Natal, (Part 2). Natal Town and Regional Planning Report 55: 631 pp. Pietermaritzburg.

Begg, G. W. 1986. The Wetlands of Natal — an overview of their extent, role and present status. Natal Town & Regional Planning Report, 68. Pietermaritzburg. 114 pp.

Begg, G. W. 1988. The Wetlands of Natal (Part 2). The distribution, extent and status of wetlands in the Mfolozi catchment. Natal Town and Regional Planning Main Studies Report 71: 1–262, Pietermaritzburg.

Begg, G. W. 1989. The Wetlands of Natal (Part 3). The location, status and function of the priority wetlands of Natal. Natal Town & Regional Planning Report 73: 1–256, Pietermaritzburg.

Begg, G. W. 1992. The future prospects for wetland conservation in South Africa: An overview of both the positive and negative aspects. In: Porter, D. J., Craven, H. S., Johnson, D. N. & Porter, M. J. (eds.) Proceedings of the First Southern African Crane Conference, held 9–10 December 1989. Southern African Crane Foundation, Durban. 91–105.

Bernacsek, G. 1985. Impact Evaluation of the Feeder Roads for Fisheries Development Project in Northern Zambia: On the Fisheries of Lake Tanganyika, Lake Mweru-Wa-Ntipa, Lake Mweru and Bangweulu Lakes/Swamp Complex (with special emphasis on stock assessment). A study commissioned by CIDA, Lusaka, Zambia. 205 pp.

Botswana Society 1976. Symposium on the Okavango Delta and its Future Utilisation, Aug. 30 to Sept. 2 1976, held at the National Museum, Gaborone. Botswana Society, Gaborone.

Breen, C. M. 1991. Are intermittently flooded wetlands of arid environments important conservation sites? In: Simmons, R. E., Brown, C. J. & Griffin, M. (eds.), The Status and Conservation of Wetlands in Namibia, Proceedings of a workshop held on 22 November 1988, Windhoek, Namibia. Madoqua 17: 61–65.

Breen, C. M. & Begg, G. W. 1989. Conservation status of southern African wetlands. In: Huntley, B. J. (ed.) Biotic Diversity in Southern Africa: Concepts and Conservation. Oxford University Press, Cape Town. 254–263.

Bullock, A. 1992a. The role of dambos in determining river flow regimes in Zimbabwe. Journal of Hydrology 134: 349–372.

Bullock, A. 1992b. Dambo hydrology in southern Africa — review and reassessment. Journal of Hydrology 134: 373–396.

Bullock, A. 1993. Personal communication. Institute of Hydrology, Crowmarsh Gifford, Wallingford, Oxford, OX10 8BB, U.K.

Burgis, M. J. & Symoens, J. J. (eds.) 1987. African wetlands and shallow water bodies. ORSTOM, Paris.

Chidumayo, E. N. 1992. The utilisation and status of dambos in southern Africa: a Zambian case study. In: Matiza, T. & Chabwela, H. N., (eds.) 1992. Wetlands Conservation Conference for Southern Africa, Proceedings of the Southern African Development Coordination Conference, Gaborone, Botswana, 3–5 June 1991. IUCN, Gland. 242 pp.

Clarke, J. F. 1973. Zambia's Wetlands. Black Lechwe 12: 14- -17.

Cowardin, L. M., Carter, V., Golet, F. C. & La Roe, E. T. 1979. Classification of wetlands and deepwater habitats of the United States. US Fish and Wildlife Service (Department of the Interior). Report FWS/OBS-79/31. US Government Printing Office, Washington, D.C. 131 pp.

Denny, P. (ed.) 1985. The Ecology and Management of African Wetland Vegetation. Kluwer Academic Publishers, Dordrecht.

Ducks Unlimited 1992. Answer to FGDC Fact Finding question 8. In: Federal Geographic Data Committee, 1992. Application of satellite data for mapping and monitoring wetlands — fact-finding report; Technical Report 1. Wetlands Sub- Committee, FGDC. US Department of the Interior, Washington, D.C.

Dugan, P. J. 1990. Wetland Conservation. A review of Current Issues and Required Action. IUCN, Gland. 96 pp.

Ellenbroek, G. A. 1987. Ecology and productivity of an African wetland system. The Kafue Flats, Zambia. Dr W. Junk, Dordrecht. 267 pp.

FAO/UNDP/GOB 1986. Soil Map of the Boro-Shorobe area. Soil mapping and Advisory Service Project BOT/80/003, FAO/UNDP/Government of Botswana. Government Printer, Gaborone.

78

Federal Geographic Data Committee 1992. Application of satellite data for mapping and monitoring wetlands — fact- finding report; Technical Report 1. Wetlands Sub-Committee, FGDC US Department of the Interior, Washington, D.C.

Geldenhuys, J. N. 1981. Classification of the pans in the Orange Free State according to vegetation structure, with reference to avifaunal communities. South African Journal of Wildlife Research 12: 55–62.

Harrison, A. D. 1959. General statement on South African hydrobiological regions. National Institute for Water Research, Report No. 1, Project 6.8H. Cited by O'Keeffe *et al.* 1989.

Heeg, J. & Breen, C. M. 1982. Man and the Pongolo floodplain. South African National Scientific Programmes Report 56, CSIR, Pretoria. 117 pp.

Heydorn, A. E. F. & Tinley, K. L. 1980. Estuaries of the Cape, Part 1 — Synopsis of the Cape Coast. Natural Features, Dynamics and Utilisation. CSIR Research Report 380, Stellenbosch.

Heydorn, A. E. F. & Grindley, J. R. (eds) 1981–1985. Estuaries of the Cape: Part II: Synopses of available information on individual systems. CSIR Research Reports, Stellenbosch.

Howard, G. W. 1985. The Kafue Flats of Zambia — a wetland ecosystem comparable with floodplain areas of northern Australia. Proceedings of the Ecological Society of Australia 13: 293–306.

Howard, G. W. & Williams, G. J. (eds) 1982. The consequences of hydroelectric power development on the utilisation of the Kafue Flats. Kafue Basin Research Committee, University of Zambia, Lusaka. 159 pp.

Hughes, R. H. & Hughes, J. S. 1992. A Directory of African Wetlands. IUCN, Gland and Cambridge/UNEP, Nairobi/WCMC, Cambridge. 820 pp.

IUCN Wetlands Newsletter, No. 5, June 1992, p. 2. IUCN, Gland.

Jeffery, R. C. V., Chabwela, H. N., Howard, G. W. & Dugan, P. J. (eds) 1992. Managing the Wetlands of Kafue Flats and Bangweulu Basin. Proceedings of the WWF — Zambia Wetlands Project Workshop, November 1986. IUCN, Gland. 113 pp.

Jones, T. A. (compiler) 1993. A Directory of Wetlands of International Importance, Part One: Africa. Ramsar Convention Bureau, Gland.

Marshall, B. E. 1978. Aspects of the ecology of the benthic fauna in Lake McIlwaine, Rhodesia. Freshwater Biology 8: 241-249.

Marshall, B. E. 1979. Fish populations and fisheries potential of lake Kariba. South African Journal of Science 75: 485–488.

Matiza, T. & Chabwela, H. N. (eds) 1992. Wetlands Conservation Conference for Southern Africa. Proceedings of the Southern African Development Coordination Conference held in Gaborone, Botswana, 3–5 June 1991. IUCN Gland. 224 pp.

McCarthy, T. S. 1992. Physical and Biological Processes Controlling the Okavango Delta — A Review of Recent Research. Department of Geology, University of the Witwatersrand, Johannesburg.

Meadows, M. E. 1985. Dambos and environmental change in Malawi, Central Africa. Zeitschrift Geomorphologie N.F. 52: 147–169.

Muyanga, E. D. & Chipungu, P. M. 1982. A short Review of the Kafue Flats Fishery from 1968 to 1978. pp. 105–113 in: Howard & Williams (eds) 1982, *ibid.*

Mzembe, C. P. 1990. Malawi: Wetland Development and Management. In: FAO Workshop on Technical Cooperation Network in Wetland Development and Management, Bajul, Gambia, April 1990. FAO Rome. pp. 128–139.

Nakayama, M. 1992. Application of remote sensing techniques to monitor lakes and wetlands in developing countries. In: Isozaki, H. M. A. & Natori, Y. (eds) Proceedings of the Asian Wetlands Symposium, 15–20 October 1992, Otsu and Kushiro, Japan.

International Lake Environment Committee Foundation, Kusatsu, Japan.

Noble, R. G. & Hemens, J. 1978. Inland Water Ecosystems in South Africa — a review of research needs. South African National Scientific Programme Report, 34. 150 pp.

O'Keeffe, J. H., Davies, B. R., King, J. M. & Skelton, P. H. 1989. The conservation status of southern African rivers. In: Huntley, B. J. (ed): Biotic Diversity in Southern Africa: Concepts and conservation. Oxford University Press, Cape Town, pp. 266–289.

Perera, N. P. 1982. The evolution of wetlands (dambos) of Zambia, and their evaluation for agriculture — a model for the management of wetlands in sub-humid eastern and southern Africa. Int. J. Ecol. Environ. Sci. 8: 27–38.

Schulze, R. E. 1979. Hydrology and water resources of the Drakensberg. Natal Town and Regional Planning Report, 42: 179 pp. Pietermaritzburg.

Schwabe, C. A. 1989. The assessment, planning and management of wetlands in the Maluti/Drakensberg mountain catchments. Institute of Natural Resources, Pietermaritzburg, Investigation Report 38: 1–73.

Schwabe, C. A. & Nthabane, D. K. 1989. Document No. 3, Drakensberg/Maluti Mountain Catchment Conservation Programme. Wetlands Project, Institute of Natural Resources, Pietermaritzburg.

Scoones, I. 1991. Wetlands in Drylands — key resources for agricultural and pastoral production in Africa. Ambio 20: 366–371.

Scotney, D. M. 1970. Soils and Land-use Planning in the Howick extension area. PhD thesis (Pasture Science), University of Natal, Pietermaritzburg: 361 pp (Appendix and soils maps: 362–380).

Scotney, D. M. 1978. The present situation in Natal. Proceedings of a symposium on the relationship between agriculture and environmental conservation in Natal and KwaZulu, Durban. 19–20 October 1978: 16–34.

Scotney, D. M. & Wilby, A. F. 1983. Wetlands and agriculture. Journal of Limnological Society of southern Africa 9: 134–140.

Scott, D. A. 1989. Design of Wetland Data Sheet for Database on Ramsar Sites. Report to Ramsar Convention Bureau, Gland. 41 pp.

Scudder, T. 1993. Personal communication. Division of the Humanities and Social Sciences, California Institute of Technology, Pasadena, California 91125, USA.

Scudder, T., Manley, R. E., Coley, R. W., Davis, R. K., Green, J., Howard, G. W., Lawry, S. W., Martz, D., Rogers, P. P., Taylor, A. R. D., Turner, S. D., White, G. F. & Wright, E. P. 1993. The IUCN Review of the Southern Okavango Integrated Water Development Project. IUCN, Gland. 543 pp.

Semeniuk, V. & Semeniuk, C. 1993. Cited by Finlayson, C. L. M., 1993 (personal communication).

Silberbauer, M. J. & King, J. M. 1991. The distribution of wetlands in the south-western Cape Province, South Africa. South African Journal of Aquatic Sciences 17: 65–81.

Simmons, R. E., Brown, C. J. & Griffin, M. (eds) 1991. The Status and Conservation of Wetlands in Namibia. Special edition: Madoqua, 17. 254 pp.

SMEC/Kalahari Conservation Society 1989. Ecological Zoning, Okavango Delta. Report in 2 volumes to Ngamiland District Land Use Planning Unit (Ministry of Local Government, Lands and Housing). Snowy Mountains Engineering Corporation.

Staring, G. J. 1978. Soils of the Okavango Delta, Field Document No. 14: Swamp and dryland soils of the Okavango. Project BOT 72/019, UNDP/FAO and Government of Botswana (report and maps).

Stormanns, C. H. Breen, C. M. (eds) 1986. Proceedings of the Greater Mkuze Swamp Symposium and Workshop, 26 March

1986. Natal Parks Board, Pietermaritzburg. Insitute of Natural Resources Investigation Report (22).

Subramaniam, S. P. 1992. A brief review of the status of the fisheries of the Bangweulu basin and Kafue Flats. pp. 44–55 in Jeffery *et al.* (eds), *ibid.*

Tweddle, D., Hastings, R. E. & Jones, T. 1978. The development of a floodplain fishery: Elephant Marsh, Malawi. In: Welcomme, R. (ed.), Symposium on river and floodplain fisheries in Africa. Burundi 1977. CIFA Technical Paper No. 5. FAO, Rome.

Tweddle, D. & Mkoto, B. J. 1986. A limnological bibliography of Malawi. CIFA Occasional Paper 13. FAO, Rome.

van Rensburg, H. J. 1968. Ecology and development. Multipurpose Survey of the Kafue River Basin, Zambia. Vol. 4, FAO/SF.35 ZAM, FAO, Rome.

van der Eyck, J. J., MacVicar, C. N. & de Villiers, J. M. 1969. Soils of the Tugela Basin — a study in sub-tropical Africa. Natal Town and Regional Planning Report, 15. 236 pp.

Vesey-Fitzgerald, L. D. E. F. 1965. Lechwe pastures. The Puku 3: 142–147.

Walmsley, R. D. & Boomker, E. A. (eds) 1988. Inventory and classification of wetlands in South Africa. Proceedings of a Workshop. Foundation for Research Development, CSIR, Pretoria. Occasional Report No. 34, Ecosystem Programmes. 90 pp.

Whigham, D. F., Dykyjova, D. & Hejny, S. 1993. Wetlands of the World I: Inventory, Ecology and Management. Kluwer Academic Publishers, Dordrecht.

Whitlow, J. R. 1984. A survey of dambos in Zimbabwe. Zimbabwe Agricultural Journal, 81(4): 129–138.

Whitlow, J. R. 1985. Dambos in Zimbabwe: a review. Zeitschrift Geomorphologie N. F. 52: 115–146.

Vegetatio **118**: 81–101, 1995.

A review of wetland inventory and classification in Australia

R.L. Pressey[1] & P. Adam[2]
[1]*New South Wales National Parks and Wildlife Service, PO Box 402, Armidale NSW 2350, Australia*
[2]*School of Biological Science, University of New South Wales, NSW 2052, Australia*

Key words: Australia, Classification, Inventory, Wetlands

Abstract

Studies of wetlands in Australia, as in other countries, have taken a wide variety of approaches to defining, surveying and classifying these environments. Past and current approaches in Australia are reviewed for each of the States and Territories which provide the context for much of the natural resource investigation in the country. While there are obvious advantages of national, and perhaps international, agreement on definition and types of wetlands, a variety of approaches to inventory and classification will always be necessary for particular purposes. More fundamental than general agreement on approaches is the need for wetland scientists and managers to maximise the accuracy of survey information, to test the assumptions involved in the use of classifications, and to ensure that the classifications they use are the most appropriate for their purposes. The issue of a global wetland classification scheme is discussed on the basis of a representative range of views by Australian wetland workers.

Introduction

The processes of inventory and classification of wetlands influence our perceptions of an important natural resource. Methods for inventory determine the accuracy and precision of information on wetlands and its utility in all aspects of planning, management and conservation. The choice of how to group wetlands into types with some homogeneity of features important to particular tasks is also critical. The wrong choice will produce classes which are not informative about processes or the distribution, in time and space, of the features of interest.

Scientists and managers have for many years applied a wide variety of approaches to inventory and classification of wetlands. At least in Australia, and perhaps in other parts of the world, there has been little exchange of ideas on the appropriateness of particular approaches for particular purposes. Much of the Australian literature in this field has limited circulation and meetings of specialists to discuss techniques are uncommon. A review of past approaches and current thinking therefore has two advantages: it facilitates

communication between individuals and groups; and it can identify problems to be addressed.

This review briefly describes the environmental and political setting of Australian wetlands and discusses the problems of defining wetlands and identifying their boundaries. The main review of activities is structured according to the separate States and Territories as these provide the political, administrative and geographical context for most surveys and classifications. This is followed by a section that attempts to synthesise the information from the review and finishes with a summary of Australian views on standardised global approaches.

The setting for Australian wetlands

Environmental overview

Much of the surface geology of Australia is ancient and many of the soils and landforms are millions of years old. The topography is, therefore, mainly subdued with a major system of highlands, the Great Dividing Range, close to the eastern seaboard and several range systems in the west (Fig. 1). Australia has been divided into twelve major drainage divisions (Fig. 2) comprising varying numbers of basins (Australian Water Resources Council 1976).

Australia spans 32 degrees of latitude and, although the range of climates is less than on other continents, there is still substantial variation in temperature and rainfall regimes (Linacre & Hobbs 1977). Most of the continent is arid or semi-arid. Median annual rainfall in Australia shows a concentric pattern with the highest falls around the coastal fringes (Fig. 3). Seasonality of rainfall varies strongly and is summarised for drainage divisions in Table 1. Rainfall variability in Australia is very high by world standards (Stafford Smith & Morton 1990) and is also distributed concentrically with highest values in the lowest rainfall regions (Paijmans *et al.* 1985). The trend in annual potential evaporation, too, is roughly the reverse of that for median rainfall. These patterns in rainfall and evaporation lead to large differences in average annual runoff between the twelve drainage divisions (Table 1) and a great diversity of water regimes in Australian wetlands, from permanent to seasonal at various times of year to irregularly filled at different average frequencies (Finlayson 1991).

The general aridity of the inland by no means indicates a lack of wetlands. The semi-arid areas that are penetrated by major rivers from the eastern highlands have long been recognised as containing extensive and biologically productive wetlands (Frith 1967). Evidence is also accumulating for the great extent and significance, on a world scale, of wetlands still further into the arid interior (Halse 1990; Kingsford *et al.* 1991; Kingsford & Porter 1993).

For many estuarine and nearshore marine wetlands, other factors are important in influencing their development and composition. Davies (1986) has suggested a broad categorisation of the Australian coastline into four segments – warm temperate humid, warm temperate arid, tropical arid and tropical humid. Within each segment there is a broad similarity of climate, wave energy, tidal regime, coastal landforms and biota. Around much of the coast there are micro- or mesorange tides, but in the north and north-west macrotidal ranges, up to 11 metres, prevail. The lateral extent of intertidal wetlands is therefore much greater in the tropics than on the temperate coasts.

Political divisions
The Commonwealth of Australia consists of six States – New South Wales, Queensland, South Australia,

Tasmania, Victoria and Western Australia – and two mainland Territories – the Australian Capital Territory and Northern Territory (Fig. 2). Australia administers three subantarctic islands, Macquarie Island (politically part of Tasmania) and Heard and Macdonald Islands in the Indian Ocean as well as a number of subtropical and tropical islands and reefs. In addition, the Australian Antarctic Territory is one of the largest territorial claims on the Antarctic continent.

Most issues concerning the management and conservation of land or water resources are the responsibility of the separate States and Territories which vary considerably in their legislative and bureaucratic mechanisms for planning, conservation, land use control and resource management. The dispersion of involvement with wetlands is even greater when the particular concerns of agencies are considered. Barson & Williams (1991) identified 57 State or Territory agencies with responsibilities for wetlands, either because they have charters for management of natural resources or because their activities affect wetlands. As in other continents subdivided into states or nations, the Australian State and Territory borders have little to do with natural boundaries such as geological discontinuities, catchments or river systems. Many wetlands that have similar requirements for management and conservation are therefore placed under different administrations and covered by different approaches to inventory and classification.

The national government (the Commonwealth of Australia) does, however, play a role in wetland issues. Australia has entered into several international treaties and agreements dealing with environmental matters and, under the Constitution, takes on powers in respect of these. Examples are the Ramsar Convention and the Japan-Australia Migratory Bird Agreement. The Commonwealth also has a role in coordinating and promoting initiatives in surveys of natural resources as well as their management and conservation and is responsible for developing and maintaining major environmental data bases (see Michaelis & O'Brien 1988; Bridgewater 1991 for reviews).

Concerns about wetlands
European settlement of Australia was accompanied by massive loss and disturbance of natural habitats, including wetlands. Drainage and other modifications of wetlands on the coastal floodplains and other high-rainfall areas, where the bulk of the population has always been concentrated, proceeded rapidly from the

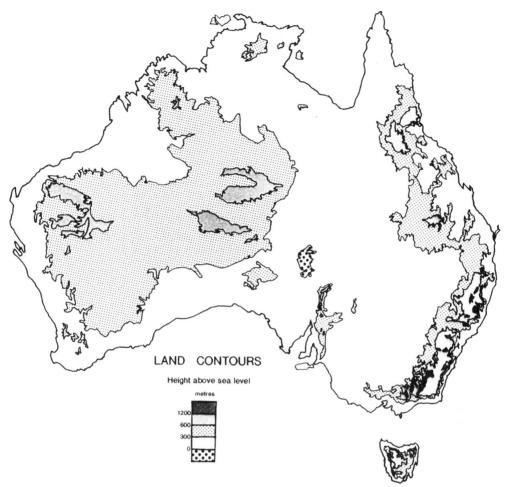

Fig. 1. Generalised relief map (from Department of Minerals and Energy 1973). This map has been derived from a Crown Copyright map and has been reproduced with the permission of the General Manager, Australian Surveying and Land Information Group, Department of Administrative Services, Canberra, ACT.

early years of settlement (Pressey & Middleton 1982; Middleton *et al.* 1985). Vast areas of wetlands were modified by domestic stock as pastoralism became the most widespread land use in the hinterland (Beadle 1948; Williams 1962; Wimbush & Costin 1979a, b). Soon afterwards, regulation of coastal and inland rivers changed the main channels and associated floodplain habitats both hydrologically and biologically (Walker 1985). Many other changes followed, including chemical pollution, burning, sedimentation, extraction of water, introduction and translocation of plants and animals, and recreation activities (see McComb & Lake 1988 for reviews).

By the late 1960s, the extent of wetland loss and damage had come to general notice, at least in the densely settled regions (Riggert 1966; Goodrick 1970).

At about the same time, the importance of estuarine wetlands for maintaining fisheries was being promoted (Pollard 1976). Research on waterfowl in the Murray-Darling Basin led to the realisation that wetlands could be damaged indirectly by changing river regimes (Frith & Sawer 1974). Campaigns by non-government organisations and increasing concern in conservation agencies over loss and alteration of wetlands led to much activity in survey and classification in the late 1970s and 1980s (Fig. 4).

The increase in survey activities and classifications of wetlands has been accompanied by moves to protect wetlands from a variety of threatening processes. The conservation of wetlands by reservation and management of processes is now a major concern in Australia. As well as formal reservation, regulatory mecha-

84

Fig. 2. Borders of States and mainland Territories and boundaries of Australian drainage divisions as defined by the Australian Water Resources Council (1976). 1: North-East Coast Division (450 705 km^2, 46 river basins); 2: South-East Coast Division (273 553 km^2, 39 basins); 3: Tasmanian Division (68 200 km^2, 19 basins); 4: Murray-Darling Division (1 062 530 km^2, 26 basins); 5: South Australian Gulf Division (82 300 km^2, 13 basins); 6: South-West Coast Division (314 090 km^2, 19 basins); 7: Indian Ocean Division (518 570 km^2, 10 basins); 8: Timor Sea Division (547 050 km^2, 26 basins); 9: Gulf of Carpentaria Division (638 460 km^2, 28 basins); 10: Lake Eyre Division (1 170 000 km^2, 7 basins); 11: Bulloo-Bancannia Division (100 570 km^2, 2 basins); 12: Western Plateau Division (2 455 000 km^2, 9 basins). This map has been derived from a Crown Copyright map and has been reproduced with the permission of the General Manager, Australian Surveying and Land Information Group, Department of Administrative Services, Canberra, ACT.

nisms and management protocols are covering increasing numbers of sites. These include listing of wetlands under the Ramsar Convention and other international agreements (Bridgewater 1991; Australian Nature Conservation Agency 1993), environmental allocations of water by bodies such as the Murray-Darling Basin Commission (Close 1990; Murphy 1990), and water management trials along the Murray River in South Australia. Recent measures to regulate potentially damaging activities in wetlands include State Environmental Planning Policy (SEPP) 14 in coastal New South Wales and the Environmental Protection (Swan Coastal Plain Lakes) Policy in Western Australia.

What is a wetland and where are its edges?

Differences in definitions of the term 'wetland' are universal. Australia has no nationally agreed definition of wetland (Barson & Williams 1991) and, even within a single State on Territory, different agencies employ different definitions for their own purposes and legislation. This leads to inconsistencies in the scope of surveys and classifications and is a source of considerable confusion to the public.

Why define wetlands?
If management and conservation were applied equally to all components of the landscape, wetlands would not

Table 1. Rainfall seasonality and runoff in drainage divisions (data from Paijmans *et al.* 1985)

	Drainage division	Rainfall seasonality	Average depth of runoff/yr (mm)
1.	North-East Coast	VMS-MS	183
2.	South-East Coast	SUW	144
3.	Tasmania	UW	730
4.	Murray-Darling	ASUW	21
5.	South Australian Gulf	AW	12
6.	South-West Coast	VMW-MW	23
7.	Indian Ocean	A	8
8.	Timor Sea	VMS	136
9.	Gulf of Carpentaria	VMS	91
10.	Lake Eyre	AS	3
11.	Bulloo-Bancannia	AS	5
12.	Western Plateau	A	0

VMS - very marked summer rainfall
VMW - very marked winter rainfall
MS - marked summer rainfall
MW - marked winter rainfall
S - summer rainfall
U - uniform rainfall
W - winter rainfall
A - arid zone rainfall

have to be separated from the other components with which they interact physically, chemically and biologically. Moreover, the many habitats often grouped together as wetlands are not a natural, homogeneous group. Many have more in common with non-wetland habitats than with each other. However, because of the history of mismanagement of many wetland types, the distinctive characteristics and management needs of most types, the specific charters of some agencies dealing with fisheries and wildlife, and the great momentum of the wetland conservation lobby, wetlands are most often surveyed and classified in isolation. The problem of definition then arises.

The planning instruments that have been applied to wetlands need lines on maps or some consistent definitions. This can limit the interpretation of wetlands to areas that can be reliably distinguished from terrestrial habitats on aerial photographs (Winning 1991a; Adam 1992), perhaps leading to the disappearance of surrounding habitats with less legal standing, regardless of conservation significance or their interactions with the wetlands themselves. In some cases, as in the Environmental Protection (Swan Coastal Plain Lakes) Policy in Western Australia, the focus has narrowed to

only a subset of areas that would typically be regarded as wetlands. Constraints on land use under such regulations mean that definitions have considerable economic and political significance, as in the United States (Bohlen 1991; Kusler 1992).

The recent emphasis on Total Catchment Management in Australia (e.g. Soil Conservation Service 1987) avoids treating wetlands in isolation. It can consider the conservation needs of other habitats and address external influences on the wetlands themselves.

Australian definitions of wetlands

'Wetlands are quirks and local aberrations of the hydrological cycle which differ from their surroundings by the persistent presence of free water' (Paijmans *et al.* 1985). These authors of the only Australian overview of wetland distribution define wetlands broadly as land permanently or temporarily under water or waterlogged, stipulating that temporary wetlands must have surface water or waterlogging of sufficient frequency and/or duration to affect the biota. They also note that inland basins that might only fill once in several decades should also be included because these areas support a distinctly aquatic biota at these times. Their

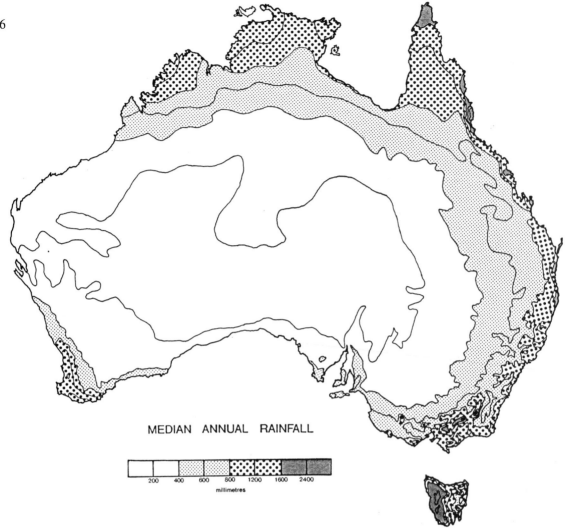

MEDIAN ANNUAL RAINFALL

Fig. 3. Distribution of median annual rainfall (from Snowy Mountains Engineering Corporation 1983). This map has been derived from a Crown Copyright map and has been reproduced with the permission of the General Manager, Australian Surveying and Land Information Group, Department of Administrative Services, Canberra, ACT.

definition includes open and vegetated basins, floodplains, stream channels, tidal flats and coastal waterbodies. Many Australian surveys of wetlands have omitted flowing waters and free-draining floodplains although, in reality, there is a spectrum of hydrological and biological conditions between flowing and still water, and many important interconnections between stream channels, floodplains and associated basins (Bren & Gibbs 1986; Mitsch & Gosselink 1986; Boulton & Lloyd 1992). Otherwise, the definition of Paijmans *et al.* (1985) broadly matches most interpretations of wetlands adopted in Australian surveys. Even so, Australian definitions differ widely in detail.

Most Australian interpretations of wetland depart substantially from the much broader Ramsar defini-

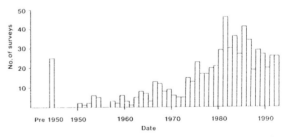

Fig. 4. Dates of completion or publication of wetland surveys and classifications in Australia.

tion (UNESCO 1971): 'areas of marsh, fen, peatland or water, whether natural or artificial, permanent or temporary, with water that is static or flowing, fresh, brackish or salt, including areas of marine water the depth of which at low tide does not exceed 6 m'. Many Australian wetland surveys have excluded artificial water bodies and most have restricted their coverage of marine areas only to the intertidal areas of soft sediments (Barson & Williams 1991; Pressey *et al.* in press). Coral reefs have seldom been regarded as wetlands in Australia and the six metre demarcation is problematical, not only for reefs but for seagrass beds which extend down to at least twenty metres depth in some areas (Coles *et al.* 1992).

Many delegates at a wetlands workshop in 1991 organised by the Bureau of Rural Resources opted to use the Ramsar definition because of its wide application overseas and for the sake of having a definition that could be promoted for consistent use. However, its general acceptance is still uncertain and a variety of definitions are still in use.

Wetland boundaries
The problem of delineating wetlands is partly solved by defining the types of environments to be considered but, even when a definition has been adopted for the purposes of survey, the edges of wetlands are not always easy to identify. This leads to problems of credibility and to uncertainties over ownership and the application of land use controls. Deciding on the exact extent of wetlands can also be expensive when consultants are needed to interpret definitions, as for Federal regulations in the United States (Kusler 1992).

The position of a wetland boundary can become crucially significant where legal instruments are applied. Knights (1991) considered that most of the existing wetland mapping in New South Wales was not precise enough to support legal protective mechanisms. For this type of regulation, a legal fiction is required – an immovable boundary sharply defined on paper and on the ground. In reality, wetlands and adjacent landward or seaward habitats are connected by a variety of processes, and wetland boundaries are often diffuse and fluctuate through time (Howard-Williams 1991; Winning 1991a). In many areas, short-term fluctuations in wetland boundaries could be insignificant compared to those that might result from climate change and rises in sea level (Bryant 1990; Burchmore 1990). While the importance of buffer zones around wetlands is often recognised, the functions and processes of wetland buffers are poorly understood, leading to arbitrary recommendations on their width and composition (Lane 1991).

Wetland inventory and classification in Australia

A review of Australian approaches to wetland inventory and classification is largely a review of activities in each separate State and Territory. Such a piecemeal approach does not give an overview of natural regions or particular management concerns (Bowman & Whitehead 1989), but the fact remains that the States and Territories are the framework within which inventory and classification of wetlands and other natural resources proceed. Some surveys and proposals for classification do transcend political borders, however, and these are reviewed in a separate section.

The review has a number of existing compilations to draw from, notably those of McComb and Lake (1988) and Donohue and Phillips (1991). The summaries below, as well as being based on many other reports and papers, also contain more recent information gathered in mid-1992. This was obtained by writing to some 68 wetland scientists and managers throughout Australia to ask for information on current activities and methods and for views on the feasibility and desirability of global standards for inventory and classification.

The summaries in the following sections refer only to some of the studies and findings on wetland inventory and classification in Australia. A much more comprehensive list of surveys and classifications is in a bibliography compiled as a basis for this paper (Pressey *et al.* in press). Terms used to describe the scale and detail of surveys follow those of Dent and Young (1981): reconnaissance – typically 1:250 000; broad ('semi-detailed') – 1:100 000 to 1:50 000, detailed – 1:25 000 to 1:10 000; and intensive – finer than 1:10 000. Wetland inventories or surveys are studies that have mapped and described the wetlands of an area, although not all have focused on wetlands specifically. Wetland classifications are any attempts, intuitive or numerical, to group wetlands with common characteristics or to identify the types of environments and biota they contain.

National or multi-State surveys and classifications
Several studies have produced overviews of the types of wetlands occurring throughout Australia. Those by Briggs (1981), Beadle (1981) and Brock (1994) are based on different definitions of vegetation types. Timms (1992a) has proposed a classification of lakes based on geomorphic origin, drawing mainly from

examples in Australia. Limnological overviews are also emerging (De Deckker & Williams 1986 and references therein). In their review of wetlands in temperate Australia, Jacobs and Brock (1993) used a framework that was primarily geographical with subcategories based on geomorphology, vegetation and water regime. Paijmans *et al.* (1985) developed a classification that was designed to be interpreted from reconnaissance-scale topographic maps which they used to map the occurrence and density of wetlands across the country. This work followed an assessment of the feasibility of a more detailed national inventory (Paijmans 1978a). Another national perspective is the recent Directory of Important Wetlands in Australia (Australian Nature Conservation Agency 1993), a compilation of information and references on 520 wetlands or wetland complexes.

Some of the surveys and classifications concentrating on estuarine wetlands have transcended State and Territory boundaries. A national inventory of Australian estuaries and enclosed marine waters was completed in 1989 (Bucher & Saenger 1989, 1991), generating a data base on 783 estuaries and embayments and their catchments. Information from this data base and the coastal components of the survey of Paijmans *et al.* (1985) have been summarised by the Resource Assessment Commission (1993). Surveys and classifications of estuarine wetlands most often focus on the three main vegetation types: seagrass, mangrove and saltmarsh, although unvegetated, hypersaline flats are extensive in northern Australia (Bucher & Saenger 1991) and might be functionally important for nutrient cycling in coastal waters (Ridd *et al.* 1988).

Australia is a major centre of diversity for seagrasses. Most seagrass beds are dominated by a single species so the widely used informal classifications recognise dominants. Multi-State studies include those on distribution, habitats and species of seagrasses (Poiner *et al.* 1989; Shepherd & Robertson 1989; West *et al.* 1989), the dynamics and decline of seagrasses in a broad context (Clarke & Kirkman 1989; Shepherd *et al.* 1989), and faunal assemblages (Bell & Pollard 1989; Howard *et al.* 1989).

Australia is also a major centre of mangrove diversity with nearly 40 species. The area of mangroves exceeds 11 000 square kilometres, the majority in the tropics (Galloway 1982; Galloway *et al.* 1984). Distinctions between mangrove types include the phytosociological classification of Bridgewater (1985), the subdivision by Saenger *et al.* (1977) of mangrove coasts into biogeographic provinces and, like

Love (1981), discussions of different zonation patterns. Galloway (1982) separated mangrove coasts into sections with relatively consistent geomorphological features.

Despite the popular perception that saltmarsh is the temperate replacement of mangrove vegetation, most of the Australian saltmarsh occurs in the tropics (Bucher & Saenger 1991). Bridgewater (1982) and Adam *et al.* (1988) have identified phytosociological types and Saenger *et al.* (1977) and Bridgewater (1982) have subdivided saltmarsh coasts biogeographically.

A few surveys of inland wetlands, aside from the national overviews, have also straddled State borders, including two studies of the Murray River system. The first of these (Pressey 1986) covered mainly floodplain basins and developed specific classifications of geomorphology and hydrology. The second concerned riparian vegetation (Margules & Partners and others 1990), using numerical classification of floristic site data and interpretation of aerial photographs for dominant species and structure.

In another multi-State survey, Paijmans (1978b) covered four extensive areas in New South Wales and Queensland, classifying wetlands broadly according to association with streams, tidal influence and physiography. Since 1983, there has been an annual aerial survey of waterbirds in eastern Australia that also provides data on the number and water area of wetlands (e.g. Kingsford *et al.* 1991). For the eastern coastline, Timms (1986) synthesised the limnological information available on dune lakes and distinguished six types based mainly on geomorphic position and hydrology. Ponder (1986) has compiled information on the mound springs of the Great Artesian Basin, covering parts of Queensland, New South Wales and South Australia, and reviewed variations in physical, chemical and biological features. Areas in both Victoria and New South Wales have been covered by Johnston and Barson (1993) in tests of the Landsat Thematic Mapper for identifying major wetland plant communities.

Antarctica and subantarctic islands
Antarctica is the driest continent, although its icemass contains some 70% of the world's fresh water. Most of its surface is ice-covered, but there are limited areas of exposed rock and ice-free ground (Walton 1984). One of these is the Vestfold Hills, site of the Australian Davis Station. Within this area of about 400 km^2 are a number of lakes, some fresh and others saline or hypersaline (Adamson & Pickard 1986).

The subantarctic islands are subject to cold, wet and windy climates. Macquarie and Macdonald Islands have no permanent snow cover whereas Heard Island, which rises to about 2750 m, has only about 15% snow-free ground (Walton 1984). The vascular flora of all the subantarctic islands is small (Wace 1960). The wetland vegetation of Macquarie Island includes a range of mires, drainage channels and lakes and is described by Selkirk *et al.* (1990). The less extensive area of wetland vegetation on Heard Island has been described by Hughes (1987).

Australian Capital Territory
Although the smallest of the States and Territories of the Australian land mass, the pattern of wetland surveys in the Australian Capital Territory is similar to most of the others – there has been no comprehensive survey of wetlands and the available information comes from a variety of studies that have addressed wetlands at different scales and for different purposes (Williams 1991). The major lentic wetlands in the region are the open waters and fringing shallows of the manmade urban lakes in Canberra and the subalpine and montane bogs and fens of the ranges. The urban wetlands are also being augmented by ongoing construction of ponds to trap sediments from eroding catchments.

Information is very incomplete on the higher altitude bogs and fens of the Territory and the neighbouring areas of New South Wales and Victoria so the ACT Parks and Wildlife Service has recently compiled a report on these habitats (Evans & Keenan 1992) as a basis for further documentation and survey.

New South Wales
The general types of wetlands in the State have been defined in several broad geographic subdivisions. Those of Goodrick (1983) and Pressey and Harris (1988) are similar to the major elements of Jacobs' (1983) more detailed breakdown of the State's wetlands according to vegetation. They were followed by the comprehensive classification of Winning (1988) who divided the State into physiographic regions, each with systems and subsystems having distinct hydrology and geomorphology. The most detailed State-wide classification of wetlands is based on geomorphology and was proposed by Riley *et al.* (1984). Roy (1984) proposed a classification of New South Wales estuaries based on two criteria: entrance conditions at the time of formation and degree of infilling.

Pressey & Harris (1988) found that the coastal wetlands of the State are relatively well known, having been covered completely in several surveys, from reconnaissance to detailed in scale, and partly by localised detailed and intensive studies, mostly north of Sydney. A long-established classification of coastal wetlands came from the survey by Goodrick (1970) and is based mainly on vegetation and water regime. Subsequent studies have used different methods for inventory and classification and focused on particular wetland types. For example, Griffith (1984) identified plant associations in the coastal dune formations, West *et al.* (1985) mapped the broad types of estuarine vegetation for the whole coast, and Pressey (1989) listed a set of attributes of wetlands larger than 0.1 ha on the Clarence floodplain. The only overall survey after Goodrick's (Adam *et al.* 1985) mapped the boundaries of certain types of coastal wetlands, without classifying or describing them, as a basis for regulating land use.

The diversity of wetlands on the ranges and tablelands reflects local and regional differences in climate, geology, landform and hydrology. Only localised areas have been surveyed in any detail and there are large gaps even in the coverage of reconnaissance and broad-scale surveys and in knowledge of the variation in wetland types. One of the few extensive surveys is that of peatlands by Hope and Southern (1983).

A general picture of the occurrence and types of wetlands in inland New South Wales has been provided by reconnaissance-scale mapping of land systems, recurring patterns of landform, soils and vegetation (Walker 1991). The land system approach has also been adapted to identify wetland systems in the northwest of the State (Goodrick 1984) and on part of the Gwydir River (State Pollution Control Commission 1978). With concern over the effects of river regulation and other environmental changes, the inland riverine wetlands are the subjects of increasing numbers of broad-scale surveys, based on interpretation of aerial photographs and selective field visits. Most of these have been undertaken by the Department of Water Resources, previously the Water Resources Commission (e.g. Water Resources Commission 1986; Green 1992), with classification derived from dominant vegetation and hydrology and a total area surveyed in the order of 1.5 million hectares. The non-riverine wetlands of the inland, apart from those in the far northwest, remain poorly known.

In other recent work, Timms (1992b) sampled 102 lakes in New South Wales to produce a summary of ecological condition, covering the range of geomorphic types roughly in proportion to abundance. Future survey work will include further inland and coastal studies by the Department of Water Resources (P. Wettin, Department of Water Resources, pers. comm.). In addition, the Fisheries Research Institute of the Department of Fisheries has plans to re-map estuarine wetlands as well as deepwater marine habitats with more emphasis on field verification than in the earlier coverage (R. West, Fisheries Research Institute, pers. comm).

Northern Territory

Finlayson *et al.* (1988a) summarised the information on the Northern Territory's wetlands up to that date, using a framework of wetland types similar to that used nationally by Paijmans *et al.* (1985). Saltmarshes were poorly known. Mangrove areas were much more thoroughly investigated with broad-scale mapping complete for the whole coastline and relationships with environmental factors identified. The mangrove areas of the Alligator Rivers region had been surveyed in more detail. Seasonally-covered floodplains are those on the lower reaches of rivers draining to the coast and include two, on the South Alligator River and Magela Creek, that had been intensively studied (e.g. Williams 1979; Finlayson *et al.* 1988b). Knowledge of the extensive and intermittent swamps and lakes of the drier parts of the Territory was relatively poor, with only very general or localised descriptions published. In these regions, as in the monsoonal north, reconnaissance-scale surveys (e.g. Mabbutt 1962) have provided overviews of the distribution, geomorphology and vegetation of wetlands.

Recent work by the Conservation Commission of the Northern Territory has involved further floristic surveys of the northern floodplains followed by numerical classification of sites into plant communities (e.g. Whitehead *et al.* 1990; Wilson *et al.* 1991). Large areas of seasonal floodplains remain unsurveyed in eastern Arnhemland. The swamp forests, dominated by *Melaleuca*, have been covered only incidentally by floodplain surveys and there are plans for a more comprehensive inventory (P. Whitehead, Conservation Commission, pers. comm.). Another gap in knowledge in the north is the nature and distribution of small ephemeral wetlands in the river headwaters and the hinterland. An ongoing project is developing methods for monitoring ephemeral wetlands across the Australian subhumid tropics using NOAA satellite imagery (M. Fleming, Conservation Commission, pers. comm.).

A problem for description and classification of the non-tidal wetlands in the Territory is temporal change, not only in the intermittent streams and basins of the inland, but in the seasonal floodplains of the north which dry and reflood annually (Fleming 1991). Further work on the floodplains might lead to a classification scheme with a stochastic element (P. Whitehead, pers. comm.) to define an envelope within which the community composition of a site would move as factors like water depth and salinity varied.

While having to deal with these year-to-year changes, inventory for management must be sensitive to medium-term (say 5–10 year) variation due to the effects of feral buffalo or salinity encroachment (B. Wilson, Queensland Department of Environment and Heritage, pers. comm.). This gives surveys a monitoring role for demonstrating trends in the features considered important for conservation.

Queensland

The first overview of wetlands in Queensland was compiled by Stanton (1975) and this is still the only overall mapping exercise of the State's wetlands. This study was a reconnaissance of 'significant' wetlands based on interpretation of aerial photographs and previous field experience. The survey identified 142 wetlands or wetland aggregations in the State, estimated to be at least 90% of the total extent and to represent all types. The wetland classification was at four hierarchical levels. Tidal wetlands were subdivided only once according to vegetation, but inland wetlands were separated according to salinity, water regime and a combination of geomorphology and rainfall. The exercise was assisted by previous reconnaissance surveys (e.g. Christian *et al.* 1953; Queensland Division of Land Utilisation 1974).

Over a decade later, Stanton's (1975) work still formed an important basis for the review of Queensland wetlands by Arthington & Hegerl (1988). These authors used a subdivision of the Queensland mainland into twelve biogeographic regions by Stanton & Morgan (1977) to structure their compilation of information and reservation status. They listed many surveys and classifications of a more localised and intensive nature than Stanton's earlier work, most of which covered parts of the coastal strip in more southern latitudes. They also listed the types of inland wetlands in Queensland according to vegetation structure and floristics.

Arthington (1988) reviewed the types of lakes in the tropics and subtropics of the State.

There is considerable ongoing work to document the State's estuarine wetlands by the Queensland Department of Primary Industries. Although extensive, some of these surveys have produced detailed (Olsen *et al.* 1980) or broad-scale maps (Danaher & Luck 1991). The more recent work has focused on seagrass areas and has involved methods ranging from point sampling by divers (Coles *et al.* 1992) to satellite imagery (Lennon & Luck 1990). Another major estuarine program is a joint project by the Australian Littoral Society and the Queensland Department of Environment and Heritage to compile a data base on intertidal wetlands by summarising all available information on each area (E. Hegerl, Australian Littoral Society, pers. comm.).

The largest current survey is by the Department of Environment and Heritage in the Gulf Plains Biogeographic Region in the north-west. This involves delineation of wetlands from classified satellite imagery, in some cases assisted by interpretation of aerial photographs, and field description of selected sites. Classification of wetlands is modified from Cowardin *et al.* (1979) and the new framework is being proposed for State-wide use (Johnston 1991; Blackman *et al.* 1992). Associated with this work is a project on the Burdekin River floodplain managed from James Cook University. Wetlands will be classified numerically when chemical and biological data are compiled, although classification is complicated by within- and between-year variation in water levels and biota (R. Congdon, James Cook University, pers. comm.). Survey work in the coastal south-east of the State by the Queensland University of Technology is concentrating on *Melaleuca* swamps (J. Davie, Queensland University of Technology, pers. comm.).

South Australia

The most comprehensive listing of wetlands in South Australia is by Lloyd & Balla (1986) who extended the work of Lloyd (1986) to identify about 1500 wetlands or wetland complexes State-wide, over 500 of which were known well enough to be rated for environmental significance. They classified inland wetlands into four types – swamps, lakes, rivers and springs – subdivided according to salinity and water regime.

Lothian and Williams (1988) provided an overview of the wetlands of the State. They adopted an informal regional framework for their compilation but, within

regions, recognised the same wetland types as Lloyd (1986). In the 80% of the State north of (drier than) the 250 mm isohyet, they identified three broad types of wetlands: mound springs, the natural surface discharge sites of artesian water, which have been well surveyed and documented (Greenslade *et al.* 1985); the vast complex of channels and basins in the far north-east that receives water from rivers rising in Queensland and which had not been surveyed in any detail at that time; and the salt lakes, many of which have been sampled limnologically (Williams 1981; De Deckker 1983). The many other wetlands in the arid part of the State, scattered through regions without major lakes or streams, were apparently poorly known, although these are numerous and extensive (Jessup 1951).

In the southern agricultural region, wetlands were generally more thoroughly surveyed than those further north. Wetlands in the south-east of the region are relatively well known (see Jones 1978). Five other wetland environments had been surveyed in some detail, although with various purposes and methodologies: the Coorong, a major lagoon complex in the south-east; the floodplain of the Murray River, the only major river in the State (Thompson 1986); the Murray's terminal lakes; small estuaries of minor coastal streams; and nine small, protected coastal embayments.

The gaps in the survey coverage of South Australian wetlands were highlighted again in Drewien's (1991) review. The uneven coverage reflects the urgency for specific information for planning decisions (A. Jensen, G. Drewien, Department of Environment and Planning, pers. comm.). Even in the best known regions there is a need for detailed or intensive, localised surveys for decisions on management and conservation. A high priority is to repeat Thompson's (1986) Murray survey and update assessments of conservation value. Intensive work is also underway in the South-East where rising groundwater will affect many wetlands. Large flows into the wetland complexes of the Cooper and Warburton Creeks in recent years have stimulated greater interest in the north-east of the State. A broadscale survey of part of the Cooper system began in 1991, based on aerial photography and ground truthing and is using the Ramsar definition and classification of wetlands.

Tasmania

Much of the knowledge of Tasmanian wetlands comes from systematic surveys between 1978 and 1980 of most of the enclosed natural wetlands of the lowlands and the Bass Strait Islands and some of the wetlands of the eastern Central Plateau (Kirkpatrick & Tyler 1988). The surveys covered some 530 wetlands and concentrated on macrophytic vegetation (see Kirkpatrick and Harwood 1981, 1983a, b for botanical details) but also included morphology, water regime, chemistry and other attributes. Data from these surveys were used to classify the wetlands similarly to those in Victoria (Corrick & Norman 1980) using water depth, permanence, salinity and vegetation structure. Kirkpatrick and Tyler (1988) produced summary maps of the occurrence of wetland classes which also include types not covered by the ground surveys of inland wetlands: artificial impoundments, natural wetlands of the glaciated high country, tidal saltmarsh (see Kirkpatrick & Glasby 1981) and shallow estuarine waters.

Tasmania has been relatively well explored limnologically and perspectives are available on salinity, nutrient status, optical properties and temperature regimes, among other attributes (e.g. Bowling *et al.* 1986, 1993; Tyler 1974, 1993). The diverse geomorphic origins of many wetlands have also been described. Wetlands of glacial origin are common, the island having been glaciated at least three times in the last 2–3 million years (Jennings & Ahmad 1957; Davies 1967). Kirkpatrick and Tyler (1988) have described informally some of the chemical, physical and geomorphological types of waterbodies in the State.

More recent inventories of selected wetlands (Blackhall 1986, 1988) involved field visits to each site and described them in terms of habitat features likely to reflect diversity or abundance of waterbirds. Several other surveys have covered specific types of wetlands. Whinam *et al.* (1989) described a survey of *Sphagnum* peatlands that was followed by numerical classification according to floristics and invertebrates. Pannell (1992) surveyed the swamp forests of the State and classified them numerically into floristic communities. Jarman *et al.* (1988) have completed a survey and classification of buttongrass moorlands. Other recent work on Tasmanian wetlands has been summarised by Blackhall (1991).

The Tasmanian Department of Parks, Wildlife and Heritage has begun a computerised database which currently holds information on about 800 sites and will be progressively updated and extended (Atkinson 1991). Several wetland types, including flowing waters, have been under-represented in surveys. Temporal variation is seen as a problem for both inventory and classification and there are plans to regularly monitor about 100 wetlands around the State (S. Blackhall & J. Atkinson, pers. comm.). All the sites in the data base have been reclassified using the Ramsar system as an interim framework pending more general agreement on a national classification scheme.

Victoria

A detailed survey of Victorian wetlands began in 1975 in the Gippsland region (Corrick & Norman 1980) and has since covered the whole State. Wetlands were located on aerial photographs and with ground surveys and information on many was collected in the field. The classification scheme was specific to the project and oriented toward use by birds. It has six categories for natural wetlands, based on water regime and salinity, with subcategories determined mainly by dominant vegetation. The distribution and other characteristics of wetlands surveyed up to December 1985 have been summarised by Norman and Corrick (1988). Additional information has come from landscape and vegetation mapping by the Land Conservation Council and Soil Conservation Council (e.g. Land Conservation Council 1987 and see Smith 1975 for early references).

Major gaps in detailed survey coverage in Victoria are the bogs and fens of the high country, although several vegetation and soil studies have described these and outlined categories (e.g. Costin 1962). Other gaps include wet heaths, flowing waters and inshore areas such as sea grass beds (P. S. Lake, Monash University, pers. comm.). As in other States, detailed data on flora, fauna, geomorphology and limnology are lacking for most wetlands. The extensive limnological work undertaken in Victoria has concentrated on lakes (see Bayly & Williams 1973; De Deckker & Williams 1986 for reviews).

A Wetlands Unit has been established within the Department of Conservation and Natural Resources with responsibilities including co-ordination of inventory, developing a wetland classification system, and establishing a data base (Shaw 1991). Further detailed surveys of wetlands have been undertaken in some regions including the western areas of Melbourne (e.g. Schulz *et al.* 1991) and parts of the Murray Valley (Lugg *et al.* 1989; Lloyd *et al.* 1991). Another recent survey, by Deakin University, was a compilation exercise to identify gaps in the types of wetlands included in National Estate listings (Deluca & Williams 1992).

The Wetlands Unit has identified a minimum data set to standardise inventory and allow a more detailed classification of wetlands (M. Beilharz, Deakin University, pers. comm.). The classification is hierarchical with several sequential criteria – geology, geomorphology, water regime, water chemistry and vegetation – and is applied within broad physiographic regions. The data on each criterion remain discrete and can be merged or rearranged to produce classes with different compositions for various purposes. Consideration is also being given to developing wetland classes that incorporate the variability inherent in many wetlands.

Western Australia

Lane and McComb (1988) concluded that 'no authoritative inventory of the total wetland resources of either the State, or any sizable region of the State, has yet been attempted'. The lakes and estuaries of the Swan Coastal Plain around Perth were the first to be surveyed by Riggert (1966), applying the classification scheme of Martin *et al.* (1953) based primarily on vegetation, water regime and salinity and emphasising use by waterfowl. Subsequent surveys have also concentrated on this region because of the ongoing impacts of urban, industrial and agricultural developments.

The Water Authority of Western Australia has made substantial progress in surveying and classifying wetlands in the south-west coastal regions, including the Perth-Bunbury area (LeProvost, Semeniuk & Chalmer 1987). New wetland maps for the whole of the Swan Coastal Plain have been prepared using the classification of Semeniuk (1987), based principally on water regime and landform. Suites of wetlands with common or inter-related features are referred to as 'consanguineous' assemblages (Semeniuk 1988) and have been mapped as wetland domains. The classification can be complemented with a system based on vegetation pattern and form (Semeniuk *et al.* 1990). Another major mapping exercise on the Swan Coastal Plain was completed by the Environmental Protection Authority to delineate lakes for protective regulation of activities. Possible further work by the Water Authority includes detailed wetland mapping in the Albany-Esperance region and the northern part of the Perth Basin, and hydrographic mapping for the wheatbelt (A. Hill, Water Authority of Western Australia, pers. comm.).

In the south-west, recent surveys by the Department of Conservation and Land Management have led to classification of wetlands according to use by waterbirds (Halse *et al.* 1993a, Storey *et al.* 1993). Similar work is proceeding with data on aquatic invertebrates

(Growns *et al.* 1992). The agreement between waterbird and invertebrate classifications varies with geographical scale and the types of wetlands being considered (S. Halse, Department of Conservation and Land Management, pers. comm.). There are ongoing aerial surveys of waterbirds and wetlands (Halse *et al.* 1992) and vegetation surveys of selected wetlands have led to informal descriptions of types (Halse *et al.* 1993b).

Estuarine wetlands have been surveyed and classified in several parts of the State (e.g. Backshall & Bridgewater 1981, Semeniuk & Wurm 1987; Semeniuk & Semeniuk 1990). The wetlands of the arid interior, while explored ornithologically, are largely unsurveyed except for isolated areas (Halse 1990) or regions covered by reconnaissance-level land inventory (Speck *et al.* 1964).

Summary of Australian studies and future directions

Number and coverage of studies

The bibliography compiled as background to this review (Pressey *et al.* in press) contains over 600 entries, most of which have been produced since the mid-1970s (Fig. 4). While some of the studies listed represent a common approach applied in stages or to different areas, the sheer number of references indicates a wide diversity of ideas on survey and classification.

The wetlands near the main population centres, scattered around the coastal fringes, have been surveyed most thoroughly and in most detail, sometimes repeatedly at different scales (Paijmans *et al.* 1985; Winning 1991b). Recognition of wetland loss and degradation in regions like the Murray-Darling Basin or north coastal Queensland has stimulated survey and conservation efforts distant from the cities but there remain vast tracts of the country in which the distribution and nature of wetlands are poorly known. In most cases, the need for knowledge is driven by perceived threat – 'community awareness about the need for conservation grows in inverse proportion to the area, status or condition of the components of our natural heritage' (Lothian & Williams 1988).

Given the inertia of bureaucracies and the perennial lack of resources for surveys of wetlands that are politically insignificant, governments and resource agencies will not necessarily respond in a timely and appropriate fashion to the need for information as a basis for management. A review mechanism is necessary to identify critical gaps in the survey coverage in relation to management needs and potential threats to wetlands. This should probably be done at both Commonwealth and State/Territory levels and should guide the allocation of funds.

Approaches to inventory

Within and between States and Territories, past, present and planned methods for inventory vary widely. One interpretation of this variety is that surveys are unco-ordinated and chaotic. While a case can certainly be made for more coordination, there are also good reasons for different approaches. First, the design of any survey is a response to several factors – the types of wetlands involved, the sorts of information required, the geographical scope of the study, the available funds and time, the type and level of expertise of the personnel, and their exposure to particular techniques and views. A wide variety of survey approaches is therefore inevitable.

Another reason for variation is that surveys of some regions span more than twenty years so differences in priorities and approaches are to be expected. Through the information they provide and the critical review they undergo, surveys give new perspectives on the way later ones should be done. A diversity of survey approaches over time, even applied to the same area, is therefore desirable if techniques are to evolve. In addition, there are many cases of reconnaissance or broad-scale surveys providing the rationale and the framework for subsequent, more detailed surveys.

It is not possible or desirable for all surveys to be standardised according to the data collected, accuracy, resolution, and precision of mapped boundaries. Indeed, in view of the gradual and fluctuating nature of wetland edges, the establishment of precise but biologically meaningless boundaries might have limited use. Pressure from agencies for neatness of coverage and uniformity of approach would stifle innovative methods and go unheeded by those facing particular management problems which could not be catered for by the 'standard' approach.

Four developments in Australian wetland survey which would be useful are:

1. regular meetings of scientists and managers involved in undertaking wetland surveys or using survey information;

2. the collation of geographical data bases of inventories that indicate survey type and quality (Knights 1991) as a basis for systematic assessments of priority areas; ideally, these collations should take a broad view of what constitutes a wetland survey;

3. establishment of a minimum data set for inventory, regardless of scope and purpose, emphasising the importance of collecting primary rather than classified data wherever possible so that options for later uses are maximised (see McComb & Lake 1988; Barson & Williams 1991);

4. most importantly, investigation of ways to make survey information more accurate: important sources of inaccuracy include temporal change, the use of untested predictive indices for attributes that are difficult to measure directly, and the collection of data without validation from aerial photographs and satellite imagery.

Approaches to classification

There is a similar diversity of approaches to classification. The earlier classifications of wetlands were primarily concerned with habitat for waterbirds (Riggert 1966; Goodrick 1970; Corrick & Norman 1980). These were followed by other schemes that attempted to categorise wetlands more generally. The Water Authority has taken the lead in wetland mapping in Western Australia and is committed to the locally derived classification scheme of Semeniuk (1987). A major project in the Gulf Country of Queensland is using the approach of Cowardin *et al.* (1979) developed in the United States. The Australian scheme of Paijmans *et al.* (1985) has been used to summarise the categories of wetlands in the Northern Territory. The Tasmanian inventory and new work in South Australia are using Ramsar types. With slight modifications, the Ramsar classes are also being used in the Directory of Australia's Important Wetlands sponsored by the Australian National Parks and Wildlife Service. In Victoria, there are plans to replace an existing State-wide classification (Corrick & Norman 1980) by a more detailed approach developed specifically for the needs of the new Wetlands Conservation Program. There is a plethora of additional classifications used in each of the States and Territories, including an increasing number of numerical analyses applied after collection of data in the field. There is also a growing recognition of the problems posed by temporal dynamics for classification schemes. New approaches incorporating temporal change could emerge in the near future.

The variety of classifications could also be used to support calls for co-ordination and uniformity of approach. This would be a simplistic view of the situation. There is no doubt that a national system of classification would be useful for communication and comparison. More important, however, is the need to see classifications in two ways: (1) as hypotheses about the way in which the features of wetlands are arranged in space and time; and (2) as responses to the need for particular types of information for particular purposes, dependent also on the geographical scale of the study and the variability of the wetlands. Time and energy spent in imposing uniform approaches would be better spent in tackling these more important issues.

Techniques are available now to measure the extent to which classifications do the jobs they are meant to do (e.g. Pressey & Bedward 1991a, b; Bedward *et al.* 1992) and these approaches will continue to improve. Advances are also being made in *exploring* wetland data rather than simply slotting wetlands into pigeonholes. Numerical classifications of plant and invertebrate data are giving new pictures of wetland types in several States and will allow established schemes to be assessed critically.

The issue of whether a national classification scheme should be adopted is still unresolved in Australia, although it has obvious benefits for overviews of wetland resources and communication about management and conservation. Despite the apparent commitment of several agencies to different classifications, there is scope for developing a broad, flexible approach for national communication. A useful one might be similar to that recently developed in Victoria with two key elements: the use of several broad criteria including geomorphology and hydrology which are recorded independently and so can be applied individually or in various combinations; and an optional framework of environmental regions that can be used to stratify the classification geographically.

Global standards for inventory and classification?

Nearly all of the responses from wetland scientists and managers on the questions of global standards concerned only classification. However, given the many factors that influence the approach taken to wetland inventory and the wide diversity of inventory methods currently used to achieve diverse goals, it is safe to conclude that a standardised inventory approach would not be feasible within any of the Australian States or Territories. The chances of a global approach being adopted

seem remote. Ongoing discussions in Australia on the issue of minimum data sets in wetland inventory, however, could lead to more national integration of survey results.

On a global classification scheme, there was a wide variety of responses, with some respondents posing arguments for and against. Positive views, less than half the responses, acknowledged that there would be benefits for international communication, comparisons, exchange of ideas on problems and solutions, use in extensive surveys, and summaries of characteristics and trends, particularly for countries linked by migratory species or conservation treaties. International consistency could, for example, allow some formal assessment of the conservation status of international flyways. Two respondents commented that a global system would have some heuristic value in understanding types of Australian wetlands and highlighting information needs. Other perceived benefits included the establishment of concise technical definitions, less time spent debating the merits of alternative approaches, the possibility of better approaches being adopted, and the improvement of global data sets for modelling climate change, for example in relation to methanogenesis.

Negative responses were slightly in the majority. An important argument was that classifications are no more than tools to understand variation for a particular management or conservation purpose and should be developed specifically – a single classification could not serve all needs and, in areas where wetlands were not well known, could confuse and hinder conservation efforts. In the same vein, others were concerned that a standard scheme would impose inappropriate or suboptimal methods on scientists and managers while stifling the development of approaches responsive to particular problems. Several people considered that classifications should be developed that specifically address the temporal variability of many Australian wetlands, as well as hydrologically variable wetlands in other countries. Another view was that classifications should be developed from data on the features of concern, perhaps iteratively as more information comes to hand.

If an attempt is made to apply a global classification in Australia, it should be as flexible as possible. Global comparability would be best achieved with a broad system based primarily on factors like climate, water regime and geomorphology, preferably recorded independently. So, if necessary, all wet tropical wetlands could be identified independently of geomorphology

96

or all floodplain wetlands could be identified independently of climate. The use of common terms like marsh or bog in any global system should be accompanied by precise definitions to facilitate consistent application.

The problem would then be to identify more detailed types which would be useful as indicators of biological characteristics and management needs. It might be possible to make a broad global subdivision useful in national, regional and local contexts by stratifying the broad types geographically, as in the biophysical approach of Stanton and Morgan (1977) in Queensland, and within major coastal environments for estuarine and marine wetlands. This would allow data on the classes to be used at several geographical levels: without stratification for global statistics, and with various levels of physical or biogeographic stratification to achieve greater homogeneity for specific purposes.

While acknowledging the potential benefits of a global classification, it is important to recognise that the real work of wetland classification is to assist with the management and conservation of the resource in many ways and at many scales. There is and will continue to be a need for various approaches to inventory and classification that are geared to particular problems. It is therefore critical that any agreement on a global scheme, if forthcoming, does not interfere with their development. Adherence to a single scheme for convenience and uniformity ignores the most fundamental fact about classifications – that they are simply ways of representing processes or patterns of selected features in time and space. To be sure that they are effective in this, and therefore reliable bases for conservation and management, scientists and managers must put much more effort into testing and refining classifications.

Acknowledgments

Travel costs for RLP to attend the IV International Wetlands Conference were generously provided by Ducks Unlimited (USA). This support made possible the symposium presentation, much valuable interchange and this paper.

We are grateful to the many people who responded to our requests by providing information and opinions for this review: Angela Arthington, Brendan Atkins, Michele Barson, John Beumer, Margie Beilharz, Stewart Blackhall, Gavin Blackman, Matt Bolton, David Bowman, Sue Briggs, Margaret Brock, Bob Congdon, Andrew Corrick, Jim Davie, Jenny Davis, Gary Drewien, Mike Fleming, Bill Freeland, Stuart Halse, Eddie Hegerl, Alan Hill, Bob Humphries, Surrey Jacobs, Anne Jensen, Peter Johnston, Richard Kingsford, Sam Lake, Greg Laughlin, Jim Lane, Mark Lintermans, Lance Lloyd, Mike Maher, Arthur McComb, David Mitchell, Marianne Newton, Mike O'Brien, Jane Roberts, Peter Saenger, Vic Semeniuk, Peter Stoddart, John Sutton, Andrew Taplin, Brian Timms, Keith Walker, Ron West, Paul Wettin, Peter Whitehead, Bill Williams, Bruce Wilson and Geoff Winning. Some of these people assisted further by commenting on the draft of the paper.

Sharon Tully gave us invaluable help by preparing the figures for the paper and corresponding with the people listed above.

References

Adam, P. 1992. Wetlands and wetland boundaries: problems, expectations, perceptions and reality. Wetlands (Australia) 11: 60–67.

Adam, P., Urwin, N., Weiner, P. & Sim, I. 1985. Coastal wetlands of New South Wales – a survey prepared for the Coastal Council of New South Wales. Department of Environment and Planning, Sydney.

Adam, P., Wilson, N. C. & Huntley, B. 1988. The phytosociology of coastal saltmarsh vegetation in New South Wales. Wetlands (Australia) 7: 35–85.

Adamson, D. A. & Pickard, J. 1986. Physiography and geomorphology of the Vestfold Hills pp. 99–139. In: Pickard, J. (ed), Antarctic oasis: terrestrial environments and history of the Vestfold Hills. Academic Press, Sydney.

Arthington, A. H. 1988. The distribution, characteristics and conservation status of lakes in the wet tropics and subtropics of Queensland. Proceedings of the Ecological Society of Australia 15: 177–189.

Arthington, A. H. & Hegerl, E. J. 1988. The distribution, conservation status and management problems of Queensland's athalassic and tidal waters. pp. 59–109. In: McComb, A. J. & Lake, P. S. (eds), The conservation of Australian wetlands. Surrey Beatty and Sons, Sydney.

Atkinson, J. 1991. An inventory system for Tasmanian wetlands. Tasmanian Department of Parks, Wildlife and Heritage, Hobart.

Australian Nature Conservation Agency 1993. A directory of important wetlands in Australia. ANCA, Canberra.

Australian Water Resources Council 1976. Review of Australia's water resources (1975). Australian Government Publishing Service, Canberra.

Backshall, D. J. & Bridgewater, P. B. 1981. Peripheral vegetation of Peel Inlet and the Harvey Estuary, Western Australia. Journal of the Royal Society of Western Australia 64: 5–11.

Barson, M. M. & Williams, J. E. 1991. Wetland inventory – towards a unified approach. Bureau of Rural Resources, Canberra.

Bayly, I. A. E. & Williams, W. D. 1973. Inland waters and their ecology. Longman, Melbourne.

Beadle, N. C. W. 1948. The vegetation and pastures of western New South Wales. Government Printer, Sydney.

Beadle, N. C. W. 1981. The vegetation of Australia. Cambridge University Press, Cambridge.

Bedward, M., Keith, D. A. & Pressey, R. L. 1992. Homogeneity analysis: assessing the utility of classifications and maps of natural resources. Australian Journal of Ecology 17: 133–139.

Bell, J. D. & Pollard, D. A. 1989. Ecology of fish assemblages and fisheries associated with seagrasses. pp. 597–609. In: Larkum, A. W. D., McComb, A. J. & Shepherd, S. A. (eds), Biology of seagrasses: a treatise on the biology of seagrasses with special reference to the Australian region. Elsevier, Amsterdam.

Blackhall, S. A. 1986. A survey to determine waterbird usage and conservation significance of selected Tasmanian wetlands. Tasmanian National Parks and Wildlife Service Occasional Paper No. 14.

Blackhall, S. A. 1988. A survey to determine waterbird usage and conservation significance of selected Tasmanian wetlands, Stage II. Tasmanian Department of Lands, Parks and Wildlife Occasional Paper No. 16.

Blackhall, S. A. 1991. Wetlands in Tasmania: a brief review of their conservation and management. pp. 39–41. In: Donohue, R. & Phillips, B. (eds), Educating and Managing for Wetlands Conservation. Australian National Parks and Wildlife Service, Canberra.

Blackman, J. G., Spain, A. V. & Whiteley, L. A. 1992. Provisional handbook for the classification and field assessment of Queensland wetlands and deep water habitats. Department of Environment and Heritage, Townsville.

Bohlen, C. C. 1991. Controversy over federal definition of wetlands. BioScience 41: 139.

Boulton, A. J. & Lloyd, L. N. 1992. Flooding frequency and invertebrate emergence from dry floodplain sediments of the River Murray, Australia. Regulated Rivers: Research and Management 7: 137–151.

Bowling, L. C., Banks, M. R., Croome, R. L. & Tyler, P. A. 1993. Reconnaissance limnology of Tasmania. 2. Limnological features of Tasmanian freshwater coastal lagoons. Archiv fur Hydrobiologie 126: 385–403.

Bowling, L. C., Steane, M. S. & Tyler, P. A. 1986. The spectral distribution and attenuation of underwater irradiance in Tasmanian Inland Waters. Freshwater Biology 16: 313–335.

Bowman, D. M. J. S. & Whitehead, P. J. 1989. Book review – the conservation of Australian wetlands. Conservation Biology 3: 209–210.

Bren, L. J. & Gibbs, N. L. 1986. Relationships between flood frequency, vegetation and topography in a river red gum forest. Australian Forest Research 16: 357–370.

Bridgewater, P. B. 1982. Phytosociology of coastal saltmarshes in the Mediterranean climatic region of Australia. Phytocoenologia 10: 257–296.

Bridgewater, P. B. 1985. Variation in the mangal along the west coastline of Australia. Proceedings of the Ecological Society of Australia 13: 243–256.

Bridgewater, P. 1991. Wetland conservation challenges in Oceania. pp. 53–60. In: Donohue, R. & Phillips, B. (eds), Educating and managing for wetlands conservation. Australian National Parks and Wildlife Service, Canberra.

Briggs, S. V. 1981. Freshwater wetlands. pp. 335–360. In: Groves, R. H. (ed), Australian Vegetation. Cambridge University Press, Cambridge.

Brock, M. A. 1994. Aquatic vegetation of inland wetlands pp. 437–466. In: Groves, R. H. (ed). Australian Vegetation. 2nd Edition. Cambridge University Press, Cambridge (in press).

Bryant, E. A. 1990. Sea level change and greenhouse: implications for wetlands. Wetlands (Australia) 10: 7–13.

Bucher, D. & Saenger, P. 1989. An inventory of Australian estuaries and enclosed marine waters. Report for the Australian Recreational and Sport Fishing Confederation and the Australian National Parks and Wildlife Service by the Centre for Coastal Management, University of New England, Lismore.

Bucher, D. & Saenger, P. 1991. An inventory of Australian estuaries and enclosed marine waters: an overview of results. Australian Geographical Studies 29: 370–381.

Burchmore, J. J. 1990. Implications of the greenhouse effect for native freshwater fishes in New South Wales. Wetlands (Australia) 10: 30–32.

Christian, C. S., Paterson, S. J., Perry, R. A., Slatyer, R. O., Stewart, G. A. & Traves, D. M. 1953. Survey of the Townsville-Bowen region, north Queensland, 1950. CSIRO Australian Land Research Series No. 2.

Clarke, S. M. & Kirkman, H. 1989. Seagrass dynamics. pp. 334–345. In: Larkum, A. W. D., McComb, A. J. & Shepherd, S. A. (eds), Biology of seagrasses: a treatise on the biology of seagrasses with special reference to the Australian region. Elsevier, Amsterdam.

Close, A. 1990. The impact of man on the natural flow regime. pp. 61–74. In: Mackay, N. & Eastburn, D. (eds.), The Murray. Murray-Darling Basin Commission, Canberra.

Coles, R. G., Lee Long, W. J., Miller, K. J., Vidler, K. P. & Derbyshire, K. Y. 1992. Seagrass beds and juvenile prawn and fish nursery grounds between Water Park Point and Hervey Bay, Queensland. Queensland Department of Primary Industries Information Series QI92011.

Corrick, A. H. & Norman, F. I. 1980. Wetlands of Victoria. I. Wetlands and waterbirds of the Snowy River and Gippsland Lakes catchment. Proceedings of the Royal Society of Victoria 91: 1–15.

Costin, A. B. 1962. Ecology of the high plains. Proceedings of the Royal Society of Victoria 75: 327–330.

Cowardin, L. M., Carter, V., Golet, F. C. & LaRoe, E. T. 1979. Classification of wetlands and deepwater habitats of the United States. US Fish and Wildlife Service Report 79/31.

Danaher, K. F. & Luck, P. E. 1991. Mapping mangrove communities in Moreton Bay using Landsat thematic mapper imagery. Proceedings of a conference on remote sensing and GIS for coastal and catchment management.

Davies, J. L. 1967. Tasmanian landforms and Quaternary climates. In: Jennings, J. N. & Mabbutt, J. A. (eds.), Landform studies from Australia and New Guinea. Australian National University Press, Canberra.

Davies, J. L. 1986. The coast. pp. 203–222. In: Jeans, D. N. (ed.), Australia: a geography. Volume 1. The natural environment. Sydney University Press, Sydney.

De Deckker, P. 1983. Australian salt lakes: their history, chemistry and biota – a review. Hydrobiologia 105: 231–244.

De Deckker, P. & Williams, W. D. 1986. Limnology in Australia. CSIRO, Melbourne/Junk, Dordrecht.

Deluca, S. & Williams, C. A. 1992. Wetlands of the National Estate in Victoria. Part one: statewide inventory and update of significance for south eastern Victoria. Deakin University Department of Heritage and Resource Management.

Dent, D. & Young, A. 1981. Soil Survey and Land Evaluation. George Allen and Unwin, London.

Department of Minerals and Energy 1973. Atlas of Australian resources. Landforms. Australian Government Publishing Service, Canberra.

Donohue, R. & Phillips, B. (eds) 1991. Educating and managing for wetlands conservation. Australian National Parks and Wildlife Service, Canberra.

98

Drewien, G. 1991. Wetlands of South Australia: a management overview. pp. 33–38. In: Donohue, R. & Phillips, B. (eds), Educating and Managing for Wetlands Conservation. Australian National Parks and Wildlife Service, Canberra.

Evans, L. & Keenan, C. 1992. Directory of important wetlands in Australia: wetlands in high altitude areas of the Australian Capital Territory. Report to the Australian National Parks and Wildlife Service.

Finlayson, C. M. 1991. Australasia and Oceania. pp. 179–208. In: Finlayson, M. & Moser, M. (eds), Wetlands. International Waterfowl and Wetlands Research Bureau, Oxford.

Finlayson, C. M., Bailey, B. J., Freeland, W. J. & Fleming, M. R. 1988a. Wetlands of the Northern Territory. pp. 103-126. In: McComb, A. J. & Lake, P. S. (eds), The Conservation of Australian Wetlands. Surrey Beatty and Sons, Sydney.

Finlayson, C. M., Bailey, B. J. & Cowie, I. D. 1988b. Macrophyte vegetation of the Magela Creek Floodplain, Alligator Rivers Region, Northern Territory. Supervising Scientist for the Alligator Rivers Region Research Report.

Fleming, M. R. 1991. Northern Territory wetlands: an overview of policy and priorities. pp. 17–25. In: Donohue, R. & Phillips, B. (eds), Educating and Managing for Wetlands Conservation. Australian National Parks and Wildlife Service, Canberra.

Frith, H. J. 1967. Waterfowl in Australia. Angus and Robertson, Sydney.

Frith, H. J. & Sawer, G. (eds.) 1974. The Murray waters: man, nature and a river system. Angus and Robertson, Sydney.

Galloway, R. W. 1982. Distribution and physiographic patterns of Australian mangroves. pp. 31–54. In: Clough, B. F. (ed.), Mangrove Ecosystems in Australia. ANU Press, Canberra.

Galloway, R. W., Story, R., Cooper, R. & Yapp, G. A. 1984. Coastal lands of Australia. CSIRO Division of Water and Land Resources, Natural Resources Series 1.

Goodrick, G. N. 1970. A survey of wetlands of coastal New South Wales. CSIRO Division of Wildlife Research Technical Memorandum No. 5.

Goodrick, G. 1983. A description of wetlands in New South Wales. pp. 10–13. In: Haigh, C. (ed.), Parks and wildlife – wetlands in New South Wales. National Parks and Wildlife Service, Sydney.

Goodrick, G. N. 1984. Wetlands of northwestern New South Wales. New South Wales National Parks and Wildlife Service Occasional Paper No. 6.

Green, D. L. 1992. Survey of wetlands of the Warrego River. NSW Department of Water Resources Technical Services Report 92.081.

Greenslade, J., Joseph, L. & Reeves, A. (eds.) 1985. South Australia's mound springs. Nature Conservation Society of South Australia, Adelaide.

Griffith, S. J. 1984. A survey of the vegetation of Yuraygir National Park. Report to the New South Wales National Parks and Wildlife Service. (unpublished).

Growns, J. E., Davis, J. A., Cheal, F., Schmidt, L. G., Rosich, R. S. & Bradley, S. J. 1992. Multivariate pattern analysis of wetland invertebrate communities and environmental variables in Western Australia. Australian Journal of Ecology 17: 275-288.

Halse, S. A. (ed.) 1990. The natural features of Lake Gregory: a preliminary review. Western Australian Department of Conservation and Land Management Occasional Paper 2/90.

Halse, S. A., Vervest, R. M., Munro, D. R., Pearson, G. B. & Yung, F. H. 1992. Annual waterfowl counts in south-west Western Australia – 1989/90. Western Australian Department of Conservation and Land Management Technical Report 29.

Halse, S. A., Williams, M. R., Jaensch, R. P. & Lane, J. A. K. 1993a. Wetland characteristics and waterbird use of wetlands in south-western Australia. Wildlife Research 20: 103-126.

Halse, S. A., Pearson, G. B. & Patrick, S. 1993b. Vegetation of depth-gauged wetlands in nature reserves of south-western Western Australia. CalmScience.

Hope, G. & Southern, W. 1983. Organic deposits of the Southern Tablelands region. NSW National Parks and Wildlife Service internal report.

Howard, R. K., Edgar, G. J. & Hutchings, P. A. 1989. Faunal assemblages of seagrass beds. pp. 558–564. In: Larkum, A. W. D., McComb, A. J. & Shepherd, S. A. (eds), Biology of seagrasses: a treatise on the biology of seagrasses with special reference to the Australian region. Elsevier, Amsterdam.

Howard-Williams, C. 1991. Dynamic processes in New Zealand land-water ecotones. New Zealand Journal of Ecology 15: 87-98.

Hughes, J. M. R. 1987. The distribution and composition of vascular plant communities on Heard Island. Polar Biology 7: 153-162.

Jacobs, S. 1983. Vegetation. pp. 14–19. In: Haigh, C. (ed), Parks and wildlife – wetlands in New South Wales. New South Wales National Parks and Wildlife Service, Sydney.

Jacobs, S. W. L. & Brock, M. A. 1993. Wetlands of Australia: southern (temperate) Australia. pp. 244–304. In: Whigham, D. F., Dykyjova, D. & Henjy, S. (eds.), Wetlands of the world. 1. Inventory, ecology and management. Kluwer Academic Publishers, Dordrecht.

Jarman, S. J., Kantvilas, G. & Brown, M. J. 1988. Buttongrass moorland in Tasmania. Tasmanian Forest Research Council Research Report No. 2.

Jennings, J. N. & Ahmad, N. 1957. The legacy of an ice cap: the lakes in the western part of the central plateau of Tasmania. Australian Geographer 7: 62–75.

Jessup, R. W. 1951. The soils, geology and vegetation of northwestern South Australia. Transactions of the Royal Society of South Australia 74: 189–273.

Johnston, P. 1991. Wetlands in Queensland: an overview of issues, policies and priorities. pp. 26–32. In: Donohue, R. & Phillips, B. (eds), Educating and managing for wetlands conservation. Australian National Parks and Wildlife Service, Canberra.

Johnston, R. & Barson, M. 1993. Remote sensing of Australian wetlands: an evaluation of Landsat TM data for inventory and classification. Australian Journal of Marine and Freshwater Research 44 (in press).

Jones, W. 1978. The wetlands of the south-east of South Australia. Nature Conservation Society of South Australia, Adelaide.

Kingsford, R. T., Porter, J. L., Ferster Levy, R., Smith, J. D. B. & Holland, P. 1991. An aerial survey of wetland birds in eastern Australia – October 1990. New South Wales National Parks and Wildlife Service Occasional Paper No. 10.

Kingsford, R. T. & Porter, J. L. 1993. Waterbirds of Lake Eyre. Biological Conservation 65: 141–151.

Kirkpatrick, J.B. & Glasby, J. 1981. Salt marshes in Tasmania: distribution, community composition and conservation. University of Tasmania Department of Geography Occasional Paper 8.

Kirkpatrick, J. B. & Harwood, C. E. 1981. The conservation of Tasmanian wetland macrophytic species and communities. Report to the Australian Heritage Commission from the Tasmanian Conservation Trust Inc.

Kirkpatrick, J. B. & Harwood, C. E. 1983a. Plant communities of Tasmanian wetlands. Australian Journal of Botany 31: 437–451.

Kirkpatrick, J. B. & Harwood, C. E. 1983b. The conservation of Tasmanian macrophytic wetland vegetation. Papers and Proceedings of the Royal Society of Tasmania 117: 5–20.

Kirkpatrick, J. B. & Tyler, P. A. 1988. Tasmanian wetlands and their conservation. pp. 1–16. In: McComb, A. J. & Lake, P. S. (eds), The conservation of Australian wetlands. Surrey Beatty and Sons, Sydney.

Knights, P. 1991. New South Wales wetlands: current conservation issues, policies and programs. pp. 14–16. In: Donohue, R. & Phillips, B. (eds), Educating and managing for wetlands conservation. Australian National Parks and Wildlife Service, Canberra.

Kusler, J. 1992. Wetlands delineation: an issue of science or politics? Environment 34: 7–37.

Land Conservation Council 1987. Report on the mallee area review. LCC, Melbourne.

Lane, J. 1991. The wise use of wetlands – managing wildlife habitat. pp. 151–155. In: Donohue, R. & Phillips, B. (eds), Educating and managing for wetlands conservation. Australian National Parks and Wildlife Service, Canberra.

Lane, J. A. K. & McComb, A. J. 1988. Western Australian wetlands. pp. 127–146. In: McComb, A. J. & Lake, P. S. (eds), The conservation of Australian wetlands. Surrey Beatty and Sons, Sydney.

Lennon, P. & Luck, P. 1990. Seagrass mapping using Landsat TM data: a case study in southern Queensland. Asian-Pacific Remote Sensing Journal 2: 1–6.

LeProvost, Semeniuk & Chalmer 1987. Environmental significance of wetlands in the Perth to Bunbury region. Western Australian Water Resources Council Publication Vol. 1.

Linacre, E. & Hobbs, J. 1977. The Australian climatic environment. John Wiley and Sons, Brisbane.

Lloyd, L. N. 1986. Regional wetland survey. Department of Environment and Planning, Adelaide.

Lloyd, L. N., Atkins, B. P., Beovich, E. K. & Warner, A. C. 1991. Hydrological management of mid-Murray floodplain for ecological conservation. Department of Conservation and Environment, Shepparton.

Lloyd, L. N. & Balla, S. A. 1986. Wetlands and water resources of South Australia. Department of Environment and Planning, Adelaide.

Lothian, J. A. & Williams, W. D. 1988. Wetland conservation in South Australia. pp. 147–166. In: McComb, A. J. & Lake, P. S. (eds), The conservation of Australian wetlands. Surrey Beatty and Sons, Sydney.

Love, L. D. 1981. Mangrove swamps and salt marshes. pp. 319–334. In: Groves, R. H. (ed), Australian vegetation. Cambridge University Press, Cambridge.

Lugg, A., Heron, S., Fleming, G. & O'Donnell, T. 1989. Conservation value of the wetlands of the Kerang Lakes area – report 1 to Kerang Lakes Area working group. Department of Conservation, Forests and Lands.

Mabbutt, J. A. 1962. Geomorphology of the Alice Springs area. CSIRO Australian Land Research Series 6: 163–184.

Margules & Partners Pty Ltd, P. & J. Smith Ecological Consultants & Victorian Department of Conservation, Forests and Lands 1990. River Murray riparian vegetation study. Report to the Murray-Darling Basin Commission.

Martin, A. C., Hotchkiss, N., Uhler, F. M. & Bourn, W. S. 1953. Classification of wetlands of the United States. United States Fish and Wildlife Service Special Scientific Report on Wildlife No. 20.

McComb, A. J. & Lake, P. S. 1988. The conservation of Australian wetlands. Surrey Beatty and Sons, Sydney.

Michaelis, F. B. & O'Brien, M. 1988. Preservation of Australia's wetlands: a Commonwealth approach. pp. 167–177. In: McComb, A. J. & Lake, P. S. (eds), The conservation of Australian wetlands. Surrey Beatty and Sons, Sydney.

Middleton, M. J., Rimmer, M. A. & Williams, R. J. 1985. Structural flood mitigation works and estuarine management in New South Wales – a case study of the Macleay River. Coastal Zone Management Journal 13: 1–23.

Mitsch, W. J. & Gosselink, J. G. 1986. Wetlands. Van Nostrand Reinhold, New York.

Murphy, J. 1990. Watering the Millewa forest. pp. 245–248. In: Mackay, N. & Eastburn, D. (eds.), The Murray. Murray-Darling Basin Commission, Canberra.

Norman, F. I. & Corrick, A. H. 1988. Wetlands in Victoria: a brief review. pp. 17–34. In: McComb, A. J. & Lake, P. S. (eds), The conservation of Australian wetlands. Surrey Beatty and Sons, Sydney.

Olsen, H. F., Dowling, R. M. & Bateman, D. 1980. Biological resources survey (estuarine inventory) – Round Hill Head to Tannum Sands, Queensland, Australia. Queensland Fisheries Service Research Bulletin No. 2.

Paijmans, K. 1978a. Feasibility report on a national wetland survey. CSIRO Division of Land Use Research Technical Memorandum 78/6.

Paijmans, K. 1978b. A reconnaissance of four wetland pilot study areas. CSIRO Division of Land Use Research Technical Memorandum 78/3.

Paijmans, K, Galloway, R. W., Faith, D. P., Fleming, P. M., Haantjens, H. A., Heyligers, P. C., Kalma, J. D. & Loffler, E. 1985. Aspects of Australian wetlands. CSIRO Division of Water and Land Resources Technical Paper No. 44.

Pannell, J. R. 1992. Swamp forests of Tasmania. Tasmanian Forestry Commission, Hobart.

Poiner, I. R., Walker, D. I. & Coles, R. G. 1989. Regional studies – seagrasses of tropical Australia. pp. 296–303. In: Larkum, A. W. D., McComb, A. J. & Shepherd, S. A. (eds), Biology of seagrasses: a treatise on the biology of seagrasses with special reference to the Australian region. Elsevier, Amsterdam.

Pollard, D. A. 1976. Estuaries must be protected. Australian Fisheries 35: 1–5.

Ponder, W. F. 1986. Mound Springs of the Great Artesian Basin. pp. 403–420. In: De Deckker, P. & Williams, W. D. (eds), Limnology in Australia. CSIRO, Melbourne/Junk, Dordrecht.

Pressey, R. L. 1986. Wetlands of the River Murray below Lake Hume. River Murray Commission Environmental Report 86/1.

Pressey, R. L. 1989. Wetlands of the lower Clarence floodplain, northern coastal New South Wales. Proceedings of the Linnean Society of New South Wales 111: 143–155.

Pressey, R. L., Adam, P. & Bowen, P. F. A bibliography of Australian surveys and classifications of lentic and tidal wetlands. Wetlands (Australia) (in press).

Pressey, R. L. & Bedward, M. 1991a. Mapping the environment at different scales: benefits and costs for nature conservation. pp. 7–13. In: Margules, C. R. & Austin, M. P. (eds), Nature conservation: cost effective biological surveys and data analysis. CSIRO, Melbourne.

Pressey, R. L. & Bedward, M. 1991b. Inventory and classification of wetlands: what for and how effective? pp. 190–198. In: Donohue, R. & Phillips, B. (eds), Educating and managing for wetlands conservation. Australian National Parks and Wildlife Service, Canberra.

Pressey, R. L. & Harris, J. H. 1988. Wetlands of New South Wales. pp. 35–57. In: McComb, A. J. & Lake, P. S. (eds), The conservation of Australian wetlands. Surrey Beatty and Sons, Sydney.

Pressey, R. L. & Middleton, M. J. 1982. Impacts of flood mitigation works on coastal wetlands in New South Wales. Wetlands (Australia) 2: 27–44.

Queensland Division of Land Utilisation 1974. Western arid region land use study – part 1. DIL Technical Bulletin No. 12.

Resource Assessment Commission 1993. RAC Coastal Zone Inquiry Information Paper No. 3. Australian Government Publishing Service, Canberra.

Ridd, P., Sandstrom, M. & Wolanski, E. 1988. Outwelling from tropical tidal salt flats. Estuarine, Coastal and Shelf Science 26: 243–253.

Riggert, T.L. 1966. Wetlands of Western Australia 1964–66 – a study of the wetlands of the swan coastal plain. Department of Fisheries and Fauna, Perth.

Riley, S. J., Warner, R. F. & Erskine, W. 1984. Classification of waterbodies in New South Wales. Water Resources Commission, Sydney.

Roy, P. S. 1984. New South Wales estuaries: their origin and evolution. pp. 99–121. In: Thom, B. G. (ed), Coastal geomorphology in Australia. Academic Press, Sydney.

Saenger, P., Specht, M. M., Specht, R. L. & Chapman, V. J. 1977. Mangal and coastal saltmarsh communities in Australasia. pp. 293–345. In: Chapman, V. J. (ed), Ecosystems of the world. 1. Wet coastal ecosystems. Elsevier, Amsterdam.

Schulz, M., Beardsell, C. & Sandiford, K. 1991. Sites of faunal significance in the western wetlands of Melbourne. Department of Conservation and Environment, Melbourne.

Selkirk, P. M., Seppelt, R. D. & Selkirk, D. R. 1990. Subantarctic Macquarie Island: environment and biology. Cambridge University Press. Cambridge.

Semeniuk, C. A. 1987. Wetlands of the Darling System – a geomorphic approach to habitat classification. Journal of the Royal Society of Western Australia 69: 95–112.

Semeniuk, C. A. 1988. Consanguineous wetlands and their distribution in the Darling System, south western Australia. Journal of the Royal Society of Western Australia 70: 69–87.

Semeniuk, C. A., Semeniuk, V., Cresswell, I. D. & Marchant, N. G. 1990. Wetlands of the Darling System, southwestern Australia: a descriptive classification using vegetation pattern and form. Journal of the Royal Society of Western Australia 72: 109–121.

Semeniuk, C. A. & Semeniuk, V. 1990. The coastal landforms and peripheral wetlands of the Peel-Harvey estuarine system. Journal of the Royal Society of Western Australia 73: 9–21.

Semeniuk, V. & Wurm, P. A. S. 1987. The mangroves of the Dampier Archipelago, Western Australia. Journal of the Royal Society of Western Australia 69: 29–87.

Shaw, J. 1991. Victorian wetlands: current conservation programs and policies. pp. 42–46. In: Donohue, R. & Phillips, B. (eds), Educating and managing for wetlands conservation. Australian National Parks and Wildlife Service, Canberra.

Shepherd, S. A. & Robertson, E. L. 1989. Regional studies – seagrasses of South Australia, western Victoria and Bass Strait. pp. 225–229. In: Larkum, A. W. D., McComb, A. J. & Shepherd, S. A. (eds), Biology of seagrasses: a treatise on the biology of seagrasses with special reference to the Australian region. Elsevier, Amsterdam.

Shepherd, S. A., McComb, A. J., Bulthuis, D. A., Neverauskas, V., Steffensen, D. A. & West, R. 1989. Decline of seagrasses. pp. 387–393. In: Larkum, A. W. D., McComb, A. J. & Shepherd, S. A. (eds), Biology of seagrasses: a treatise on the biology of seagrasses with special reference to the Australian region. Elsevier, Amsterdam.

Smith, A. J. 1975. A review of literature and other information on Victorian wetlands. CSIRO Division of Land Use Research Technical Memorandum 75/5.

Snowy Mountains Engineering Corporation 1983. An assessment of the quality and quantity of Australia's surface water resources. Water 2000 consultants' reports 1. Australian Government Publishing Service, Canberra.

Soil Conservation Service 1987. Total catchment management: a State policy. SCS, Sydney.

Speck, N. H., Wright, R. L., Rutherford, G. K., Fitzgerald, F., Thomas, F., Arnold, J. M., Basinski, J. J., Fitzpatrick, E. A., Lazarides, M. & Perry, R. A. 1964. General report on the lands of the West Kimberley area, W.A. CSIRO Australian Land Research Series 9.

Stafford Smith, D. M. & Morton, S. R. 1990. A framework for the ecology of arid Australia. Journal of Arid Environments 18: 255–278.

Stanton, J. P. 1975. A preliminary assessment of wetlands in Queensland. CSIRO Division of Land Use Research Technical Memorandum 75/10.

Stanton, J. P. & Morgan, M. G. 1977. The rapid selection and appraisal of key and endangered sites: the Queensland case study. University of New England School of Natural Resources Report No. PR4.

State Pollution Control Commission 1978. Water hyacinth environmental study: background review. SPCC, Sydney.

Storey, A. W., Vervest, R. M., Pearson, G. B. & Halse, S. A. 1993. Wetlands of the Swan Coastal Plain. Volume 7. Waterbird usage of wetlands of the Swan Coastal Plain. Water Authority of Western Australia and Environmental Protection Authority, Perth.

Thompson, M. B. 1986. River Murray wetlands: their characteristics, significance and management. Nature Conservation Society of South Australia and Department of Environment and Planning, Adelaide.

Timms, B. V. 1986. The coastal dune lakes of eastern Australia. pp. 421–432. In: De Deckker, P. & Williams, W. D. (eds), Limnology in Australia. CSIRO, Melbourne/Junk, Dordrecht.

Timms, B. V. 1992a. Lake geomorphology. Gleneagles, Adelaide.

Timms, B. V. 1992b. The conservation status of athalassic lakes in New South Wales, Australia. Hydrobiologia 243/244: 435–444.

Tyler, P. A. 1974. Limnological studies. pp. 29–61. In: Williams, W. D. (ed.), Biogeography and ecology of Tasmania. Junk, The Hague.

Tyler, P. A. 1993. Reconnaissance limnology of Tasmania. 1. The Picton Massif. Archiv fur Hydrobiologie 126: 257–272.

UNESCO 1971. Article 1, part 1, Convention on wetlands of international significance especially as waterfowl habitats. Published in Australia, 1976, for Department of Foreign Affairs by Australian Government Publishing Service, Treaty Series 1975, No. 48.

Wace, N. M. 1960. The botany of the southern oceanic islands. Proceedings of the Royal Society of London (B) 152: 475–490.

Walker, K. F. 1985. A review of ecological effects of river regulation in Australia. Hydrobiologia 125: 111–129.

Walker, P. J. 1991. Land systems of western New South Wales. Soil Conservation Service of New South Wales Technical Report No. 25.

Walton, D. W. H. 1984. The terrestrial environment. pp. 1–60. In: Laws, R. M. (ed), Antarctic ecology, volume 1. Academic Press, London.

Water Resources Commission 1986. Wetlands and water management strategy for the Lachlan River Valley. WRC, Sydney.

West, R. J., Larkum, A. W. D. & King, R. J. 1989. Regional studies – seagrasses of south eastern Australia. pp. 255–260. In: Larkum, A. W. D., McComb, A. J. & Shepherd, S. A. (eds), Biology of seagrasses: a treatise on the biology of seagrasses with special reference to the Australian region. Elsevier, Amsterdam.

West, R. J., Thorogood, C. A., Walford, T. R. & Williams, R. J. 1985. Estuarine inventory for New South Wales, Australia. NSW Department of Agriculture and Fisheries Bulletin No. 2.

Whinam, J., Eberhard, S., Kirkpatrick, J. & Moscal, T. 1989. Ecology and conservation of Tasmanian *Sphagnum* peatlands. Tasmanian Conservation Trust Inc., Hobart.

Whitehead, P. J., Wilson, B. A. & Bowman, D. M. J. S. 1990. Conservation of coastal wetlands of the Northern Territory of Australia: the Mary River floodplain. Biological Conservation 52: 85–111.

Williams, A. R. 1979. Vegetation and stream pattern as indicators of water movement on the Magela floodplain, Northern Territory. Australian Journal of Ecology 4: 239–247.

Williams, K. D. 1991. Australian Capital Territory wetlands: an overview of priorities and policies. pp. 9–13. In: Donohue, R. & Phillips, B. (eds), Educating and managing for wetlands conservation. Australian National Parks and Wildlife Service, Canberra.

Williams, O. B. 1962. The Riverina and its pastoral industry. In: Barnard, A. (ed.), The simple fleece: studies in the Australian wool industry. Melbourne University Press, Melbourne.

Williams, W. D. (ed.) 1981. Salt lakes. Junk, The Hague.

Wilson, B. A., Brocklehurst, P. S. & Whitehead, P. J. 1991. Classification, distribution and environmental relationships of coastal flood plain vegetation, Northern Territory, Australia. Conservation Commission of the Northern Territory Technical Memorandum 91/2.

Wimbush, D. J. & Costin, A. B. 1979a. Trends in vegetation at Kosciusko. II. Subalpine range transects, 1959–1978. Australian Journal of Botany 27: 789–831.

Wimbush, D. J. & Costin, A. B. 1979b. Trends in vegetation at Kosciusko. III. Alpine range transects, 1959–1978. Australian Journal of Botany 27: 833–871.

Winning, G. 1988. Classification of New South wetlands. Graduate Diploma in Environmental Management thesis, Mitchell College of Advanced Education.

Winning, G. 1991a. Some problems in determining the boundaries of SEPP 14 wetlands. Wetlands (Australia) 11: 10–20.

Winning, G. 1991b. A review of wetland surveys in New South Wales. Report to the NSW National Parks and Wildlife Service (unpublished).

Vegetatio **118**: 103–124, 1995.
© 1995 *Kluwer Academic Publishers. Printed in Belgium.*

A geomorphic approach to global classification for inland wetlands

C.A. Semeniuk & V. Semeniuk
V & C Semeniuk Research Group, 21 Glenmere Rd., Warwick, W.A., 6024, Australia

Key words: Climate, Wetland classification, Wetland geomorphology, Wetland hydroperiod

Abstract

A geomorphic classification of inland wetlands on criteria other than vegetation is proposed, based on their host landform and degree of wetness. Thus, the classification addresses the underlying structure of most inland wetlands, i.e. their landform setting and their various types of hydroperiod. Landforms host to wetlands include: basins, channels, flats, slopes and hills/highlands. Degrees of wetness include: permanent, seasonal or intermittent inundation, and seasonal waterlogging. From combining the landform type with hydroperiod, thirteen primary types of common wetlands are recognized: 1. permanently inundated basin = *lake*; 2. seasonally inundated basin = *sumpland*; 3. intermittently inundated basin = *playa*; 4. seasonally waterlogged basin = *dampland*; 5. permanently inundated channel = *river*; 6. seasonally inundated channel = *creek*; 7. intermittently inundated channel = *wadi*; 8. seasonally waterlogged channel = *trough*; 9. seasonally inundated flat = *floodplain*; 10. intermittently inundated flat = *barlkarra*; 11. seasonally waterlogged flat = *palusplain*; 12. seasonally waterlogged slope = *paluslope*; and 13. seasonally waterlogged highlands = *palusmont*. Water, landform and vegetation descriptors can augment the nomenclature of the primary units: e.g. salinity of water; size and shape of landform; and organisation, structure and floristics of vegetation. The classification can be used in many settings, regardless of climate and vegetation types. Using the approach adopted in this classification, in principle, more landform types and degrees of wetness, if necessary, can be added to the system to define additional wetland types.

Introduction

Wetlands have been difficult natural systems to classify because the term is imprecise, there have been confusing concepts of what constitutes a wetland, and there have been differing criteria used in their classification. For instance, some wetlands have been classified solely on their vegetation structure or floristics (e.g. salt marshes, and meadows), some according to their vegetation combined with associated soil/substrate and water types (e.g. peatlands, bogs, fens), and some on their water permanence, sometimes in association with vegetation (e.g. lakes, swamps). However, to date, no classifications have successfully encompassed the full range of climatic and physiographic wetland types. In addition, the use of many wetland terms, founded on vegetation types characterising wetland conditions, has been inconsistently applied in areas outside where they were first defined, with the result that 'meadow'

or 'swamp' are terms used for wetlands with widely differing water permanence and landform setting. For example, both a wetland basin and a wetland flat, that are either seasonally inundated with fresh water, or seasonally waterlogged (but without water ever free-standing on the surface), could be vegetated by appropriate rushes or grasses such that all the types of wetlands developed under these varying conditions of landform and water permanence would be termed a wetland 'meadow'. Yet the vegetation unit that comprises the 'meadow' is colonising widely different wetland systems. Another example of this is the use of the term 'marsh' which is a vegetation term applied to such varying wetland systems as tidal marshes and freshwater marshes which have fundamentally different settings and water regimes. A further example is provided by those wetland basins, flats, or hill slopes that under appropriate climatic and vegetation conditions support mosses and develop peat substrates – these would all be termed mires.

In recent years, there has been a consensus amongst scientists as to what constitutes a wetland in Australia and globally (UNESCO 1971; Ramsar Convention Bureau 1991; Briggs 1981; for review see Semeniuk 1987). Classification of these wetlands encompasses a wide range of systems, but as yet, there has been no consensus reached on what constitutes a workable scheme, and on whether, indeed, a classification scheme that is globally applicable is practical or even achievable (Pressey 1992).

Recently, Semeniuk (1987) proposed a geomorphic approach to classify land-based wetland systems. Landform/geomorphic and water/wetness attributes of a wetland are non-genetic criteria, and this approach to the classification of wetlands is based fundamentally on the two major factors which determine the existence of all wetlands, i.e. landform and water, regardless of the climatic setting, soil type, vegetation cover, geomorphic setting, or origin. The classification therefore, has the potential to bring out the underlying, unifying feature of wetlands that occur in diverse physiographic and climatic settings.

The classification was designed for the area of the Darling System in south-western Australia, within which there were a limited range of basic landforms that were host to wetlands, and a limited range of conditions of water permanence that determined the occurrence of a 'wet land' (i.e. a terrain that is wet). However, if the geomorphic approach was to be adopted for areas outside the region of south-western Australia [for instance, in the wetter climates where more landform types other than basins, channels and flats are host to wetlands, and in the drier climates where landforms are only intermittently flooded], the classification of Semeniuk (1987) needed to be expanded.

The purpose of this paper is to propose such a classification – one that can be applied to other parts of the world generally. The paper explains the philosophy of the geomorphic approach to wetland classification, the criteria to be used, the categories of wetland derived from the expanded classification, and, finally, the practical application of the classification.

Wetlands included in this study are all those land-based closed systems such as basins, and land-based open systems such as riverine channels, flats, slopes and highlands (Fig. 1). The open systems of deltaic and estuarine wetlands and marine wetlands are not considered.

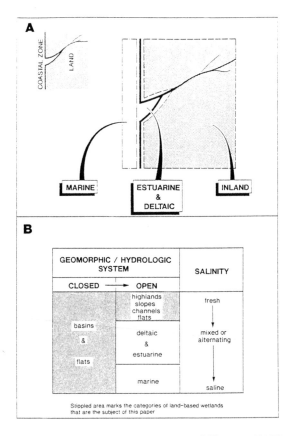

Fig. 1. Schematic diagram showing scope of this paper. (A) The relationship of inland wetlands to estuarine/deltaic wetlands and marine wetlands; (B) Conceptual summary of types of broad wetland categories and their inter-relationships in terms of geomorphic/hydrologic setting and their salinity.

Review of wetland definitions and classifications

Definition of wetland

The definition of wetland in recent literature includes permanent lakes, small to large seasonal lakes, small to large areas of seasonally waterlogged soils, fluviatile systems, estuarine systems, and marine systems (UNESCO 1971; Bayly & Williams 1973; Cowardin *et al.* 1979; Adam *et al.* 1985). However, in the older literature, the concept of wetland was more rigidly confined to encompass only lakes and water-saturated basins. The main wetland classifications developed over the past 20 years are briefly outlined below.

The Ramsar Convention, held to secure wetlands of international importance to migratory water birds, defined wetlands as 'areas of marsh, fen, peatland or water, whether natural or artificial, permanent or temporary, with water that is static or flowing, fresh, brackish or salt, including areas of marine water the depth of which at low tide does not exceed 6 m' (UNESCO 1971). This definition covers all categories of marine, estuarine and inland wetlands. Hill (1978) identified wetlands by the presence of hydrophilic vegetation typically adapted to life in areas inundated or saturated by water with the appropriate duration and frequency to promote that vegetation. This definition is problematic for those wetlands where there is no vegetation. Cowardin *et al.* (1979) defined wetlands as lands transitional between terrestrial and aquatic systems where the water table is usually at or near the surface or the land is covered by shallow water. In this context, the areas of permanent inundation are viewed as aquatic environments. It was stipulated that wetlands should have one or more of the following attributes: (1) at least periodically the land supports hydrophytes; (2) the substrate is predominantly undrained hydric soil; (3) the substrate is nonsoil and is saturated with water or covered by shallow water at some time during the growing season of each year. The criteria of Cowardin *et al.* provide useful guidelines for identifying wetlands, and, in particular, wetland margins.

Bates & Jackson (1980) defined wetlands as a general term for a group of wet habitats and included areas that are permanently wet and/or intermittently water covered, such as coastal marshes, tidal swamps and flats, and associated pools, sloughs and bayous. Zoltai and Pollet (1983) defined wetlands as areas where wet soils are prevalent, having a water table near or above the mineral soil for most of the thawed season, supporting a hydrophilic vegetation, and pools of open water (< 2 m deep), including shallow open water. They did not include areas that become temporarily flooded, but remain relatively well drained for most of the growing season.

The definition in this paper, as used by Semeniuk (1987), adopted from the Wetlands Advisory Committee (1977) is that of 'areas of seasonally, intermittently or permanently waterlogged soils or inundated land, whether natural or otherwise, fresh or saline'. This would include areas with waterlogged soils, ponds, billabongs, lakes, swamps, rivers and their tributaries, as well as marine and coastal wetlands such as tidal flats and estuaries. It is important to note that wetlands may be distinguished by the occurrence of water, or waterlogged soils, or vegetation typical of water conditions (e.g. swamp trees, reed beds), or hydric soils (i.e. formed in response to prevailing water inundation or waterlogging, and including peats, peaty sands, car-

bonate muds, etc.). This paper only deals with natural, inland (non-marine, non-estuarine) wetlands.

It is also important to note that the boundary of a wetland should be drawn to encompass all features diagnostic of 'wet lands'. There has been a tendency, at least in Western Australia, in the past to include only the permanently inundated parts of a wetland in the defined area of the wetland. The permanently inundated parts of a lake are viewed as a 'wetland', and the waterlogged soils of the outer margin viewed as a buffer zone to the wetland, or are not viewed as a wetland at all. This is not strictly correct. The boundary of a wetland should encompass all that terrain that conforms to the definition of a wetland (Figs. 2 and 3). Thus, within a given wetland there may be parts that are: 1. permanently inundated, 2. seasonally inundated, or 3. seasonally waterlogged by water table rise. All these parts would be viewed as being 'wet lands' if they had occurred in isolation, and indeed, individual basins showing any of the 'wetness' characteristics listed above would be defined as wetlands on any of the international or local definitions. These zones should also be viewed as being part of a wetland if they occur as concentric or zonal parts, or subcomponents of a large wetland. Indeed, given the general slope of wetland shores, and the fluctuation in water table or water levels, there will always be zones of progressively less wet terrain within a given wetland.

This aspect of zones of progressively less wet terrain within a wetland has been addressed in the literature by recognition of its littoral zone (Hutchinson 1957; Wetzel 1983; Ruttner 1953; Cowardin *et al.* 1979). Thus, wetlands with a central zone of permanent water will have a second zone that is inundated by seasonal water level fluctuations, and an outer zone where the soils are seasonally waterlogged by water table rise and capillary water rise. Similarly, a wetland with a central zone of seasonal inundation will have an outer zone that is seasonally waterlogged by water table rise and capillary water rise. In this study, the boundary of wetland is drawn at the outside of the area that has the characteristics of dampness, or hydric soils, or vegetation indicative of wet conditions.

Classification schemes

Previous wetland classifications have been based on biological criteria (e.g. mires, swamps; Martin *et al.* 1953; Golet & Larson 1974; Ivanov 1981; Gore 1983); physical criteria and genetic criteria (e.g. riverine lakes, glacial lakes, tectonic lakes; Hutchinson

106

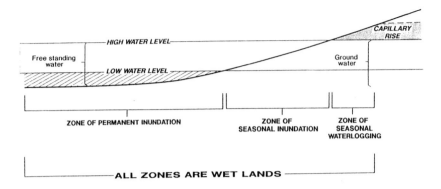

Fig. 2. The boundary of a wetland should encompass all that terrain that conforms to the definition of a 'wet land'.

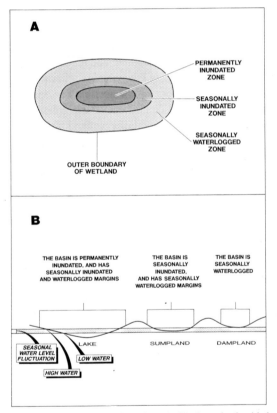

Fig. 3. (A) Zones of wetness around a typical basin wetland, with the boundary of the wetland as used in this paper; (B) The development of the 3 basin wetlands, which in this example are related to landform position with respect to fluctuating water level – the extent of 'wet' land in each is shown.

1957); biological/chemical criteria (e.g. fens, bogs; Tansley 1939; Gorham 1957; Reeves 1968; Moore & Bellamy 1974; Ruttner 1953; Gore 1983); chemical criteria (e.g. minerotrophic, ombrotrophic; Hakanson & Jansson 1983; Wetzel 1983); ontogenetic criteria,

and others. Many of the classifications are based on wetlands occurring in northern Europe, and do not always account for the wide variety of wetland types occurring in various other physiographic and climatic regimes of the world. Moreover, the approach used to date has tended to emphasise the vegetation of wetlands, or the products of the wetland environment (e.g. peats).

Semeniuk (1987), in SW Australia, presented a simple classification of land-based, mainly freshwater wetland systems, based on the two main attributes of wetlands: their landform shape and water permanence (or hydroperiod). Semeniuk (1987) recognised three basic landform types, viz., basins, channels and flats, that were host to wetlands, and recognised three types of water permanence that determined wetland conditions, viz., permanently inundated areas, seasonally inundated areas, and seasonally waterlogged areas. In combination, these attributes of landform and water permanence resulted in the recognition of 7 common types of wetlands:

1. permanently inundated basin = lake
2. seasonally inundated basin = sumpland
3. seasonally waterlogged basin = dampland
4. permanently inundated channel = river
5. seasonally inundated channel = creek
6. seasonally inundated flat = floodplain
7. seasonally waterlogged flat = palusplain

As a result, this classification, firstly, identified wetlands in south-western Australia that had previously either been unrecognised, or uncategorised (dampland and palusplain), and secondly, more consistently defined the various types of wetlands, some of which had been ambiguously and inconsistently named using terms from previous classifications. For instance, the term 'swamp' had been applied to areas of true swamps (cf. Tansley 1939; Spence 1967; Holmes 1944; Golet & Larson 1974; Ivanov 1981; Gore 1983), as well as any wetland that had the characteristics of 'swampiness' or 'marshiness'.

The classification of Semeniuk (1987) intentionally did not bring into consideration, at a primary level, aspects such as type of vegetation cover, or soil types, that occur in the wetlands. It was considered that the geomorphic approach would best bring out the underlying similarity of wetlands across a wide range of climatic, geomorphic, soil, and vegetation settings. The rationale was that landform and water characteristics would be the dominant and/or common feature for all wetlands, regardless of their setting. A descriptor for vegetation and soils could be added at a secondary or tertiary level to highlight the differing vegetation complexes and soil complexes that may occur in different types of wetlands. For this purpose, a series of terms to describe the landform, water and vegetation features of the primary wetlands was also provided by Semeniuk (1987) and Semeniuk et al. (1990). Landform was further described by scale and shape, water by its salinity and its consistency, and vegetation by its organisation, structure and floristics.

The significance of climatic setting

At a site specific level, local variation in soils, stratigraphy and hydrology can result in the development of wetlands as a consequence of ponding and/or seepage, and this can override the climatic setting as a major factor to develop wetlands at the local scale. However, the overall control of the distribution and abundance of wetlands globally is climate. Wetlands are more numerous in humid environments, and become less frequent as the climate becomes drier. Indeed, the various climates of the earth develop different types of wetlands according to the endemic landform type and the availability of water. Thus, in this context, the concept of what is a wetland should be a relative one. A wetland, at a local scale, may be viewed as that portion of the earth's surface that is permanently, seasonally or intermittently wet or damp, in relation to adjoining dry areas. The relatively wetter areas, over geological time, develop either a distinctive hydric soil that reflects this prevailing difference, or a vegetaton complex indicative of wetter conditions.

In south-western Australia, where the geomorphic classification was first developed, the climate is subhu-

mid to humid, such that the main landform host to wetlands are basins, channels and flats, and the categories of water permanence that develop wetlands are 'permanent inundation', 'seasonal inundation', and 'seasonal waterlogging'. Any areas that are 'intermittently flooded' do not tend to become prevailing wetlands, in that they do not develop wetland vegetation complexes, or hydric soils. Wetlands developed on slopes and highlands are uncommon or absent, although local zones of marked seepage can develop some small scale wetland slopes.

In more humid areas, in contrast, the whole of a regional landscape may retain water, and become a 'wetland'. Wetlands may encompass highlands, slopes, flats, channels and basins, with the basins, channels and flats being wetter than the other landscape units. The various wetland types may be differentiated on their different landscape attributes and degree of wetness. For example, in northern Europe, the climate is humid enough to ensure that there is an excess of precipitation over evaporation, such that landforms in addition to basins, channels and flats can develop into wetlands. Wetlands such as highland mires and moors are developed on slopes where there are seeps and springs, or on hills and highlands.

At the other extreme, in very arid areas, the majority of the terrain will be dry, but the lower parts of the landscape will accumulate water on an intermittent basis following any rain that results in substantial runoff. The prevailing, or long term condition is that rainfall intercepted by highlands and slopes will preferentially flow into low areas. There will always be insufficient rainfall to saturate the highlands and slopes enough to develop wetland soils or wetland vegetation. In the long term, the lower parts of arid zone landscapes may develop soils and vegetation complexes reflecting this fundamental difference between 'intermittently wet land' versus 'permanently dry land'. For example, in the arid zones of Australia, the landforms of slopes and highlands are either rocky highlands or linear dune fields. Although they may receive torrential rainfall very intermittently, inundation or waterlogging will be for a very short period. The resulting runoff will accumulate in the low zones of the topography – i.e. the basins, channels and flats. Such accumulations are not seasonal, but intermittent, following aperiodic rainfall in the region. However, since low-lying landforms tend to accumulate water more than the slopes and uplands, they have tended to develop soils and vegetation responding to this intermittent wetting. In this context, the wetter basins, channels and flats in

108

arid areas are intermittent wetlands when compared to the surrounding terrain.

A wetland classification must account for these differences, in that in some humid areas there will be development of landscape units as wetlands which would be drylands in a more arid setting. As the climate becomes more humid, more of the terrain will be subject to wet conditions, and hence will be captured as 'wetlands'. At the other extreme, as the climate becomes more arid, less water is available, and less of the terrain will be subject to wetting, be it seasonal or intermittent wetting.

Proposed geomorphic classification for wetlands

Landform and water characteristics

Landforms are the 'containers' to wetlands, or are 'host' to wetlands. They determine the size and shape of the wetland, and in addition, specifically for basins, the depth of a wetland. There are a variety of basic landform types that potentially may contain water, or retain water, and hence may become a wetland. Experience in areas outside of south-western Australia where the geomorphic classification was first developed, i.e. arid central Australia, humid northern Europe, arid to humid South Africa, and arid to humid Northern America, indicates that there are at least 5 basic landform types that determine the occurrence of wetlands. These are listed below in geometrically gradational series from concave basins to convex hills (Fig. 4):

1. basins
2. channels
3. flats
4. slopes, and
5. highlands, or hills.

These landforms, of course, are intergradational. For example, flats grade into undulating flats and then into series of basins; slopes grade into flats; basins may be isolated, or may become more and more interconnected and hence grade into channels.

When classifying a landform it is important to classify the full landform and not just parts of it. Thus, within a basin which may have steep sloping banks and a flat foor, the individual parts of this basin should not be classed as 'slopes' and 'flats'. The landform types of hills, creeks, and basins may have sub-components of slopes and flats, whereas the landform types of flats and slopes occur in isolation.

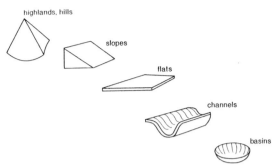

Fig. 4. Landforms that are host to wetlands.

Water is a major component that distinguishes the wetland habitat from other terrestrial habitats, and also the factor which influences biological response by its presence, depth, chemistry, and movement. The period of water availability in a wetland (= hydroperiod), i.e. its permanence or intermittency, is directly related to precipitation and evaporation, mechanisms of recharge and discharge, permeability of underlying sediments, and shape of the wetland. Four types of permanence are distinguished:

1. permanently inundated
2. seasonally inundated
3. intermittently inundated
4. seasonally waterlogged.

The term 'waterlogged' is equivalent to 'saturated'. Some authors use the term waterlogged to refer to soils that are saturated with water when inundated to shallow depth. We use the term waterlogged as distinct from inundated in the following way:

waterlogged: those soils that are saturated with water, but where the water does not inundate the soil surface;

inundated: those soils that are covered with free-standing water; the soil below the surface in these situations is also saturated (waterlogged).

The term 'water level' is used in this paper to refer to the level of the free standing water and to the level of groundwater.

In order to avoid a plethora of hydroperiod categories, and hence a profusion of wetland types, only four critical types of water permanence are used. This approach differs from that of previous workers, such as the seven-fold system of nontidal water regimes of Cowardin *et al.* (1979). Some categories of Cowardin *et al.* (1979) have been combined (Table 1). For example, the categories of 'permanently flooded throughout

Table 1. The categories of hydroperiod of this paper correlated with the seven nontidal hydroperiod categories of Cowardin *et al.* 1979.

	Categories of Cowardin *et al.* 1979	Category in this paper
1.	Permanently flooded – water covers land surface throughout year in all years	Permanently inundated
2.	Intermittently exposed – surface water present throughout year except in years of extreme drought	Permanently inundated
3.	Semipermanently flooded – surface water persists throughout growing season in most years. When surface water is absent, water table is at or near surface	Seasonally inundated
4.	Seasonally flooded – surface water is present for extended periods, especially in early growing season but is absent by the end of the season	Seasonally inundated
5.	Saturated – substrate is saturated for extended periods during growing season but surface water is seldom present	Seasonally waterlogged
6.	Temporarily flooded – surface water is present for brief period during growing season but water table is otherwise well below the soil surface	Intermittently inundated
7.	Intermittently flooded – substrate is usually exposed but surface water is present for variable periods with no seasonal periodicity	Intermittently inundated

the year in all years', and 'intermittently exposed only during periods of extreme drought' have an underlying similarity of permanent inundation, and the occasional drying out of these wetlands during droughts may not be a prevailing feature. Thus, these two categories of water regime can be combined, because in the classification proposed here, the prevailing year to year condition is used to categorise the hydroperiod. To illustrate with another example, consider those areas that are seasonally waterlogged on a prevailing year to year basis. Even if they are rarely or infrequently inundated they do not qualify as inundated areas in terms of prevailing water permanence. The correlation of hydroperiod used in this paper to those of Cowardin *et al.* (1979) is illustrated in Fig. 5, using situations where watertable rise maintains the wetland. Note that the watertable example used in Fig. 5 is not intended to imply that wetland hydroperiod is controlled by this mechanism alone. Other mechanisms of recharge can maintain a wetland, and similar correlation of hydroperiods could be devised for these other types.

The characteristic of the recurring hydroperiod should be the deciding factor in determining the classification of a wetland by its 'wetness'. Occasional or infrequent flooding of terrain can take place anywhere during torrential rainfall, or by sheet flooding, for instance, in a car park after heavy rains, but such phenomena do not categorise temporarily inundated land surfaces as wetlands because many of the other criteria to indicate prevailing wetland conditions are absent. In this context, arid zone basins that are intermittently inundated can be termed wetlands, where they have developed soils and vegetation reflecting the long term difference between them and the surrounding uplands.

In some regions, the hydroperiod may be weakly seasonal, with rainfall distributed throughout the year, but with more increased rainfall in one part of the year. Wetlands in these regions again should be classified on the prevailing condition. Also, there are some situations, as in South Africa, where basins are permanently inundated from year to year, with some water level fluctuation, and only in a five-year to ten-year period, for instance, do they dry out. Again, these wetlands would be classified on the prevailing condition of permanent inundation.

110

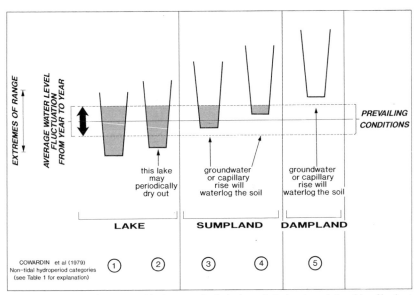

Fig. 5. Schematic diagram, for situations where watertable rise controls the hydroperiod, showing relationship of basins to average and extreme water level fluctuations that result in the various hydroperiod categories of this paper [viz., permanently inundated, seasonally inundated & seasonally waterlogged], and those of Cowardin *et al.* (1979) (see Table 1 for explanation of the five categories of water regime of Cowardin *et al.* 1979). This illustration does not imply that these categories of hydroperiod are related solely to rise and fall of watertable.

Table 2. The thirteen basic wetland categories formed from combining landform and hydroperiod attributes.

Water longevity	Landform				
	Basin	Channel	Flat	Slope	Highland
Permanent inundation	Lake	River	–	–	–
Seasonal inundation	Sumpland	Creek	Floodplain	–	–
Intermittent inundation	Playa	Wadi	Barlkarra	–	–
Seasonal waterlogging	Dampland	Trough	Palusplain	Paluslope	Palusmont

The basic wetland types

The classification of wetlands proposed here is developed by combining the various types of landform with the various types of water permanence (or hydroperiod). Combining water permanence with types of landform provides 13 main wetland types (Table 2). Thus, basins may be permanently inundated, seasonally inundated, intermittently inundated, or seasonally waterlogged. Channels may be permanently inundated, seasonally inundated, or intermittently inundated. Flats may be seasonally inundated, intermittently inundated, or seasonally waterlogged. Slopes tend to be only

seasonally waterlogged, and highlands or hills also tend to be only seasonally waterlogged. These basic categories require nomenclature so that they may form the primary units of wetland differentiation. In order not to create clumsy binomial or polynomial nomenclature, terms such as 'intermittently inundated flat' or 'seasonally waterlogged basin', were avoided, and single-word terms were used. The proposed terms for the basic wetland units are as follows:

Lake : permanently inundated basin
Sumpland : seasonally inundated basin
Dampland : seasonally waterlogged basin

Playa : intermittently inundated basin
River : permanently inundated channel
Creek : seasonally inundated channel
Wadi : intermittently inundated channel
Trough : seasonally waterlogged channel
Floodplain : seasonally inundated flat
Barlkarra : intermittently inundated flat
Palusplain : seasonally waterlogged flat
Paluslope : seasonally waterlogged slope
Palusmont : seasonally waterlogged highlands and
hills.

Some of the terms are established and well defined previously in the literature and hence have been re-utilised. Others (i.e. sumpland, dampland, palusplain) were coined in Semeniuk (1987), or have been coined in this paper. The rationale for, the definition of, and the origin of the terms are provided in Table 3. A comparison of selected proposed wetland terms with other previously established terms is provided in Table 4. It should be noted that the terms river, creek and wadi are used to denote permanence, seasonality, or intermittency of water flow in channels, respectively; the size of the river, creek or wadi is indicated by the use of a scale descriptor. Some typical wetland types occurring in isolation, or adjacent to other wetlands, are illustrated in Fig. 6.

Some combinations of categories of course are not possible as prevailing wetlands. For instance, any slopes that are inundated, would remain inundated for only a very short period before runoff would remove any free standing water, and hence, slope wetlands with hydroperiods of permanent, seasonal or intermittent inundation cannot be developed in practice. The same rationale applies to the category of permanently, seasonally, or intermittently inundated highlands or hills, and permanently inundated flats.

Lakes that have a wide seasonally exposed littoral zone need to be distinguished from sumplands, and sumplands that have a wide seasonally waterlogged outer zone need to be distinguished from damplands. A cut-off of 10% is used as the boundary to distinguish between the types (Fig. 7). Thus, if a basin that has a seasonally fluctuating water level dries out such that there is still more than 10% of water by area in the basin at the driest stage, then it is a lake, but if there is less than 10% water by area in the basin, then it is a sumpland.

Use of descriptors

Water, landform and vegetation descriptors are used to augment the nomenclature of the primary units. Large oval, water filled basins that remain fresh throughout the year can be termed macroscale, freshwater, stasohaline, ovoid lakes. The full image of descriptors is presented in Figs. 8 & 9, and described below. Soil type has not been incorporated at this stage as a descriptor into the classification. However, use of terms such as peaty, calcareous, gypseous, diatomaceous, or quartzose (sandy) could readily be added to the system if necessary.

Descriptors of water
Water in a wetland may be further described in terms of its salinity, the consistency of salinity, other chemical features of its water quality, and its source. In this paper, water in wetlands is further described only in terms of salinity and consistency of salinity. Other descriptors can be added if necessary, dependent on the need and the type of study.

Salinity may be subdivided into categories of: fresh, brackish (or mixosaline), saline and hypersaline. In the literature, definitions vary for categories such as brackish, saline, hypersaline (Davis & DeWiest 1966; Drever 1982; Cowardin *et al.* 1979; Hammer 1986). The category terms and boundaries adopted in this paper are after Hammer (1986).

Wetlands that are seasonally variable in salinity are categorised by the salinity state in which the wetland exists for the major part of each year. A wetland that ranges from freshwater for most of the year, to brackish during the season of reduced water supply, would be classified as freshwater. However, a term is introduced to denote whether salinity is constant or variable. Water salinity that is consistent throughout the year, remaining totally within a given salinity field, is termed *stasohaline*. Water quality that markedly fluctuates throughout the year is termed *poikilohaline*.

Descriptors of landform
The cross sectional geometry was initially used to subdivide a wetland into hills/highlands, slopes, flats, channels and basins. The landform host to a wetland, however, can be further categorised on the basis of plan geometry and size. In plan, encompassing the limnetic and littoral zones (Hutchinson 1957; Cowardin *et al.* 1979), wetland shapes may be described as linear, elongate, irregular, fan-shaped, ovoid or round, for

Table 3. Definition and origin of terms.

Wetland term	Definition	Defined by	Origin of term	Usage in this paper
Lake	Permanently inundated basin of variable size and shape	Mill (1900–1910) Monkhouse (1965) Bates & Jackson (1980) Fairbridge (1968) Ruttner (1953)	Established term, from Latin *lacus*, a hollow	The usage in this paper does not distinguish between shallow lakes and deep lakes
Sumpland	Seasonally inundated basin of variable size and shape	Semeniuk (1987)	After 'sump' meaning site of water retention or ponding or accumulation; the term is fortuitously similar to 'sumpf' the German term for swamp	As defined
Dampland	Seasonally waterlogged basin of variable size and shape	Semeniuk (1987)	After 'damp' meaning moist or wet. Thus it refers to a dampness or waterlogging of soils of some basin wetlands	As defined
Playa	Intermittently flooded basin of variable size and shape	Bates & Jackson (1980)	Established term referring to intermittently flooded basin; usually in arid environments such basins are 'salt lakes'	As defined
River	Permanently inundated channel of variable size and shape	Swayne (1956) Trowbridge (1962) Morisawa (1968)	Established term from Latin *rivus*, a stream (Shipley 1982)	This usage conforms with the concept authors that river is defined as channelled water flow, but is different to most authors in its necessity for permanence of water. The permanence of water, also generally implies a channel of large rather than small size
Creek	Seasonally inundated channel of variable size and shape	Whittow (1984) Monkhouse (1965) Trowbridge (1962) Bates & Jackson (1980)	Established term	This usage generally conforms with that of Australia and southwestern U.S.A.
Trough	Seasonally waterlogged channel of variable size and shape	Bates & Jackson (1980)	Established term	This usage generally conforms with that of U.S.A.
Wadi	Intermittently flooded channel of variable size and shape	Bates & Jackson (1980)	Arabic term referring to drainage channels in desert environments that flash-flood during the occasional storm	This usage generally conforms with that of U.S.A.

Table 3. Continued.

Wetland term	Definition	Defined by	Origin of term	Usage in this paper
Floodplain	Seasonally inundated flat	Mill (1900–1910) Monkhouse (1965) Moore (1949) Wolman & Leopold (1957) Ward (1978)	Established term	This differs from other authors in that inundation of the plain or flat need not be linked to a river; in general, however, a floodplain is associated with a river or creek
Palusplain	Seasonally waterlogged flat	Semeniuk (1987)	After latin *palus* meaning 'marshy'; thus the term refers to flats which are similar in wetness to dampland basins	As defined
Barlkarra	Intermittently flooded flat	This paper	NW Australian aboriginal word referring to grassy flats/plains that are flooded from time to time	As defined
Paluslope	Seasonally waterlogged slope	This paper	After latin *palus* meaning 'marshy'; thus the term refers to slopes which are similar in wetness to dampland basins, i.e., wetland slopes	As defined
Palusmont	Seasonally waterlogged highlands and hills	This paper	After latin *palus* = 'marshy, and *montanus* = 'mountain'; hence the term refers to hills and highlands that are seasonal wetlands	As defined
Stasohaline	Water of relatively constant salinity remaining in a given salinity field	Semeniuk (1987)	After staso (Greek) meaning constant	As defined
Poikilohaline	Water of variable salinity fluctuating from one salinity field to another	Originally defined Dahl (1956)	After poikilo (Greek) meaning variable	As defined
Waterlogged or saturated	Area in which groundwater stands near, or at the land surface		Established term	Usage conforms with Golet and Larson (1974), Martin *et al.* (1953) and most other authors

Table 4. Comparison of selected wetland terms used in this paper with established classifications.

This paper	Martin *et al.* (1953)	Cowardin *et al.* (1979)	Golet & Larson (1974)	Paijmans *et al.* (1985)	General European	General N. American
Lake	Open fresh water Deep fresh marshes Open saline water	Lacustrine	Open water Shrub swamp Deep marsh	Lakes Swamp Coastal water bodies	Lake Swamp	Lake Swamp
Sumpland	Wooded swamp Seasonally flooded Basins Shallow fresh marshes Deep fresh marshes Saline marshes Open saline water	Palustrine	Deep marsh Shallow marsh Shrub swamp Wooded swamp Open water	Lakes Swamp	Marsh	Marsh Meadow
Dampland	Fresh meadows Wooded swamp	Palustrine	Meadow	–	–	Meadow
River	–	Riverine	–	River and creek Channels	River Stream Creek Brook	River Stream Creek Brook
Creek	–	Riverine	–	River and creek Channels	–	In part, Arroyo
Wadi	–	–	–	–	–	In part, Arroyo
Floodplain	Shrub swamp Wooded swamp	–	Seasonally flooded flats	Land subject to inundation	Floodplain	Floodplain Seasonally flooded flat
Palusplain	Wooded swamp Saline flat Salt meadow?	Palustrine	–	–	–	–
Palusmont	–	–	–	–	Equivalent in part to highmire (see Gore 1983)	–

basins, slopes and hills/highlands; and straight, sinuous, anastomosing, or irregular for channels.

Wetlands may be further categorised according to scale. For hills/highlands, slopes, basins and flats the categories of geomorphic scale for wetlands developed therein are (after Semeniuk 1987):

Megascale: Very large scale wetlands larger than a frame of reference 10 km × 10 km;

Macroscale: Large scale wetlands encompassed by a frame of reference 1000 m × 1000 m to 10 km × 10 km;

Mesoscale: Medium scale wetlands encompassed by a frame of reference 500 m × 500 m to 1000 m × 1000 m;

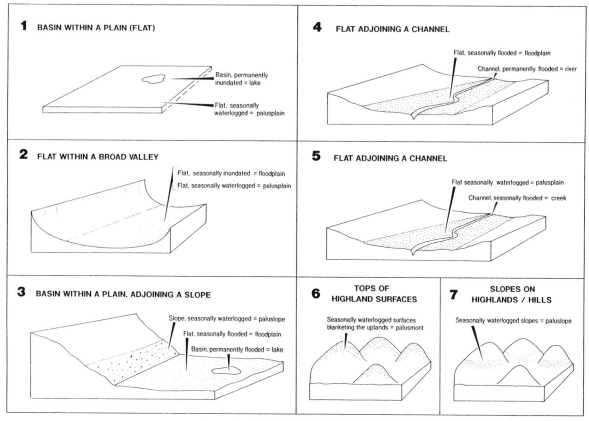

Fig. 6. Schematic diagram showing types of landforms and the wetlands developed therein. Some wetlands occur in isolation and some wetlands occur adjoining others.

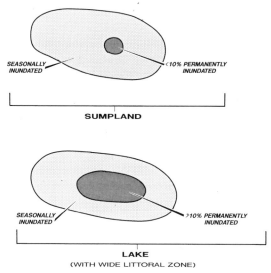

Fig. 7. Criteria to determine whether a basin is a lake or a sumpland dependent on the proportion of area of permanent inundation within the wetland.

Microscale: Small scale wetlands encompassed by a frame of reference 100 m × 100 m to 500 m × 500 m;

Leptoscale: Very small scale wetlands encompassed by a frame of reference < 100 m × 100 m.

Thus, all basins that are permanently inundated are lakes; those that are smaller than 100 m × 100 m are leptoscale lakes, those that fit in a frame of 1 km × 1 km are mesoscale lakes, and those that are of a size greater than 10 km × 10 km are megascale lakes.

In the case of channels, a definitive width to length relationship is used to separate size of channel wetlands:

Macroscale: Large scale channels 1 km and greater wide, by several to tens of kilometres long;

Mesoscale: Medium scale channels hundreds of metres wide, by thousands of metres long;

Microscale: Small scale wetlands tens of metres wide, hundreds of metres long;

Leptoscale: Fine scale channels several metres wide, tens of metres long.

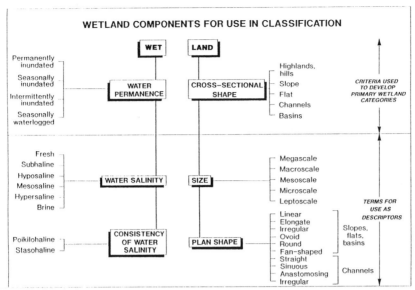

Fig. 8. Components or attributes of wetlands and the terminology used in the proposed classification.

Thus, all channels that are seasonally inundated, for instance, are creeks, but those that are less than several metres wide are leptoscale creeks, and those that are of a size greater than 1 km wide are macroscale creeks.

Descriptors of vegetation

Semeniuk *et al.* (1990) proposed a classification for wetland vegetation based on the areal extent and pattern of distribution of vegetation cover over the wetland, the internal organization of that vegetation in plan, the predominant vegetation structure, or the range of structural types in zones, and the details of floristics. Vegetation cover was divided into 3 intergradational classes: peripheral, mosaic and complete. Complexity of wetland vegetation was divided into 3 classes: homogeneous, zoned and heterogeneous. The combination of cover and internal organization resulted in the recognition of 9 basic wetland vegetation categories: periform, paniform, latiform, zoniform, gradiform, concentriform, bacataform, heteroform, and maculiform (Fig. 9). These terms are the primary part of a binary terminology, which forms the core of the vegetation classification. Established structural terms are adopted to describe the structure of wetland vegetation, and this forms the second part of the binary terminology. This approach provided a systematic way to describe and compile an inventory of wetland vegetation units. The classification provides a conceptual picture of the wetland vegetation, and the diversity

and complexity of specific wetlands then become obvious.

Semeniuk *et al.* (1990) also suggested that, if required, the vegetation terms, outlined above, could be used as descriptors to existing wetland classifications. If only the overall vegetation pattern of a wetland needs to be noted as a descriptor, the vegetation classification presented above may be modified to adjectival form, with the substitution of '... form' in the nomenclature by '... phytic'. Thus, a lake with periform forest, a sumpland with gradiform heath/sedgeland, and a dampland with zoniform forest/heath/sedgeland could be termed periphytic lake, gradiphytic sumpland and zoniphytic dampland, respectively. In these cases, the emphasis is on the classification of the wetland type, and the vegetation adjectival qualifier simply augments the nomenclature of the wetland.

Application of the proposed classification to mapping

Several areas in different physiographic and climatic regions of Western Australia have been selected to indicate the practical use of the proposed classification, and to display the range of wetlands with respect to their variability of geometry, size, and salinity. Maps from three different geomorphic and climate settings illustrating a range of wetland types in each selected area are presented in Fig. 10. The areas selected are:

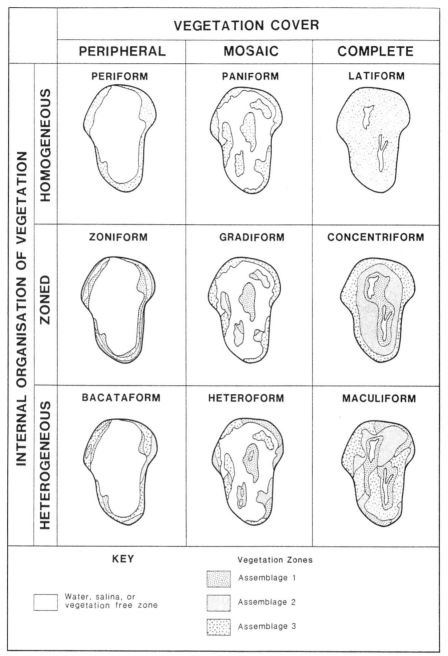

Fig. 9. The nine categories of vegetation organisation within basin wetlands as proposed by Semeniuk *et al.* (1990).

1. Scott River area – wet climate
2. Rockingham area – intermediate climate
3. Wiluna area – dry climate.

The Scott River area, located within the Scott Coastal Plain (Playford et al. 1976), is in a humid climate. The southern part of the terrain is a sandplain adjoining a coastal barrier dune system, and it is dominated by palusplain, with scattered macroscale to microscale lakes, mesoscale to microscale sumplands and damplands, floodplains, creeks and microscale rivers. The northern and eastern part of the terrain is a dissected plateau and it is dominated by leptoscale creeks and microscale rivers.

118

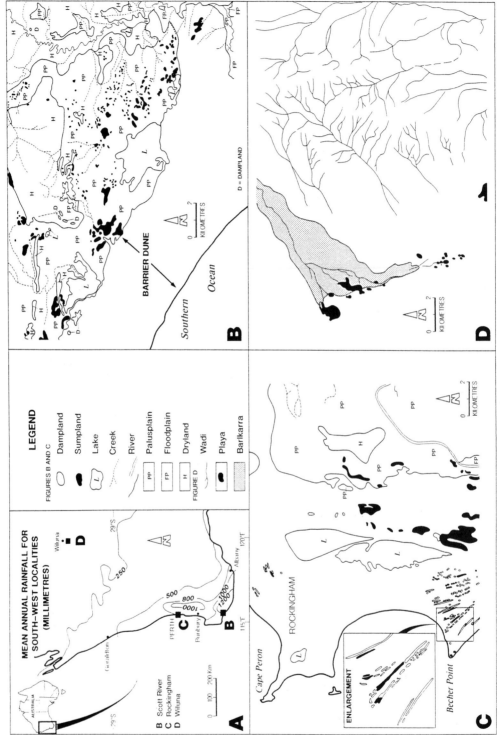

Fig. 10. Maps showing types and distribution of wetland categories for three selected areas in south-western Australia. Scott River area is in a wet climate, Wiluna is in a very dry climate, and Rockingham is in an intermediate climate.

In contrast, the Rockingham area set in the Swan Coastal Plain (McArthur & Bettenay 1960; Playford *et al.* 1976) occurs in a less humid environment. The eastern part of the area is a fluvial plain that contains a microscale, straight and sinuous, freshwater, poikilohaline river with its associated microscale floodplains which are poikilohaline, and macroscale palusplain which is freshwater, poikilohaline. The central part of the area is composed of limestone rides and contains a chain of mesoscale to macroscale, hyposaline, poikilohaline sumplands, and macroscale hyposaline, poikilohaline lakes. The western part of the area is a Holocene beachridge plain and contains one mesoscale freshwater stasohaline lake, and numerous leptoscale to microscale freshwater stasohaline sumplands and damplands (Semeniuk 1988).

The Wiluna area is in the arid Yilgarn Block and is underlain by sandplain, dunes and rocky outcrops (Glassford 1987). The area selected for mapping has the following wetlands: microscale to mesoscale hypersaline, poikilohaline playas, a barlkarra and leptoscale wadis.

Discussion and conclusions

The geomorphic classification, first developed for sub-humid to humid areas of south-western Australia, where most wetlands are basins, channels or flats, has been expanded to include a wider range of landform types that are host to wetlands under a wider variety of climatic conditions. Hence, the present classification is more globally applicable. The expanded classification has been used to classify wetlands in selected areas of arid Australia, humid south-western Australia, northern Europe, and South Africa.

Wetlands are inherently complex ecological habitats, but their analysis is simplified somewhat by a classification which brings into prominence the important underlying features of all wetlands, i.e. land and water. The classification has the advantage of separating wetlands initially on criteria other than vegetation. In current classification schemes, on the basis of vegetation, seasonally inundated basins that are meadows, sedgelands, forested swamps, or fens (depending on climatic setting, species pool and soils), would all be classified differently at a primary level (Fig. 11). In this classification they could be classified primarily as sumplands on the basis of underlying similarity of landform and water. The actual vegetation, water salinity and soil differences between these wetlands

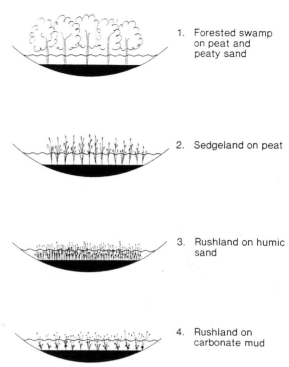

1. Forested swamp on peat and peaty sand

2. Sedgeland on peat

3. Rushland on humic sand

4. Rushland on carbonate mud

Fig. 11. A variety of sumplands that have variable soil and vegetation cover that would be classified as vastly different wetland types – the underlying, unifying feature of all these types is that they are basins with seasonal inundation.

would be systematically highlighted by the addition of descriptors to the primary classification.

It is apparent also, that with the proposed classification, a wetland can still be placed in its appropriate category even if it has been substantially altered by clearing of vegetation and disturbance of soils. As long as the hydroperiod and basic landform geometry have not been destroyed, then the inherent geomorphic wetland entity remains and can be identified and named.

The remainder of this discussion centres on nine features of the geomorphic classification of wetlands: 1. non-genetic nature of the classification, 2. addition of extra primary units, 3. basic wetland units with potential for overlayering additional information, 4. significance of the classification to conservation, 5. the wetland categories as ecological units, 6. usefulness to mapping, 7. the categories as framework to other studies, 8. the effect of climate setting and 9. the category of self-emergent wetlands.

The first feature of the wetland classification is that it is non-genetic (Fig. 12). It is based on the fundamental factors of landform types and permanence of water,

120

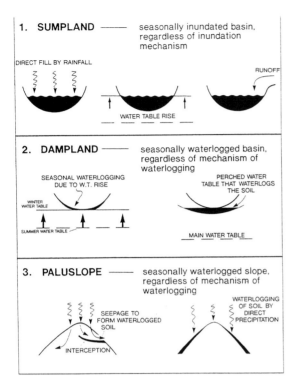

1. **SUMPLAND** —— seasonally inundated basin, regardless of inundation mechanism

DIRECT FILL BY RAINFALL

RUNOFF

WATER TABLE RISE

2. **DAMPLAND** —— seasonally waterlogged basin, regardless of mechanism of waterlogging

SEASONAL WATERLOGGING DUE TO W.T. RISE

PERCHED WATER TABLE THAT WATERLOGS THE SOIL

WINTER WATER TABLE

SUMMER WATER TABLE

MAIN WATER TABLE

3. **PALUSLOPE** —— seasonally waterlogged slope, regardless of mechanism of waterlogging

WATERLOGGING OF SOIL BY DIRECT PRECIPITATION

SEEPAGE TO FORM WATERLOGGED SOIL

INTERCEPTION

Fig. 12. The non-genetic nature of the proposed classification.

and not on such aspects as source of water, mechanism of recharge, or origin of landform (glacial, interdunal, or fluvial, etc.). As such, a basin may be described as a sumpland regardless of how the water is recharged into that sumpland (e.g. by water table rise, or ponding of rainwater, or seepage of groundwater discharge, or runoff).

The philosophy of approach developed in this classification means that, in principle, additional primary wetland units can be described and named if other host landforms and other types of water permanence are found. For instance, there is no category for either 'intermittent' or 'permanent' waterlogging. If there are wetlands where such regimes occur, then the wetland types developed under these situations can be separated and named. Also, there is no category of wetlands occupying large valleys (as distinct from wetland flats within a large valley, cf. Fig. 6.2 and 6.3). We recommend that wetlands occupying the full setting of a large valley situation be named 'palusvales'.

Even without detailed studies, or with a minimum number of field surveys, or by using only aerial photographs, a given wetland still can be readily classified into landform categories such as basin, flat,

channel, etc. The classification has the advantage that when additional detailed information becomes available, further discrimination of the basic wetland types is possible on the bases of hydroperiod, water salinity and other water properties, soil, and vegetation. Thus, variation between wetlands, in terms of size and shape (which reflects both the amount of interface with upland vegetation, and length of shoreline), water salinity and its seasonal consistency, and the type of vegetation is accented by this systematic approach to classification.

The classification is useful for management, resource allocation, and conservation in that it systematically brings out the differences between wetlands. In the first instance, by applying the classification only at a primary level, distinction can be made between various wetland types. Thus, sumplands, which have different ecological functions to lakes and damplands, and different hydrologic and geochemical roles in the landscape to floodplains, palusplains, rivers and creeks, would be clearly separated as distinct wetlands from these others. Previously, there was a tendency to aggregate all such systems into the common term 'wetlands', with the result that sometimes one system was chosen for conservation in lieu of another. When more detailed information is added to the primary categories, with addition of descriptors of scale, shape, water chemistry, and vegetation, the classification facilitates comparison within a single wetland category. Thus, a sumpland, separated at the first level from lakes, damplands and other wetlands, can be further highlighted as being similar, or dissimilar to other sumplands. For example, consider the differing characteristics of the following wetlands in south-western Australia, all of which are sumplands (Fig. 13):

1. megascale, ovoid, freshwater, stasohaline, maculiphytic sumpland (= Lake Pinjar; Lat 31° 38′ Long 115° 48′);
2. macroscale, linear to irregular, subhaline, poikilohaline, maculiphytic sumpland (= Stakehill Swamp; Lat 32° 23′ Long 115°47′);
3. mesoscale, ovoid, freshwater, stasohaline, heterophytic sumpland (= Lake Banganup; Lat 32° 10′ Long 115° 50′);
4. microscale, ovoid, freshwater, poikilohaline, baccataphytic sumpland (= Mt Brownman Swamp; Lat 32° 10′ Long 115° 48′);
5. leptoscale, ovoid, freshwater, stasohaline, concentriphytic sumpland (= Becher Swamp; Lat 32° 22′ Long 115° 45′).

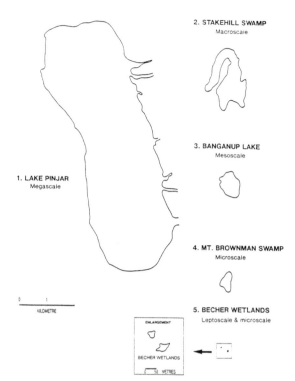

1. LAKE PINJAR
Megascale

2. STAKEHILL SWAMP
Macroscale

3. BANGANUP LAKE
Mesoscale

4. MT. BROWNMAN SWAMP
Microscale

5. BECHER WETLANDS
Leptoscale & microscale

Fig. 13. Various size and shapes of sumplands on the Swan Coastal Plain, south-western Australia. Location of these wetlands in terms of latitude and longitude is given in the text. Note that the name 'lake' given to some of these wetlands is a misnomer; they are historic locality terms and not scientific ones.

HIGHLANDS	VERY WET CLIMATES	INTERMEDIATE	VERY DRY CLIMATES
highlands, hills			
slopes			
flats			
channels			
basins LOWLANDS			

Fig. 14. The various landform types that are captured as wetlands in relationship to climate.

These differences, listed above, can have important implications in recognising intrinsically diverse wetland systems reflecting different ecological functions, such as variable use of the wetland by invertebrate and vertebrate fauna, and the variable types of vegetation cover. Thus, the proposed classification illustrates that there are 'wetlands and wetlands', with implication that preservation of diversity warrants conservation of each of the recognised types.

Since there are only thirteen basic geomorphic wetland types, the proposed classification also can provide useful mapping units since the various wetland types may be readily identified and mapped as categories. The regional differences and similarities between wetlands also emerge more clearly and may be linked to other types of mapping parameters, such as soils, geology, contours, climate. The use of the classification for mapping purposes results in an additional resource for planning in that it provides a regional picture of land capability within any given area.

Various wetland categories have a range of multifarious ecological functions and the proposed classification may parallel these functional delineations. For instance lakes, damplands and creeks are each utilised by fauna in different ways because each of these wetlands are essentially different habitats as determined by the longevity and source and mechanism of recharge of water. For exaple, some avifauna and reptiles use open water lakes for a specific range of purposes, whereas the vegetated sanctuary of many typical damplands may be utilised by mammals and other species of avifauna in a different capacity. Nutrient pathways and trophic inter-relationships also may be fundamentally distinct between these various primary wetland types. Botanists, zoologists, educationalists, recreation and land-use planners may be able to make preliminary assessments of the diversity, dependence, complexity of wetlands from the class to which it belongs. The advantage of the proposed classification is that it can be used as a basis for any wetland study regardless of the ultimate discipline of the study be it hydrology, stratigraphy, botany, zoology, and so circumvents the problem of a proliferation of nomenclature arising from specific applications/studies (evaluation of wetlands, and micro-invertebrate studies). Thus, it provides a non-genetic framework upon which to base further detailed work.

As discussed earlier, the various climatic regions of the earth develop different types of wetlands according to the availability of water and to the landform. The classification scheme proposed here accounts for this global variability. Thus, the basic landforms that are host to wetlands in very arid, and intermediate climates, are the same landforms that are host to wetlands in more humid climates. In effect, as the climate becomes more humid, more of the landforms are cap-

Table 5. Subdivision of inland wetlands on non-emergence or self-emergence

Subdivision on non-emergence/ self-emergence	Landform	Combining landform & hydroperiod
Non-emergent	basin	lake
		sumpland
		dampland
		playa
	channel	river
		creek
		trough
		wadi
	flat	floodplain
		palusplain
		barlkarra
	slope	paluslope
	hill/highland	palusmont
Self-emergent	selected examples:	
	raised bogs, mound springs, sinter mounds	

tured as wetlands (Fig. 14). As the climate becomes more arid, the hydroperiod characteristics of wetlands change from generally permanent and seasonal inundation to generally intermittent inundation, with effects on the soils and vegetation.

But while we contend that a majority of wetlands globally are captured in the wetland classification using the geomorphic approach, there are a minority of categories not so readily or easily classified. These include wetlands where organic and/or mineral deposits which have accumulated within the system have assumed sufficient relief to become *substantially* geomorphically self-emergent. Thus, while a wetland formed by spring water issuing from a fissure in granite and simply maintaining the sloping surface of bare rock in a wet state clearly could not be viewed as self-emergent, another wetland in a similar situation but where seepage promoted growth of peat that then accumulated to some thickness above the original granite surface to form a *positive mound*, could be viewed as self-emergent.

Mound springs, some raised bogs, and some geothermal wetlands are the main wetlands that fall into the category of self-emergent, and of the three, wetlands that accumulate mineral deposits to form resistant emergent mounds are the most problematic. At incipient stages, or if located within a definitive landform setting, such wetlands are still easy to classi-

fy according to the geomorphic classification system, but if the physical emergence of the wetland becomes the dominant landform feature in the area, then classification becomes more difficult. For instance, mound springs (consisting of self-emergent accumulations of mineral precipitates) form where there is seepage from springs and precipitation of mineral precipitates; with continued seepage the assemblage of water and the emergent mineral deposit forms a type of wetland. If located on a slope, and the mineral deposits are only thin, then the wetland could be classed as a paluslope. But if the deposit builds up a substantial mound, it strictly could be viewed as a local small hill, and hence a palusmont according to the geomorphic approach. The problem particularly relates to examples where springs emerge from a fault that intersects the ground surface on a dry featureless plain. The seepage of water from such faults can produce, over the millennia, a 2-3 m high mound of mineral precipitates. Whether on a plain, or on the side of a hill, where such mound springs consist of a low knoll or rounded hillock of, for instance, carbonate precipitates, covered in moss, algae and cyanobacteria, and a permanently damp surface, the use of palusmont in such situations is unacceptable, because the conspicuous landforms *have been built by wetland processes, rather than the wet-*

land being formed within the landform setting. In these cases, the wetland is a self-emergent structure.

Similarly, some raised bogs would be difficult to classify. Most bogs, and many of the raised bogs, could be classed on their landform setting, viz., basins, slopes, plains, or highlands, but where the raised bog has substantial local relief above its initial geomorphic setting, it ceases to be a part of, or dependent on its geomorphic setting. Geothermal wetlands, or hydrothermal wetlands, maintained by seepage or escape of heated water at the earth's surface, produce a similar range of wetlands to springs and other basin wetlands, except that the waters are geothermally heated. Such waters, in fact, can have a range of origins. They may be magmatic in derivation, or true springs heated by a magmatic source, or heated deep groundwaters driven to the surface by magmatic processes, or locally heated groundawter, or locally melted ice deposits. Stricty, geothermal waters and their associated wetlands will grade into springs and their related wetlands, and into groundwater-maintained wetlands. However, notwithstanding the problem of the complex origin of heating, or the relationship and gradation of geothermal waters into springs and normal groundwater systems, geothermal systems result in a range of wetland types that include: paluslopes, with thin layers of mineral precipitates; local small lakes or mud deposits maintained by escaping geothermal waters; mounds of sinter or mineral precipitates having substantial relief, whose surfaces are kept damp/wet by seepage of geothermal waters; and local small lakes or ponds that are sited *within* and built by the sinter deposits precipitated from the geothermal waters (in effect, these are small basins within the mound deposit. Of these categories, many are relatively easy to classify, and the term 'geothermal' can be added as a genetic descriptor to the wetland category. Only the mounds of sinter or mineral precipitates (like the mound springs), and those mounds associated with small lake/ponds, i.e., an assemblage of 'mounds with small lake/ponds', would be difficult to classify.

In the three categories listed above, we propose the use of the term 'self-emergent wetland' to encompass these types, with the caveat that such wetlands will grade into the more common 'non-emergent wetlands' (Table 5).

It should be noted that simple accumulation of transported, precipitated, or organic material, per se, should not be the criterion for recognising self-emergent wetlands. Generally being areas of natural depression, the majority of wetlands have accumulated material as allochthonous sedimentary material (e.g., flood deposits), autochthonous organic material (e.g., peat, diatomite, calcareous mud), or autochthonously precipitated material (e.g., calcareous mud, gypsum crystals). Many basin wetlands merely accumulate sedimentary and organic material to the level of high water, the deposits being termed basin-fills. Similarly, floodplains may accumulate sediments to form aggrading sequences, but (with the exception of natural levee bank deposits) the nature of fluvial settings is that such deposits generally do not accrete vertically sufficiently to form locally self-emergent structures, otherwise they would soon be eroded by subsequent floods. Floodplains set in tectonic subsiding areas will more actively aggrade sediments than those in non-tectonic areas, and in this sense they can be viewed as 'basin-fills'. Also, rivers, though channel-form and more prone to flushing of their sedimentary deposits, in the long term still accumulate sediments as semi-emergent structures in the repositories of the channel, point bars and mid-channel shoals. The gradational spectrum of sedimentary, mineral precipitate and organic deposits in wetlands could be viewed as a three-end-member system, as follows:

1. nil to negligible accumulation;
2. depression-, channel-, and basin-fill accumulations;
3. significant accumulation to form positive mounds.

We consider that the 3rd category only should be viewed as self-emergent deposits.

Note should be taken that, traditionally, for wetlands use of the term 'emergent' generally means 'emerged vegetation' as distinct from 'submerged vegetation' (Cowardin *et al.*, 1979). Geomorphically, sedimentologically, and organically, the term 'emergent' can refer to terrains that have become 'self-emergent', assuming positive relief above their surroundings. To avoid this potential confusion, use of the term 'emergent' should not be used in isolation, but should always be accompanied by a vegetation term, e.g., 'emergent vegetation', 'emergent hydrophytes', or 'emergent sedges', if it is to refer to status of vegetation, and the terms 'self-emergent' and 'non-emergent' be used as prefixes to wetland.

Acknowledgement

We would like to thank Dr A. Dench, Centre for Linguistics, for helpful discussion on the aboriginal word 'barlkarra'.

References

Adam, P., Unwin, N., Weiner, P., Sim, I. & Winer, P. 1985. Coastal Wetlands of New South Wales. Report for Department of Environment and Planning.

Bayly, J. A. E. & Williams, W. D. 1973. Inland Waters and Their Ecology . Longman, Cheshire.

Bates, R. L. & Jackson, J. A. 1980. Glossary of Geology. American Geological Institute.

Briggs, S. V. 1981. Freshwater Wetlands. In: Groves, R. H. (Ed.) Australian Vegetation. pp. 335–360. Cambridge University Press, London.

Cowardin, L. M., Carter, V., Golet, F. C. & LaRoe, E. T. 1979. Classification of wetlands and deepwater habitats of the United States. U.S. Dept. of the Interior; Fish & Wildlife Service Dec.

Dahl, E. 1956. Ecological salinity boundaries in poikilohaline waters. Oikos 7: 1–21.

Davis, S. N. & DeWiest, R. J. 1966. Hydrogeology. John WIley & Sons Inc., London, New York.

Drever, J. I. 1982. The Geochemistry of Natural Water. Prentice-Hall, Inc.

Fairbridge, R. W. (Ed.). 1968. The Encyclopedia of Geomorphology. Vol. III. Hutchinson & Ross Inc., Dowden.

Glassford, D. K. 1987. Cainozoic stratigraphy of the Yeelirrie area, northeastern Yilgarn Block, Western Australia. Journal Royal Society Western Australia 70: 1–24.

Golet, F. C. & Larson, J. S. 1974. Classification of Freshwater Wetlands in the Glaciated Northeast. Resource Publication 116. U.S. Fish and Wildlife Service, Washington, D.C.

Gore, A. J. P. 1983. Introduction. In: Gore, A. J. P. (Ed.). Ecosystems of the World. 4A. Mires: Swamp, Bog, Fen and Moor. General Studies. pp. 1–34. Elsevier Scientific Publishing Co., Amsterdam.

Gorham, E. 1957. The Development of Peatlands. Quarterly Review Biology 32: 147–164.

Hakanson, L. & Jansson, M. 1983. Principles of Lake Sedimentology. Springer-Verlag, Berlin, New York.

Hammer, U. T. 1986. Saline Lake Ecosystems of the World. Dr W. Junk Publishers, Dordrecht.

Hill, J. R. Jr. 1978. Corps of Engineering Efforts Related to Wetland Protection. In: Montanari, J. H. and Kusler, J. A. (Ed.) Proceedings of the National Wetlands Protection Symposium, 1977, Reston, Va. U.S. Dept. of the Interior Fish and Wildlife Service FWS/OBS-78/79. Washington, D.C.

Holmes, A. 1944. Principles of Physical Geology. Nelson.

Hutchinson, G. E. 1957. A Treatise on Limnology. Vol. 1. Wiley & Sons, London, New York.

Ivanov, K. E. 1981. Water Movement in Mirelands. Academic Press.

MacArthur, W. M. & Bettenay, E. 1960. The Development and Distribution of the Soils of the Swan Coastal Plain Western Australia. CSIRO Soil Publication No. 16.

Martin, A. C., Hotkiss, N., Uhler, F. M. & Bourn, W. S. 1953. Classification of Wetlands of the United States. U.S. Fish Wildl. Service. Spec. Sci. Rep. Wildl. No. 20.

Mill, H. R. 1900–1910. In: Sir Dudley Stamp (Ed.) (1966) A Glossary of Geographical Terms.

Monkhouse, F. J. 1965. A Dictionary of Geography. E. Arnold and Chicago Aldine.

Moore, W. G. 1949. A Dictionary of Geography. Penguin Books, Hammondsworth, Middlesex.

Moore, P. D. & Bellamy, D. J. 1974. Peatlands, London, Elek Science.

Morisawa, M. 1968. Streams: Their dynamics and morphology. McGraw-Hill Book Co., New York.

Paijmans, K., Galloway, R. W., Faith, D. P., Fleming, P. M., Haantjens, H. A., Heyligers, P. C., Kalma, J. D. & Loffler, E. 1985. Aspects of Australian Wetlands. CSIRO Australia Division Water & Land Resources Technical Paper No. 44, 1–71.

Playford, P. E., Cockbain, A. E. & Low, G. H. 1976. Geology of the Perth Basin Western Australia. Geological Survey of Western Australia Bulletin 124, 311 pp.

Pressey, R. L. & Adams, P. 1995. A review of wetland inventory and classification in Australia. Vegetatio (this volume).

Ramsar Convention Bureau. 1991. Proceedings of the 4th Meeting of the Conference of Contracting Parties. Montreux, Switzerland. Published by Ramsar Convention Bureau, Gland, Switzerland.

Reeves, C. C. Jr. 1968. Introduction to Palaeolimnology. Elsevier, Amsterdam.

Ruttner, F. 1953. Fundamentals of Limnology. University of Toronto Press.

Semeniuk, C. A. 1987. Wetlands of the Darling System – a geomorphic approach to habitat classification. Journal Royal Society Western Australia 69: 95–112.

Semeniuk, C. A. 1988. Consanguineous wetlands and their distribution in the Darling system, southwestern Australia. Journal Royal Soceity Western Australia 70: 69–87.

Semeniuk, C. A., Semeniuk, V., Cresswell, I. D. & Marchant, N. G. 1990. Wetlands of the Darling System, SW Australia: a descriptive classification using vegetation pattern and form. Journal Royal Society Western Australia 72: 109–121.

Shipley, J. T. 1982. Dictionary of Word Origins. Adams & Co., Littlefield.

Spence, D. H. N. 1967. Factors controlling the distribution of freshwater macrophytes with particular reference to the lochs of Scotland. Journal of Ecology 55: 147–169.

Swayne. 1956. A Concise Glossary of Geographic Terms. In: Sir D. Stamp (Ed.) (1966) A Glossary of Geographic Terms.

Tansley, A. G. 1939. The British Islands and Their Vegetation. Cambridge University Press, Cambridge.

Trowbridge, A. C. (Ed.). 1962. Dictionary of Geological Terms. American Geological Institute, Washington, D.C.

UNESCO. 1971. Article 1, part 1, Convention on wetlands of international significance especially as waterfowl habitats. Published in Australia, 1976, for Department of Foreign Affairs by Australian Government Publishing Service, Canberra, Treaty Series 1975, No. 48.

Ward, R. 1978. Floods, a geographic perspective. John Wiley & Sons, New York.

Wetlands Advisory Committee, 1977. The Status of Wetland Reserves in System 6. Report of the Wetlands Advisory Committee to the Environmental Protection Authority. Perth, Western Australia.

Wetzel, R. G. 1983. Limnology. Saunders College Publishing.

Whittow, J. B. 1984. Penguin Dictionary of Physical Geography. Penguin Books, Hammondsworth, Middlesex.

Wolman, M. G. & Leopold, L. B. 1957. River flood plains: some observations on their formation. U.S. Geological Survey Professional Paper 282-C.

Zoltai, S. C. & Pollett, F. C. 1983. Wetlands in Canada: Their Classification, Distribution and Use. In: Gore, A. J. P. (Ed.) Ecosystem of the World. 4B. Mires: Swamp, Bog, Fen and Moor. Regional Studies. Elsevier Scientific Publishing Co., Amsterdam.

Vegetatio **118**: 125–129, 1995.

An evaluation of the first inventory of South American wetlands

L.G. Naranjo
Departamento de Biología, Universidad del Valle, A. A. 25360, Cali, Colombia

Key words: Anthropogenic impact, Conservation, Neotropics, South America, Wetland inventory, Wetland losses

Abstract

Thanks to the efforts of the International Waterfowl Research Bureau, during 1982–1984, geographic and biotic data from 368 wetlands across South America were gathered. However, the conservation impact of this inventory has not been as striking as expected because of two reasons. First, since most contributors did not work for governmental agencies, the major points stressed in the document have been overlooked by local authorities. Second, because of logistic limitations of the inventory, large wetlands (i.e. those across the Amazon basin), were not inventoried using the IWRB criteria and thus considered as one large wetland. A critical, comparative, review of the results of the inventory revealed that in addition, the reliability of the inventory is questionable because of the differential effort put into the compilation of the information. Several countries appear to contain only a small percentage of South American wetlands, while in fact they have as many wetlands of international importance as countries that have many times their estimated area of wetlands.

Introduction

Eleven years ago, at the XXII Executive Meeting of the International Waterfowl Research Bureau (IWRB) at Edmonton, for the first time in history a group of Neotropical wetland scientists gathered to discuss the possibility of compiling a comprehensive inventory of wetlands across the Neotropics. One year later, and continuing this enterprise, the IWRB expanded the size of the first group at the meeting held at Huelva (Spain), and officially launched the project of inventorying the Neotropical wetlands under the direction of Derek Scott and his assistant, Montserrat Carbonell.

Those were times of big hopes. At the time, our expectations of this first inventory of wetlands for the entire biogeographic region included: (a) to have the basic framework for sound conservation action for wetlands and waterfowl; (b) to expand the number of neotropical countries signataries of the Ramsar convention; and (c) to create a network of people and institutions responsible for continuing the task of monitoring the status of the inventoried wetlands.

After two more years, the basic information was put together. Hundreds of data sheets were collect-

ed across Central and South America, and in spite of some differences in the apparent intensity of the coverage among countries, the data base was sufficient to produce what can be considered without any doubt a seminal volume for wetland scientists. Even recognizing the inevitable limitations of this inventory, users of Scott & Carbonell (1986) took it for granted that the information it contained was reliable enough to use it for the first attempts to develop wetland management and conservation plans in the region.

The purpose of this paper is then to evaluate the reliability of the inventory, and to assess its real impact on wetland conservation in South America during the last six years.

Methods

For the preparation of their inventory, Scott & Carbonell (1986) created a network of national coordinators from all the Neotropical countries, in charge of compiling information from local sources. In an attempt to make uniform the national reports, Scott & Carbonell prepared a standard data sheet containing

126

16 descriptive variables (Appendix 1), similar to those used for previous inventories (e.g. Scott 1980), and instructed the coordinators in their use.

In order to determine the consistency and reliability of the information among the South American countries, I examined the following variables for every wetland included in the inventory: size, elevation, biogeographic province, legal protection, and threats. Using these data, I calculated the proportion of each country represented by wetlands, the ratio of protected wetlands to total wetland area per country, and finally, ranked the importance of each threat for all the wetlands considered to be affected.

The impact of the inventory on wetland conservation was determined from two main sources: first, information provided by the regional office of the Western Hemisphere Shorebird Network in Argentina, and the UICN offices in Quito, and second, direct answers to a questionnaire sent to people involved with the original inventory in all the countries except the Guyanas. Items included in this questionnaire are given on Appendix 2.

Results

Coverage of the inventory
A total of 368 wetlands were included in the inventory for South America, for a total of 120 739 000 ha distributed among 12 countries, and ranging from subantartic glacial lakes to high elevation paramo and puna peat bogs and lakes, including floodplains, tropical marshes, estuaries and coastal swamps. It is clear the extreme variation of the total inventoried wetland area represented by the different countries. More than 95% of the South American inventoried wetlands belong to six countries (Fig. 1), and Brazil, by far the largest nation in the continent has 50% of the total wetland area. South American nations vary widely in size, so one would expect a large proportion of the wetlands to be located in the larger countries, which is in fact the case (Fig. 2).

Altitudinal distribution of wetlands
The altitudinal distribution of South American wetlands is presented on Table 1. While most of the wetlands are located below 1000 m of elevation, as anybody would expect, a comparison between countries with similar topography, size, and type of climate, revealed noteworthy differences. For instance,

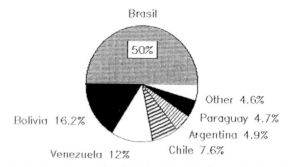

Fig. 1. Percent area of wetlands by country in South America out of a total estimated area of 120 739 000 ha.

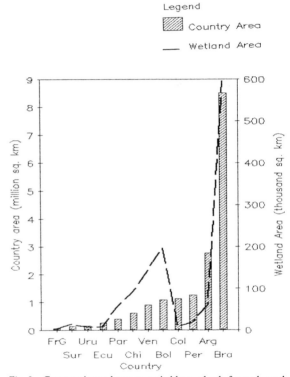

Fig. 2. Country size and area occupied by wetlands for each south american country.

Venezuela has 14.4% of the South American wetlands below 1000 m, while Colombia is included among the six countries that combined represent only 1.8% of the total. A close review of the data base revealed that the Venezuelan inventory included the entire eastern llanos as a single wetland, while their Colombian equivalent, roughly similar in area, are mentioned only as isolated wetlands rather small. Similarly, Brazilian low elevation wetlands, which represent 60.4% of the continental total, include the entire Amazon basin as a wetland, without any application of the criteria required by

Table 1. Altitudinal distribution of South American wetlands included in the IWRB (Scott & Carbonell 1986) neotropical inventory.

Elevation range (m)	Area (ha)	Percentage
0–1000	100,029,440	96.7
1000–2000	2,660	0.002
2000–3000	317,475	0.3
3000–4000	2,565,517	2.48
>4000	466,214	0.45

the IWRB inventory system (see Appendix 1). Applying these criteria, the Brazilian total gets reduced by c. 20 000 000 ha, equivalent to about 30% of the figure presented in the inventory.

Conservation of South American wetlands

The next logical question about the IWRB inventory of neotropical wetlands deals with the conservation status of the different areas at the time of publication of the results of that survey. The first striking observation is the fact that only about 15% of the inventoried wetlands (17 898 526 ha) had at least nominal protection in 1986. But most important, the percentage of protected areas among the different countries is again uneven. Five countries were responsible in 1986 for the protection of 94.4% of those areas (in decreasing order, Chile with 40.3%, Brazil 24.4%, Bolivia 20.4%, Argentina 4.7%, and Venezuela 4.5%), which clearly does not bear any relationship with country area, total wetland area/country, or any measure of economic wealth nor technological development.

But looking at the causes of wetland loss and/or degradation throughout South America, some consisting patterns clearly emerge from the inventory. Five factors (in decreasing order, industrial pollution, deforestation, drainage, grazing, and domestic pollution), are responsible for anthropogenic, negative impacts affecting more than 10% of the inventoried wetlands. Minor causes (less than 10% of inventoried wetlands affected by each factor) of wetland degradation include (in decreasing order): navigation, agriculture, construction of dams and levees, erosion (siltation), fires, construction of roads, disturbance of aquatic vegetation, and eutrophication. Unfortunately, the inventory does not provide estimates of the impact of any of the factors already mentioned, and thus makes it impossible to use it as a data-base to assess the rate of degradation of any wetland.

Discussion

Despite the apparently vast coverage of the IWRB Neotropical Wetland Inventory, it is apparent even from a superficial review of the data sheets compiled that they represent only a selected subset of the total wetland area of the continent. Although this was to be expected from the original objective of the inventory being a catalog of wetlands of international importance for waterbirds, it is unfortunate, for other purposes, to find impossible to determine how much of the total wetland area was actually covered.

The comparisons of wetland areas among countries of similar geographic features revealed striking disparities, and consequently indicated a bias in the inventory due to differential effort invested in the compilation of the information. Clear examples of discrepancies in effort in the inventory are those of Bolivia vs. Peru, and Venezuela vs. Colombia. While Bolivia has c. 18% of its total area covered by wetlands, Peru has less than 2%, and the same is true for Venezuela (c. 16%) vs. Colombia (< 1%).

Despite these inconveniences, the IWRB inventory has been a milestone for wetland conservation efforts throughout the continent. For the first time in history, wetland issues can be supported by a document compiled with the cooperation of individuals and institutions of recognized prestige. But it is legitimate to ask how successful have been wetland conservation efforts after the publication of the IWRB inventory. Unfortunately, the answer to this question is ambiguous. Seven countries have signed the Ramsar convention after the inventory was carried out, and of these, five did so after the results were published. This would mean quite a positive impact, if one take into account that before those dates, only Chile had signed the Convention in South America. However, most of the Ramsar sites dedicated after 1984 already had some protection before the IWRB launched the inventory, and according to the reports of wetland scientists from Uruguay, Brazil, Argentina and Peru, their governments commitment to sign the convention had no direct relationship with that document. In addition, the loss of wetlands across the continent has continued and it is thus very likely that the current proportion of protected wetlands in South America remains stable.

An additional point that deserves comment, is the public concern about wetlands in the Continent. Although the IWRB inventory has been frequently cited by conservation biologists and non-governmental organizations in Latin America during the last few years, few governmental agencies in charge of natural resources management and protection are even aware of its existence. The lack of a follow-up action after the completion and publication of the inventory has implied that the general public remains ignorant about the necessity of preserving wetlands.

Given the ever increasing rate of degradation of these ecosystems in South America, it is of the utmost importance to design and implement conservation actions on an international scale. With this idea in mind, I would recommend:

1. To complete and update the inventory of South American wetlands, following the IWRB protocol. This procedure would allow a comparative approach to the definition of current and future threats to wetlands.
2. To re-define the impact of the different causes of wetland degradation and loss in the continent.
3. To involve and assist governmental agencies in the design of conservation strategies related to wetlands.
4. Help local governments to identify the potential impact of large scale development plans on major wetlands.
5. To create an effective network of non-governmental organizations to promote public awareness and concern about wetlands based on wise use of their natural resources.

Acknowledgments

I want to thank Arnold Van der Walk and Max Finlayson, for their keen interest in my participation in this symposium. Thanks also to the efforts made by the organizers, Iowa State University kindly provided a travel grant allowing my attendance. In Colombia, COLCIENCIAS in Bogotá and the Universidad del Valle in Cali granted additional financial support. Several friends in different South American countries, kindly provided useful information. In this respect, I want to thank Pablo Canevari and Manuel Nores from Argentina, Inés Santos de Lima from Brazil, Víctor Pulido from Peru, and Francisco Rilla from Uruguay. Last, but not the least, I want to thank Beatriz Torres who welcomed my stay in Columbus, consequently expanding significantly my restricted budget, and two anonymous reviewers for their careful review of a first draft of this paper.

References

Atkinson-Willes, G. L., Scott D. A. & Prater A. J. 1982. Criteria for selecting wetlands of international importance: proposed amendments and guidelines on use. pp. 1017-1042 In: Spagnesi M. (ed.) Atti della Conferenza sulla conservazione delle zone umide di importanza internazionale specialmente come habitat delle uccelli aquatici, Cagliari 24–29 novembre 1980. Suppl. Ricerche di Biologia della Selvaggina, VIII.

Scott. D. A. 1980. A preliminary inventory of wetlands of international importance for waterfowl in west Europe and northwest Africa. IWRB Special Publication No. 2. Slimbridge: IWRB.

Scott, D. A. & Carbonell M. (eds). 1986. Inventario de Humedales de la Región Neotropical. Slimbridge: IWRB-UICN.

Udvardy, M. D. F. 1975. A classification of the Biogeographical Provinces of the World. IUCN Occ. Paper No. 18.

Appendix 1. Variables contained in the IWRB data sheet used for the inventory of neotropical wetlands

1. Name of Wetland.
2. Coordinates: from the Operational Navigation Charts (1: 1 000 000) of the Defense Mapping Agency of the USA.
3. Size: in hectares.
4. Elevation.
5. Biogeographic province: following Udvardy (1975).
6. Type: based on the following codes:
   01: shallow bays and coves
   02: estuaries, deltas
   03: small islands close to shore
   04: rocky coasts, clifs
   05: sea shores
   06: mudflats
   07: coastal salt marshes
   08: mangroves
   09: slow flowing rivers and creeks (lowlands)
   10: fast flowing rivers and creeks (mountains)
   11: fresh water lakes (including oxbows)
   12: fresh water marshes
   13: fresh water ponds (less than 8 ha)
   14: salt lakes (continental)
   15: dams and reservoirs
   16: seasonally flooded savannas
   17: rice paddies and aquaculture ponds
   18: swamp forests
   19: peatlands
7. Description
8. Vegetation: including land plant communities surrounding the wetland
9. Land ownership
10. Legal protection
11. Exploitation
12. Water birds
13. Other fauna: mainly species included in the WWF Red Data Book.
14. Threats: current and potential.
15. Research and conservation: a review of research projects (finished and in progress), conservation and management plans, and a list of research priorities for the site.
16. References: list of literature related to the wetland.
17. Sources: names of local experts and institutions providing information about the wetland for this inventory.
18. Criteria for inclusion: following the categories established by the Ramsar Convention in 1980 (Atkinson-Willes *et al.* 1982).

Appendix 2. Questionnaire of assessment of the impact of the IWRB Neotropical Inventory on conservation of South American wetlands

1. For your country, please give an estimate (number and size) of major wetlands not covered by the IWRB Neotropical Inventory.
2. After the publication of the IWRB inventory, have other wetland inventories been carried on your country? If so, how do they differ from the IWRB inventory? (coverage, methods, etc.).
3. Have any conservation and/or management plans related to wetlands, implemented in your country after 1986, been derived from the IWRB inventory?
4. Can you give an estimate (in %) of wetland losses in your country after 1986?

Vegetatio **118**: 131–137, 1995.

Canadian wetlands: Environmental gradients and classification

S.C. Zoltai[1] & D.H. Vitt[2]

[1]*Forestry Canada, Northern Forestry Centre, Edmonton, Alberta, Canada T6H 3S5*
[2]*Devonian Botanic Garden, The University of Alberta, Edmonton, Alberta, Canada T6G 2E1*

Key words: Bogs, Ecological gradients, Fens, Marshes, Peatlands, Shallow lakes, Swamps, Vegetation classification, Water chemistry

Abstract

The Canadian Wetland Classification System is based on manifestations of ecological processes in natural wetland ecosystems. It is hierarchical in structure and designed to allow identification at the broadest levels (class, form, type) by non-experts in different disciplines. The various levels are based on broad physiognomy and hydrology (classes); surface morphology (forms); and vegetation physiognomy (types). For more detailed studies, appropriate characterization and subdivisions can be applied. For ecological studies the wetlands can be further characterized by their chemical environment, each with distinctive indicator species, acidity, alkalinity, and base cation content. For peatlands, both chemical and vegetational differences indicate that the primary division should be acidic, *Sphagnum*-dominated bogs and poor fens on one hand and circumneutral to alkaline, brown moss-dominated rich fens on the other. Non peat-forming wetlands (marshes, swamps) lack the well developed bryophyte ground layer of the fens and bogs, and are subject to severe seasonal water level fluctuations. The Canadian Wetland Classification System has been successfully used in Arctic, Subarctic, Boreal and Temperate regions of Canada.

Wetland classification

Classification is a process of organizing information by forming groups within which the individual elements are similar in some respects, but are distinctly different from the elements of the other groups. However, the purpose of classification must be clearly defined, as the same objects, in this case wetlands, can be grouped differently, depending on the needs of the intended user. The perceptions of a wildlife or waterfowl biologist, an agronomist, a peat producer, a hydrologist, or an engineer can be vastly different when considering the same wetland. It was natural, therefore, that a number of wetland classification systems have been developed in Canada by different interest groups.

Some ecologists developed a phytosociological classification of peatland vegetation (cf. Dansereau & Segadas-Vianna 1952; Gauthier & Grandtner 1975). Meanwhile, the mire classification developed in Europe (Osvald 1925) has been used to include vegetation indicators and environmental gradients in Canada

(cf. Sjörs 1952, 1969; Vitt *et al.* 1975; Gauthier 1980; Damman 1986). The same principles were incorporated into a classification of Ontario wetlands (Jeglum *et al.* 1974). A classification of the wetlands of the Canadian prairies was developed (Millar 1976) primarily to fulfil the needs of waterfowl ecologists. An organic soil classification system was developed by the Canada Soil Survey Committee (1978), characterizing the peatlands by their soil properties. Yet another classification system was developed to serve the engineering community (Radforth 1969), based on various surface morphology and peat structure features.

This variety of classification approaches resulted in restricted communication between the various user groups. The Canadian Wetland Classification System (CWCS) has been developed to facilitate communication between different disciplines concerned about this segment of the natural landscape (Tarnocai 1988). This classification system is based on categorizing important ecosystem processes, such as water budget (amount and seasonality), carbon budget (peat accumulation), and environmental parameters, such as water quality (dissolved solids) and quantity, which interact to allow the development of distinctive ecosystems. Such a classification is synthetic and was chosen because this places the emphasis on environmental processes that determine the development of different types of wetlands. The genesis and dynamics of wetlands are implied at the highest levels of the classification by grouping similar wetland functions. This aspect is very important for all wetlands, but especially for peatlands that literally create their own environment over a long period of time.

The classification system has a hierarchical structure where the various categories represent different levels of complexity. The higher levels have been assigned multispectral (ecosystemic) definitive and descriptive criteria, the lower levels having more specific criteria (Zoltai *et al.* 1975). The three higher categories have readily recognizable definitive criteria which allow their use by non-specialists. The formal classification ends at a level which is fairly specific, but still too generalized for specialized purposes. Specialists are encouraged to develop their classification that is compatible with the CWCS, based on their interests: ecosystems, floristics, hydrology, soils, engineering qualities, etc. The CWCS provides a common platform that allows the exchange of data and results between different disciplines, using a common language.

The CWCS differs in its basic philosophy from the classification system developed for the US Fish and Wildlife Service (Cowardin *et al.* 1979). The CWCS is based chiefly on wetland functions: interrelationships of the biotic and abiotic components of the wetland ecosystems. The resulting units have implications of the hydrology, water quality, climate, vegetation interactions and chronology (genesis) of the particular wetlands. The USFWS system, on the other hand, is an objective approach that relies largely on observable features which require few process oriented decisions. It is designed to accommodate a wide range of environments, from deep waters to bogs and from rocky substrates to peatlands. This system answers the questions of what and where, but does not provide a framework for understanding the rationale of wetland development.

In the CWCS a wetland is defined as land that is saturated with water long enough to promote wetland or aquatic processes as indicated by poorly drained soils, hydrophytic vegetation, and various kinds of biological activity that are adapted to a wet environment (Tarnocai *et al.* 1988). Wetlands include organic

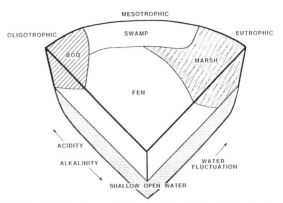

Fig. 1. The five classes of the Canadian Wetland Classification System in relation to the important chemical gradients. The vertical axis is a qualitative expression of the influence of water depth.

wetlands (peatlands) and mineral wetlands (generally non-peaty).

The broadest category of wetlands, wetland classes are recognized on the basis of overall genetic origin of the wetland ecosystem. Five wetland classes have been recognized: bogs, fens, swamps, marshes and shallow open waters (Fig. 1). They are defined and described on the basis of distinctive abiotic parameters such as hydrologic regime, water chemistry, or mineral material, which interact with the biota to form characteristic vegetation cover, and in some classes, peat.

In the next lower category are the wetland forms, which are based on surface morphology and pattern, water quality, relationship to open water, and morphology of the underlying mineral soil, as expressed by the ecosystems that are established on the wetlands. This is an open-ended category, with new wetland forms being described and defined as needed. The third category, wetland types, is based on vegetation physiognomy. The wetland is then named to identify all three categories, for example: a wooded (type) raised (form) bog (class); or shrubby (type) patterned poor (form) fen (class).

Wetland characteristics

Quantitative and qualitative differences in ecological processes, operating within given climatic conditions, are primarily responsible for differences in wetlands. Whereas it is often difficult to measure ecological processes themselves, it is relatively easy to determine the manifestations of these processes in terms of ecosystem characteristics. These characteristics in turn can be used to define a synthetic classification system. The classification itself serves to reflect the fundamental ecological processes.

The classification of wetlands should be based on the factors that control wetland processes. Most of these factors are present as gradients, and these are often non-linear in form. Although these gradients are often correlated with one another, they do not necessarily vary at similar rates or quantities. Despite these complications, wetlands can be grouped into fundamental types that reflect basic ecological processes (Figures 1, 2). The Canadian System recognizes five classes of wetlands that form nodes along several complex hydrological, chemical, and biotic gradients.

Shallow open waters are wetlands that exist whenever the water levels are sufficient to create habitats for aquatic and floating vegetation. Although seasonal fluctuations in water level may expose the bottom substrate, aquatic processes characterize this wetland. Chemistry of the water does not differentiate this wetland class from the remaining four (Fig. 1), but may be important in determining the type of shallow water vegetation that is present. These wetlands form a transition to truly aquatic ecosystems and are largely influenced by the adjoining aquatic system (Vitt & Slack 1975).

Marshes are treeless wetlands that are subject to relatively large seasonal water level fluctuations. These wetlands are strongly influenced by either ground water or surface water and have relatively high amounts of water flow. As a result, available nutrients (here considered only as nitrogen and phosphorus) are abundant and vascular plant production is high, however decomposition rates are also high. Bryophytes are nearly lacking, owing to their inability to compete with vascular plants under eutrophic conditions and when fluctuating water levels are present. The influence of surface and ground waters generally serves to increase base cations and HCO_3^-, thus creating a well buffered wetland. Trees are lacking due to either regional climatic conditions and/or the seasonally wet conditions. As a result, marshes are wetlands that reflect eutrophic, temporally variable conditions that prohibit a well developed bryophyte layer, largely due to high vascular plant production. Rapid decomposition, along with the poorly developed ground layer, inhibits significant peat accumulation. Chemical differences greatly influence the vegetation composition of marshes. Freshwater marshes are dominated by calcium and bicarbonate (and are alkaline systems), while saline marshes

have sodium and sulphate as their dominant ions (and are thus saline and non-alkaline). Tidal marshes, with very different vegetation, are dominated by sodium and chloride ions (and thus are also saline and non-alkaline wetlands) (Fig. 2).

Swamps have many similarities to marshes in that swamps have strong seasonal water level fluctuations and relatively strong water flow. Like marshes, the bryophyte ground layer is poorly developed or lacking in most swamps. Both production and decomposition is generally high, and the seasonal lowering of the water level permits the establishment of a well-developed tree or shrub layer. When peat is produced in some quantity, it is well decomposed; however, in general, peat accumulation is small. Tree development often inhibits water flow and some swamps can become acidic, especially in areas of acidic ground water. These acidic treed wetlands rarely develop a bryophyte dominated ground layer due to the fluctuating water levels, but stable water levels under dense conifer cover allow the formation of a moss carpet. Peat accumulation is small owing to rapid decomposition. Vegetationally, swamps can be quite diverse, from deciduous alder-ash, to evergreen white cedar, to densely forested black spruce, to cypress swamps farther south. All of these wetlands, however, are characterized by the presence of trees or tall shrubs, generally little peat accumulation and extreme seasonal water level fluctuation.

Peatlands are distinct from non-peat forming wetlands by a combination of hydrological and biotic factors that function together to create conditions suitable for decreased plant production and decreased decomposition. The stability of seasonal water levels restricts total water flow through a wetland and allows a ground layer of bryophytes to develop. Available nutrient levels are low, as nutrients are accumulated in non-available forms by this ground layer. As a result, production of the vascular plant component is reduced, but production of the ground layer increases due to active nutrient sequestering by the bryophyte layer (Bayley *et al.* 1987). Stable water levels and decreased nutrient availability lead to a decrease in decomposition and peat accumulation increases.

Fens are always bryophyte dominated wetlands that are influenced by the chemistry of the surrounding mineral soil deposits. Fens having bicarbonate (thus they are alkaline) as their dominant anion and calcium as their dominant cation are rich fens. These wetlands are characterized by brown mosses largely of the

134

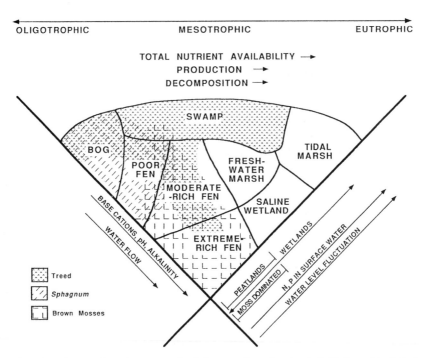

Fig. 2. The relationship of bog, fen, marsh, and swamp wetland classes to the major chemical, biotic, and hydrologic gradients. Fens and marshes are divided into several wetland forms in order to illustrate the pattern of variability of these classes. Saline wetlands may be either fens or marshes.

family Amblystegiaceae, and an abundance of sedges. Extreme-rich fens have pH above 7.0 (and are basic) and often have deposits of marl in their wetter areas, while moderate-rich fens have pH above 5.5 and below 7.0 (and are acid). These rich fens are strongly influenced by water from surrounding uplands and from the larger watershed, and generally have sufficient flow to allow the wetlands to be mesotrophic. Although data are scarce, it appears that production of the ground layer in rich fens is similar to that in poor fens and bogs (Vitt 1990), while production of the vascular plant component may be greater than in these latter wetlands due to somewhat greater decomposition allowing more nutrients to be made available. Poor fens (and bogs) are acidic, non-alkaline wetlands that are dominated by the genus *Sphagnum*. *Sphagnum*, through its abilities to acidify and hold large quantities of water, restricts both water flow and nutrient availability, which in turn decrease vascular plant production and decomposition. Poor fens, as rich fens, are influenced by geogenous water. The presence of a well developed ground layer of *Sphagnum* serves to stabilize the wetland and increase the acidity. From both vegetational and chemical viewpoints, poor fens are more similar to bogs than they are to rich fens (Malmer *et al.* 1992); however,

the hydrology of poor fens has greater similarities to that of rich fens.

Bogs are *Sphagnum*-dominated, ombrotrophic wetlands. They are acidic, largely due to humic acid production during decomposition processes (Hemond 1980) and to cation exchange due to uronic acid production from *Sphagnum*, and non-alkaline. The base cation content is limited and ombrotrophy greatly restricts water flow. Nutrients are largely tied up in *Sphagnum* and accumulated peat and mineralization processes are reduced. Thus the rate of production in the ground layer remains relatively similar to that of fens seemingly due to internal cycling of nutrients within the *Sphagnum* plants, while vascular plant production is limited.

Wetlands in Canada form distinct types along complex gradients. Especially important are: (1) Hydrology (in particular, seasonal water level fluctuation and amount of water flow through wetlands); (2) Chemistry (in particular nutrient [N, P] availability, acidity [hydrogen ion content], alkalinity [bicarbonate content], and base cation content [Ca, Mg, Na, K]). Nutrient availability leads to classification of wetlands as oligotrophic, mesotrophic or eutrophic; and (3) Biotic (in particular the development of a ground layer dominated by either brown mosses or *Sphagnum* and the presence of a tree layer). Proper classification of wetlands will lead to proper management and any classification system must recognize the complex interrelationships between the gradients formed from these controlling hydrological, chemical, and biotic factors. The CWCS has been used as the framework for evaluating wetlands for conservation or use in Canada (Bond *et al.* 1992).

The needs of particular disciplines can be accommodated by combining the CWCS with parameters that are important to that discipline. It is evident from wetlands of west-central Canada that data sets of these characteristics can be used to formulate wetland sites into an ecologically relevant classification (Tables 1 and 2).

Inventorying wetlands

It must be clearly understood that wetland classification and the inventorying of wetlands are two distinct and separate operations. Although the wetland classification should be used as an inventory and mapping criterion, the purpose of the inventory, and the desired detail and scale of the inventory map will dictate the kinds of wetlands that will be recognized. Therefore, a universal, global inventory of wetlands is possible only if these parameters are determined in advance. This is not an easy task, as circumstances vary greatly around globe. In some countries with limited wetland resources, the few wetlands may be very well documented, whilst in other countries, such as Canada, vast tracts of wetlands are virtually untouched and never studied.

The wetlands of Canada have not been accurately inventoried. A small scale map was generated (National Wetlands Working Group (1986) from available published and unpublished information. In recent years, however, a number of wetland inventories were initiated by various provinces that provide more accurate information. The mapping units are generally compatible with or were inspired by the CWCS. The wetland resources of New Brunswick were inventoried at a scale of 1:50 000 (Airphoto Analysis Associates Consultants Ltd. 1975). The Ontario portion of the Hudson Bay Lowlands were mapped at a scale of 1:250 000 (Pala & Boissonneau 1982). The peatlands of southern Quebec have been mapped and an atlas published (Buteau 1989) at a scale of 1:250 000; the inventorying

of the wetlands of northern Quebec is under way. The wetlands of Alberta (with the exclusion of southern parts of the province) have been mapped at 1:250 000 scale, and a summary map was published at a scale of 1:1 000 000 (Vitt 1992). The mapping of the wetlands of Manitoba is currently under way. In addition to wetland inventory maps, the surficial geology maps of the northern Mackenzie River valley show the extent of fens and bogs at a scale of 1:250 000 (Duk-Rodkin & Hughes 1992).

In our experience, technology must be combined with field work in order to obtain reliable data on the extent and kind of wetlands. Remotely sensed data in the form of aerial photographs or spectral data (passive or active systems) are necessary to cover areas of poor access. Although computer-enhanced remotely sensed information shows promise (Pala & Boissonneau 1982; Anon 1989), the evaluation and validation of this information is necessary before confidence can be established. This rapidly developing technology, however, dictates that this option must be monitored for possible future use.

The use of aerial photos is a well-established and tried method of wetland inventory (Buteau 1989; Vitt 1992). Wetland types can be readily identified to a detail limited only by the scale of the photos. The thickness of the peat cannot be determined from air photos with confidence, although relationships between peat depth and topography can be established locally within climatic regions. Ground truthing is necessary to determine the vegetation types, cation and nutrient levels, in addition to peat thickness and stratigraphy measurements. Ideally, a multi-tiered approach should be applied: (1) field work limited to a small portion of the wetlands; (2) expansion of this knowledge to large areas through air photo interpretation; (3) extension of the results of air photo study to a regional scale through interpretation of remotely sensed information.

We feel confident that the CWCS can be applied to Boreal, Subarctic, and Arctic regions of the Northern Hemisphere. The definition of the wetland classes is compatible with that of northern European wetland ecologists. The wetland forms are especially useful for air photo interpretation, and new forms can be defined, as needed. The broad vegetation physiognomy, used to identify wetland types, should not pose any difficulty. However, we lack the necessary experience to judge the relevance of CWCS to wetlands of lower latitudes.

Table 1. Variability in surface water chemistry in the wetland classes from a small region (Elk Island National Park) of central Alberta, Canada (Nicholson 1992; n = 331).

	Bog	Fen	Swamp	Marsh
pH	3.5–3.6	4.0–6.2	5.9–6.1	5.2–6.4
Reduced Conductivity* (μS cm^{-1})	16–27	40–160	230–330	160–530 (900)**
Ca (mg l^{-1})	4–7	2–33	26–43	27–65
Na (mg l^{-1})	2–3	2–5	5–22	3–125 (800)**
Organic N (μg l^{-1})	2900–3000	1350–2850	2000–3000	200–2500 (6400)**
NO_3^- (μg l^{-1})	13–20	8–23	7–10	9–175
NH_4^+ (μg l^{-1})	160–250	23–80	28–146	73–130 (6400)**
P (total) (μg l^{-1})	350–480	135–400	220–650	250–520 (1100)**

* Conductivity minus the effect due to hydrogen ions (Sjörs 1952)
** ( ) Data from saline marshes.

Table 2. Average chemical composition of surface water in various peatlands in west-central Canada (Zoltai & Johnson 1987). All data are from point measurements at mid-summer. Standard error of the mean in brackets.

	Bogs	Poor fens	Moderate-rich fens	Extreme-rich fens
No. sites	71	33	147	18
Depth to water table	37	12	11	5
below surface (cm)	(1.8)	(4.0)	(1.2)	(1.6)
pH	4.5	4.8	5.8	6.5
	(.02)	(.04)	(.05)	(.12)
Reduced conductivity* (μS cm^{-1})	62	53	212	374
	(3)	(7)	(12)	(33)
Ca (mg l^{-1})	2.04	2.90	24.98	53.60
	(.18)	(.63)	(1.59)	(5.29)
Mg (mg l^{-1})	0.87	1.19	10.16	14.20
	(.12)	(.32)	(.69)	(1.24)
Na (mg l^{-1})	2.59	3.89	4.75	6.54
	(.28)	(.93)	(.36)	(.87)
P (mg l^{-1})	0.17	0.14	0.13	0.12
	(.018)	(.024)	(.046)	(.020)
K (mg l^{-1})	1.42	1.26	1.42	0.97
	(.10)	(.19)	(.11)	(.20)

* Conductivity minus the effect due to hydrogen ions (Sjörs 1952).

References

Airphoto Analysis Associates Consultants Limited. 1975. Wetlands – peatlands resources, New Brunswick. Prepared for New Brunswick Department of Natural Resources, Fredericton, New Brunswick. Unpublished Report, 188 pp.

Anon. 1989. Inventory of peatlands in northern Quebec. In: Peatlands and Remote Sensing. Québec Ministère de l'Énergie et des Ressources, Service de la cartographie & Service géologique de Québec. 4 pp.

Bayley, S. E., Vitt, D. H., Newbury, R. W., Beaty, K. G., Behr, R. & Miller, C. 1987. Experimental acidification of a *Sphagnum*-dominated peatland. Canadian Journal of Fisheries and Aquatic Sciences (Suppl.) 44: 194–205.

Bond, W. K., Cox, K. W., Heberlin, T., Manning, E. W., Witty, D. R. & Young, D. A. 1992. Wetland evaluation guide. North American Wetlands Conservation Council (Canada), Issues Paper No. 1992-1, 121 pp.

Buteau, P. 1989. Atlas des tourbières du Québec méridional. Service géologique de Québec. Gouvernement du Québec, Québec.

Canada Soil Survey Committee. 1978. The Canadian system of soil classification. Publication No. 1646, Canada Department of Agriculture, Ottawa, Canada. 164 pp.

Cowardin, L. M., Carter, V., Golet, F. C. & LaRoe, E. T. 1979. Classification of wetlands and deepwater habitats of the United States.

FWS/OBS-79/31. US Fish & Wildlife Service, Washington, D.C. 103 pp.

Damman, A. W. H. 1986. Hydrology, development, and biochemistry of ombrogenous peat bogs with special reference to nutrient relocation in a western Newfoundland bog. Canadian Journal of Botany 64: 384–394.

Dansereau, P. & Segadas-Vianna, F. 1952. Ecological study of the peat bogs of eastern North America. Structure and evolution of vegetation. Canadian Journal of Botany 30: 490–520.

Duk-Rodkin, A. & Hughes, O. L. 1992. Surficial geology, Fort McPherson-Bell River (Map 1745A), Arctic Red River (Map 1746A), Travaillant Lake (Map 1747A), Canot Lake (Map 1478A), Trail River-Eagle River (Map 1744A), Martin House (Map 1743A), Ontaratue River (Map 1742A), Fort Good Hope (Map 1741A), Sans Sault Rapids (Map 1784A). Geological Survey of Canada, Ottawa, Ontario.

Gauthier, R. 1980. La végétation des tourbières et les sphaignes du parc des Laurentides, Québec. Forest Ecology Laboratory, Laval University. Études écologiques No. 3. Québec, Québec. 634 pp.

Gauthier, R. & Grandtner, M. M. 1975. Étude phytosociologique des tourbières du Bas Saint-Laurent, Québec. Naturaliste canadien 102: 109–153.

Hemond, H. F. 1980. Biogeochemistry of Thoreau's Bog, Concord, Massachusetts. Ecological Monographs 50: 507–526.

Jeglum, J. K., Boissonneau, A. N. & Haavisto, V. F. 1974. Toward a wetland classification for Ontario. Dept. of Environment, Canadian Forestry Service, Information Report 0-X-215, 54 pp.

Malmer, N., Horton, D. G. & Vitt, D. H. 1992. Element concentrations in mosses and surface waters of western Canadian mires relative to precipitation chemistry and hydrology. Ecography 15: 114–128.

Millar, J. B. 1976. Wetland classification in western Canada. Environment Canada, Canadian Wildlife Service. Report Series No. 37. 37 pp.

National Wetlands Working Group. 1986. Canada: Distribution of wetlands. In: National Atlas of Canada. Energy, Mines and Resources Canada, Ottawa. 5th edition, Map MCR 4107.

Nicholson, B. H. 1992. The Wetlands of Elk Island National Park: Vegetation, Development and Chemistry. Ph.D. Dissertation. University of Alberta, Edmonton.

Osvald, H. 1925. Die Hochmoortypen Europas. Veroff. Geobot. Inst. Eidg. Tech. Hochsch. Volume 3. Rubel. Zürich, Switzerland 707–723.

Pala, S. & Boissonneau, A. 1982. Wetland classification maps for the Hudson Bay Lowlands. Naturaliste canadien 109: 653–659.

Radforth, N. W. 1969. Classification of muskeg. In: Muskeg Engineering Handbook. I. C. MacFarlane, editor. University of Toronto Press. Toronto, Ontario: 31–52.

Sjörs, H. 1952. On the relation between vegetation and electrolytes in North Swedish mire waters. Oikos 2(1950): 241–258.

Sjörs, H. 1969. Bogs and fens of the Hudson Bay Lowlands. Arctic 12: 3–19.

Tarnocai, C. (chairman). 1988. The Canadian Wetland Classification System. In: Wetlands of Canada. Rubec, C.D.A. (coordinator). Polyscience Publications Inc., Montréal, Québec: 413–427.

Vitt, D. H. 1990. Growth and production dynamics of boreal mosses over climatic, chemical, and topographic gradients. Botanical Journal of the Linnean Society 104: 35–59.

Vitt, D. H. (project leader). 1992. The Peatlands of Alberta: A 1:1 000 000 summary map. Nicholson, B. H. & Halsey, L. H. (eds). Alberta Forestry, Lands and Wildlife, Edmonton.

Vitt, D. H. & Slack, N. G. 1975. An analysis of the vegetation of Sphagnum-dominated kettle hole bogs in relation to environmental gradients. Canadian Journal of Botany 53: 332–359.

Vitt, D. H., Achuff, P. & Andrus, R. E. 1975. The vegetation and chemical properties of patterned fens in the Swan Hills, north central Alberta. Canadian Journal of Botany 53: 2776–2795.

Zoltai, S. C. & Johnson, J. D. 1987. Relationships between nutrients and vegetation in peatlands of the Prairie Provinces. In: Proceedings, Symposium '87: Wetlands/Peatlands. Rubec, C. D. A. & Overend, R. P. (compilers). Edmonton, August 23–27, 1987: 535–542.

Zoltai, S. C., Pollett, F. C., Jeglum, J. K., & Adams, G. D. 1975. Developing a wetland classification for Canada. In: Forest soils and forest land management. Bernier, B. & Winget, C. H. (eds). Proc. 4th North American Forest Soils Conference, Québec, August, 1973: 497–511.

Vegetatio **118**: 139–152, 1995.
© 1995 *Kluwer Academic Publishers. Printed in Belgium.*

US Fish and Wildlife Service 1979 wetland classification: A review *

Lewis M. Cowardin[1] & Francis C. Golet[2]
[1]*US Fish and Wildlife Service, Northern Prairie Wildlife Research Center, Jamestown, ND 58401, USA;*
[2]*Department of Natural Resources Science, University of Rhode Island, Kingston, RI 02881, USA*

Key words: Classification, Definition, United States, Wetland

Abstract

In 1979 the US Fish and Wildlife Service published and adopted a classification of wetlands and deepwater habitats of the United States. The system was designed for use in a national inventory of wetlands. It was intended to be ecologically based, to furnish the mapping units needed for the inventory, and to provide national consistency in terminology and definition. We review the performance of the classification after 13 years of use. The definition of wetland is based on national lists of hydric soils and plants that occur in wetlands. Our experience suggests that wetland classifications must facilitate mapping and inventory because these data gathering functions are essential to management and preservation of the wetland resource, but the definitions and taxa must have ecological basis. The most serious problem faced in construction of the classification was lack of data for many of the diverse wetland types. Review of the performance of the classification suggests that, for the most part, it was successful in accomplishing its objectives, but that problem areas should be corrected and modification could strengthen its utility. The classification, at least in concept, could be applied outside the United States. Experience gained in use of the classification can furnish guidance as to pitfalls to be avoided in the wetland classification process.

Introduction

Development of US Classification

The wetland classification in use today by the US Fish and Wildlife Service (USFWS) was developed between 1975 and 1979 (Cowardin *et al.* 1979). It originated from a need by the USFWS to inventory the wetland resources of the United States. The stated purposes of the classification were to: (1) describe ecological units that have certain homogeneous natural attributes; (2) arrange those units in a system that would aid decisions about resource management; (3) furnish units for inventory and mapping; and (4) provide uniformity in concepts and terminology throughout the United States. The history and development of the classification is closely related to the development of the National Wetland Inventory (NWI). This paper describes and evaluates the classification. The inventory is described by Wilen and Bates (this volume).

The USFWS has a long history of involvement in wetland classification and inventory. Conservation and management of migratory waterfowl are a responsibility of the USFWS based on migratory bird treaties with Canada and Mexico. Thus, conservation of wetland habitats is one of the agency's primary objectives. The USFWS conducted the first quantitative national inventory of wetlands in the mid-1950s; the results were summarized in US Fish and Wildlife Service Circular 39 (Shaw & Fredine 1956). That inventory was based on a classification, developed by Martin *et al.* (1953), which included 20 classes of wetlands. After close scrutiny, the authors found that the Martin *et al.* classification was inconsistently applied among regions. The reason was primarily lack of detail in definitions. By the mid-1970s, when the current inventory was being planned, there had been an explosion of public and professional interest in wetlands that transcended the habitat function for migratory birds. Numerous excellent regional classifications (e.g., Stewart & Kantrud 1971, Golet & Larson 1974; Jeglum *et al.*

---
* The US Government's right to retain a non-exclusive, royalty free licence in and to any copyright is acknowledged.

140

1974; Odum *et al.* 1974; Zoltai *et al.* 1975; Millar 1976) also had been developed since the Martin *et al.* classification.

In January of 1975, the USFWS convened a small number of interested individuals from various agencies and regions to formulate the skeleton of a new classification that could serve as the basis for a new national wetlands inventory. Three important points were agreed upon: (1) none of the existing classifications met the requirement for national uniformity, (2) regional classifications would not suffice because of the confusion resulting at regional boundaries, and (3) a new classification should be hierarchical in structure. Following that meeting, Cowardin and Carter (1975) prepared a tentative classification that was presented at a July 1975 national workshop, where 150 federal and state wetland management personnel were invited to comment on the proposed classification (Sather 1976). Input from that workshop resulted in major modifications of the Cowardin and Carter paper and led to the preparation of a revision, Interim *Classification of Wetlands and Aquatic Habitats of the United States* (Cowardin *et al.* 1976). The revised system was tested using both high- and low-altitude aerial photographs and field-checks at 21 sites scattered across the country. The NWI staff worked with the authors of the 1976 classification to resolve practical problems encountered during testing. At the same time, the authors tested the evolving classification at numerous locations throughout the United States. The final version was published in 1979 (Cowardin *et al.* 1979); it was reprinted in 1985 and 1992. The classification has been used by the NWI for 13 years. It has also been widely used by other federal and state regulatory and resource management agencies, as well as by wetland researchers.

Objectives of paper
This paper has four objectives: (1) to acquaint the reader with the structure of the USFWS classification system, (2) to explain the rationale for the approach to classification, (3) to review successes as well as problems encountered during its use, and (4) to evaluate the potential for use of the system on an international scale.

Overview of the classification

Definition of wetland and deepwater habitat
Under the USFWS classification (Cowardin *et al.* 1979:3), wetlands are defined as follows:

> Wetlands are lands transitional between terrestrial and aquatic systems where the water table is usually at or near the surface or the land is covered by shallow water. For the purposes of this classification wetlands must have one or more of the following three attributes: (1) at least periodically, the land supports predominantly hydrophytes; (2) the substrate is predominantly undrained hydric soil; and (3) the substrate is nonsoil and is saturated with water or covered by shallow water at some time during the growing season of each year.

In support of this definition, the USFWS has prepared a list of plants known to occur in U.S. wetlands. The first draft, released in 1977, contained 4235 species; the latest version includes more than 7000 species (Reed 1988). Each of the species in this list is placed into one of four categories according to its frequency of occurrence in wetlands: Obligate Wetland, Facultative Wetland, Facultative, or Facultative Upland (see Table 1 for definitions).

Also to supplement the USFWS wetland definition, the US Soil Conservation Service has developed a definition of hydric soil, taxonomic and hydrologic criteria for identifying hydric soils, and a list of the hydric soils of the United States. The first draft, Hydric Soils of the United Sates, was published in 1982 (Soil Conservation Service 1982) and subsequent editions (Soil Conservation Service 1985, 1987, 1991) were released. In the latest edition (Soil Conservation Service 1991:1), hydric soil is defined as follows:

> A hydric soil is a soil that is saturated, flooded, or ponded long enough during the growing season to develop anaerobic conditions in the upper part.

The aim of the NWI was to map not only those areas traditionally regarded as wetlands, but also those deeper waters which frequently are associated with wetlands. For that reason, Cowardin *et al.* (1979) made a clear distinction between wetlands and 'deepwater habitats'. The latter were defined as permanently flooded lands lying below the deepwater boundary of wetlands. In nontidal areas, the boundary between wetland and deepwater habitat was placed at a depth of 2 m below low water – the maximum depth to which rooted, emergent plants normally grow (Welch 1952; Zhadin

Table 1. Wetland indicator categories for plants that occur in US wetlands (from Reed 1988).

Category	Definition
Obligate wetland	Under natural conditions, occurs almost always (estimated probability > 99%) in wetlands.
Facultative wetland	Usually occurs in wetlands (estimated probability 67–99%), but occasionally found in nonwetlands.
Facultative	Equally likely to occur in wetlands or nonwetlands (estimated probability 34–66%).
Facultative upland	Usually occurs in nonwetlands (estimated probability 67–99%, but occasionally found in wetlands (estimated probability 1%–33%).

& Gerd 1963; Sculthorpe 1967). Marked, short-term fluctuations in the level of tidal waters called for different criteria for separating wetlands from deepwater habitats. In tidal areas the boundary was placed at the elevation of extreme low water; thus, sites that are permanently covered with tidal water are considered deepwater habitats, regardless of water depth.

Hierarchical structure of the classification
The classification structure consists of five levels, arranged in a hierarchical fashion. Proceeding from the highest to the lowest level, these are: Systems, Subsystems, Classes, Subclasses, and Dominance Types. Figure 1 illustrates the classification structure to the Class level, and Table 2 presents the distribution of Subclasses within the classification hierarchy.

The *System* is the uppermost level in the classification. It describes the overall complex of hydrological, geomorphological, physical, chemical, and biological features that certain groups of wetlands and deepwater habitats share. Five Systems are recognized: Marine, Estuarine, Riverine, Lacustrine, and Palustrine. Salinity, wave energy, basin morphology, water depth, and surface water area are some of the key features distinguishing Systems.

Systems are further divided into *Subsystems*, primarily on the basis of water depth, surface water permanence, or, in the case of the Riverine System, stream gradient and extent of tidal influence. Marine and Estuarine Systems each have two Subsystems, Subtidal and Intertidal; the Lacustrine System has two Subsystems, Limnetic and Littoral; and the Riverine System has four Subsystems, Tidal, Lower Perennial, Upper Perennial, and Intermittent. The Palustrine System is not divided into Subsystems.

The *Class* is the third level in the classification hierarchy. It may be thought of as the basic habitat

type. The Class describes the general appearance of the habitat in terms of either the dominant life form of the vegetation, in the case of vegetated habitats, or the form and general composition of the substrate, along with water regime, in the case of nonvegetated habitats. The same Class may occur within two or more Systems or Subsystems (Fig. 1). There are six Classes of nonvegetated habitats: Rock Bottom, Unconsolidated Bottom, Rocky Shore, Unconsolidated Shore, Streambed, and Reef. Vegetated Classes include: Aquatic Bed, Emergent (herbaceous) Wetland, Scrub-Shrub Wetland, Forested Wetland, and Moss-Lichen Wetland.

Each of the 11 Classes contains two or more *Subclasses* (Table 2). Subclasses are distinguished by finer differences in either vegetative life form or substrate composition. For example, Forested Wetlands are divided into five Subclasses: Broad-leaved Deciduous, Needle-leaved Deciduous, Broad-leaved Evergreen, Needle-leaved Evergreen, and Dead. Unconsolidated Bottoms have four Subclasses: Cobble-Gravel, Sand, Mud, and Organic.

Dominance Type is the lowest level in the classification hierarchy. It describes the dominant plant or sedentary or sessile animal species within a particular Subclass at a specific site. When the Subclass is based on vegetative life form, the Dominance Type is the most abundant single species, or combination of species (in the case of codominance), in the vegetation layer used to name the Subclass. Thus, in a Broad-leaved Deciduous Forested Wetland, the Dominance Type would be the most abundant broad-leaved deciduous tree species (e.g., *Acer rubrum*). When the Subclass is based on substrate composition, the Dominance Type is the predominant plant or sedentary or sessile macroinvertebrate species on the site. Dominance Types are not listed in the classification; they are determined onsite by the user.

142

Class/Subclass	Marine ST	Marine IT	Estuarine ST	Estuarine IT	Riverine TI	Riverine LP	Riverine UP	Riverine IN	Lacustrine LM	Lacustrine LT	Palustrine —
Rock Bottom											
Bedrock	X		X		X		X		X	X	X
Rubble	X		X		X		X		X	X	X
Unconsolidated Bottom											
Cobble-Gravel	X		X		X	X	X		X	X	X
Sand	X		X		X	X	X		X	X	X
Mud	X		X		X	X	X		X	X	X
Organic			X		X	X			X	X	X
Aquatic Bed											
Algal	X	X	X	X	X	X	X		X	X	X
Aquatic Moss					X	X	X		X	X	X
Rooted Vascular	X	X	X	X	X	X	X		X	X	X
Floating Vascular			X	X	X	X	X		X	X	X
Reef											
Coral	X	X									
Mollusk			X	X							
Worm	X	X	X	X							
Streambed											
Bedrock				X	X			X			
Rubble				X	X			X			
Cobble-Gravel				X	X			X			
Sand				X	X			X			
Mud				X	X			X			
Organic				X	X			X			
Vegetated								X			
Rocky Shore											
Bedrock		X		X	X	X	X			X	
Rubble		X		X	X	X	X			X	
Unconsolidated Shore											
Cobble-Gravel		X		X	X	X	X			X	X
Sand		X		X	X	X	X			X	X
Mud		X		X	X	X	X			X	X
Organic		X		X	X	X	X			X	X
Vegetated					X	X	X			X	X
Moss–Lichen Wetland											
Moss											X
Lichen											X
Emergent Wetland											
Persistent				X							X
Nonpersistent				X	X	X	X			X	X
Scrub–Shrub Wetland											
Broad-leaved Deciduous				X							X
Needle-leaved Deciduous				X							X
Broad-leaved Evergreen				X							X
Needle-leaved Evergreen				X							X
Dead				X							X
Forested Wetland											
Broad-leaved Deciduous				X							X
Needle-leaved Deciduous				X							X
Broad-leaved Evergreen				X							X
Needle-leaved Evergreen				X							X
Dead				X							X

[a]ST = Subtidal, IT = Intertidal, TI = Tidal, LP = Lower Perennial, UP = Upper Perennial, IN = Intermittent, LM = Limnetic, LT = Littoral.

Fig. 1. Classification hierarchy of wetlands and deepwater habitats, showing Systems, Subsystems, and Classes in the USFWS classification (from Cowardin *et al.* 1979: Fig. 1). The Palustrine System does not include deepwater habitats.

Table 2. Distribution of Subclasses within the USFWS classification hierarchy (from Cowardin *et al.* 1979: Table 1).

System	Subsystem	Class

WETLANDS AND DEEPWATER HABITATS

Marine
- *Subtidal*
  - Rock Bottom
  - Unconsolidated Bottom
  - Aquatic Bed
  - Reef
- *Intertidal*
  - Aquatic Bed
  - Reef
  - Rocky Shore
  - Unconsolidated Shore

Estuarine
- *Subtidal*
  - Rock Bottom
  - Unconsolidated Bottom
  - Aquatic Bed
  - Reef
- *Intertidal*
  - Aquatic Bed
  - Reef
  - Streambed
  - Rocky Shore
  - Unconsolidated Shore
  - Emergent Wetland
  - Scrub-Shrub Wetland
  - Forested Wetland

Riverine
- *Tidal*
  - Rock Bottom
  - Unconsolidated Bottom
  - Aquatic Bed
  - Streambed
  - Rocky Shore
  - Unconsolidated Shore
  - Emergent Wetland
- *Lower Perennial*
  - Rock Bottom
  - Unconsolidated Bottom
  - Aquatic Bed
  - Rocky Shore
  - Unconsolidated Shore
  - Emergent Wetland
- *Upper Perennial*
  - Rock Bottom
  - Unconsolidated Bottom
  - Aquatic Bed
  - Rocky Shore
  - Unconsolidated Shore
- *Intermittent*
  - Streambed

Lacustrine
- *Limnetic*
  - Rock Bottom
  - Unconsolidated Bottom
  - Aquatic Bed
- *Littoral*
  - Rock Bottom
  - Unconsolidated Bottom
  - Aquatic Bed
  - Rocky Shore
  - Unconsolidated Shore
  - Emergent Wetland

Palustrine
- Rock Bottom
- Unconsolidated Bottom
- Aquatic Bed
- Unconsolidated Shore
- Moss-Lichen Wetland
- Emergent Wetland
- Scrub-Shrub Wetland
- Forested Wetland

Modifiers

Besides vegetation and substrate composition, the classification addresses several other aspects of wetlands and deepwater habitats, namely water regime, water chemistry, soil type, and modification by humans or beavers (*Castor canadensis*). These features are treated as Modifiers that are applied once the habitat has been placed in the classification hierarchy.

Without long-term measurements of water levels at individual sites, it is impossible to accurately describe a site's hydrologic regime, but the authors of the classification believed that even broad categorization of hydrology can be useful. *Water Regime Modifiers* provide a gross description of a site's water regime, the frequency and duration of surface water inundation or soil saturation. Two major categories of water regimes are recognized, Tidal and Nontidal. Tidal Water Regime Modifiers describe the frequency and duration of flooding or exposure by ocean tides; four are recognized: subtidal, irregularly exposed, regularly flooded, and irregularly flooded. Nontidal Water Regime Modifiers describe hydrologic conditions at inland sites during the growing season; the eight Nontidal Modifiers are: permanently flooded, intermittently exposed, semipermanently flooded, seasonally flooded, saturated, temporarily flooded, intermittently flooded, and artificially flooded.

Water Chemistry Modifiers address two key variables, salinity (Cowardin *et al.* 1979: Table 2) and hydrogen-ion concentration or pH (Cowardin *et al.* 1979: Table 3). All habitats are classified according to salinity, and freshwater habitats (< 0.5 ppt salinity) are further classified by pH. The suffix 'haline' is used for the Marine and Estuarine Systems, in which ocean salts predominate, while the suffix 'saline' is reserved for Riverine, Lacustrine, and Palustrine Systems; however, the same prefixes and hierarchy of salinity values apply to inland and coastal habitats.

Soil Modifiers, taken directly from *Soil Taxonomy* (Soil Survey Staff 1975), are used for those wetlands in which an unconsolidated substrate is capable of supporting emergent herbs, emergent mosses, lichens, shrubs, or trees. Deepwater habitats and wetlands that are too wet to support emergent vegetation are not considered to have soil. Wetland soils are broken into two major categories, mineral and organic.

Finally, a series of Special Modifiers was developed to indicate that certain habitats have been created or modified by humans or beavers. These include: excavated, impounded, diked, partly drained, farmed, and artificial (i.e., nonvegetated substrates emplaced by humans). Special Modifiers may be used singly or in combination wherever they apply.

Regionalization

In the USFWS classification, a given taxon has no particular regional alliance; its representatives may be found in one or many parts of the United States. Yet, for the purpose of planning and for organization, retrieval, and interpretation of inventory data, it is important to be able to place habitats within a regional context. The USFWS classification adopted Bailey's (1976, 1978) classification and map of ecoregions of the United States to fill the need for regionalization inland. Bailey's hierarchical classification addresses subcontinental and regional differences in climate as well as major vegetation types. Cowardin *et al.* (1979: Fig. 7) (Fig. 2), developed 10 additional provinces for the Marine and Estuarine areas of the United States.

Key aspects of the approach and rationale

Classification by individual components

In most North American wetland classifications developed prior to Cowardin *et al.* (1979), (e.g., Martin *et al.* 1953; Stewart & Kantrud 1971; Golet & Larson 1974; Jeglum *et al.* 1974; Zoltai *et al.* 1975) traditional terms, such as marsh, swamp, bog, fen, and wet meadow played a central role. The first draft of the current USFWS classification (Cowardin & Carter 1975) also used such terms. However, after two national reviews and many months of haggling over definitions, the authors of the USFWS system agreed that the meaning of terms such as swamp and bog varies so widely across the United States that it would be a mistake to base a national classification on them. Rather than try to force users nationwide to adopt new, standardized definitions for these familiar terms – and risk failure – the authors decided to abandon this traditional terminology and, instead, to create a classification of wetland components (i.e., vegetative life form, substrate composition and texture, water regime, water chemistry, and soil). Our belief was that the latter approach to classification would be more direct, more accurate, and less likely to result in national inventory statistics that were meaningless because of regional variations in the interpretation of classification taxa. Moreover, it soon became clear that, once a wetland's separate components were classified, it was relatively easy to

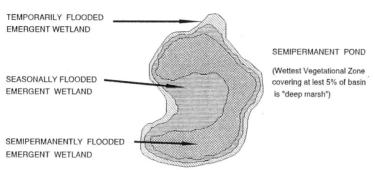

COWARDIN ET AL. (1979)   STEWART AND KANTRUD (1971)

TEMPORARILY FLOODED
EMERGENT WETLAND

SEASONALLY FLOODED
EMERGENT WETLAND

SEMIPERMANENTLY FLOODED
EMERGENT WETLAND

SEMIPERMANENT POND

(Wettest Vegetational Zone
covering at lest 5% of basin
is "deep marsh")

Fig. 2. Comparison between the zone classification of Cowardin *et al.* (1979) and the basin classification of Stewart and Kantrud (1971). In actual mapping by the National Wetlands Inventory the narrow marginal zones may be too small delineate and classify individually.

determine which of the more traditional terms, such as swamp, bog, or marsh, would be appropriate within a given region of the country.

The decision to change to a 'classification by components' was resisted strongly by some, including some of the authors themselves, but in the end, we were satisfied that what was lost in colorful mental images was more than regained in accuracy of classification. We also want to stress that we view the continued use of more traditional regional classifications as wholly appropriate if they better serve a user's particular need. The advantage of using the USFWS classification is national consistency.

Inclusion of nonvegetated habitats
Based on recommendations presented at the July 1975 classification workshop (Sather 1976), the authors of the USFWS classification made another major departure from traditional practice: the wetland definition and classification were extended to a wide variety of nonvegetated habitats such as beaches and rocky shores. The thinking was that it is hydrology, not the presence of vegetation, that determines the existence of wetland. It seemed to make little sense to ignore nonvegetated habitats such as mud flats, while calling contiguous vegetated habitats, with virtually the same water regime, wetland. Actually, the trend to include nonvegetated areas in the concept of wetland was apparent even in Martin *et al.* (1953). That classification, designed primarily for waterfowl habitat, included inland and coastal saline flats along with 14 vegetated wetland types and 4 open-water types.

Classification of habitat zones
Under the USFWS classification, homogeneous areas or zones within wetlands are delineated and classified individually, in the same manner as a forest stand classification. Thus, the number of classes within a single wetland basin may vary widely, depending on the diversity of habitats within the basin and the scale of the remote sensing imagery and final maps on which the wetlands are delineated. This approach contrasts sharply with basin classification, where a single taxon is applied to an entire basin, regardless of the habitat diversity within the basin (e.g., Stewart & Kantrud 1971). Although these two approaches are quite different, it is possible, in some cases, to generate a basin class from the separate polygons classified on an NWI map. Figure 2, for example, compares two classifications of a hypothetical Palustrine wetland: a zonal classification by Cowardin *et al.* (1979), as it might appear on an NWI map, and a basin classification by Stewart and Kantrud (1971). Under the Stewart and Kantrud approach, the wetland basin class is determined by the zone with the most permanent water regime provided that the zone occupies at least 5 percent of the basin.

In this example, the zone on the NWI map is 'semipermanently flooded emergent wetland' (equivalent to 'deep marsh' zone recognized by Stewart and Kantrud). Based on this information, the wetland basin would be classified as a 'semipermanent pond' under the Stewart and Kantrud system.

One method of classification is not inherently superior to another. They merely serve different purposes. The authors of the USFWS classification chose to construct a classification of habitat zones for several reasons: (1) A national classification must be applicable to wetlands that vary greatly in size. For Marine, Estuarine, and Riverine Systems, a basin classification has little meaning, although entire estuaries or river systems can be, and have been, classified for certain purposes (e.g., Odum *et al.* 1974). Similarly, large Palustrine-Riverine-Lacustrine complexes simply do not lend themselves to basin classification; (2) If the inventory map furnishes sufficient detail on the characteristics of individual wetland zones or polygons, it is sometimes possible to derive basin classifications, as noted in the paragraph above. Thus, NWI maps may serve the needs of various users with different purposes. However, combining mapping units to produce new taxa may be difficult, especially if an attempt is made to automate the process; (3) The inventory was to be accomplished primarily through interpretation of high-altitude aircraft imagery, and, generally, it is far easier to identify and delineate habitat zones than entire basins, particularly in areas of low topographic relief.

Inventory demands
Although the primary goal of the authors of the USFWS wetland classification was to produce an ecologically meaningful system for describing wetland habitats, some modifications were required to accommodate use of high-altitude aerial photographs. Certain important wetland characteristics or attributes that might be readily recognized in the field could not be consistently distinguished on high-altitude imagery; as a result, such characteristics were either relegated to the lowest levels of the classification hierarchy (e.g., Dominance Type) or to the Modifiers (e.g., water regime, water chemistry), or they were dropped from the classification altogether. The practical needs of the inventory thus had an influence on the final form of the classification. Inventory-related constraints are discussed in more detail below, under 'Limitations of Remote Sensing'.

Relation to regulatory programs
When the authors of the classification first invited comment from various federal and state agencies on how the classification and inventory should be structured, we were frequently asked to align both the definition of wetland and the structure of the classification with state and federal laws and regulations. We chose not to do this. Laws and regulations vary significantly among agencies. They are drafted for various purposes, and

they may or may not have an ecological basis. We were convinced that, without separation from the problems of wetland regulation, the classification would not have been completed, and the inventory would have been hopelessly mired. Each NWI map has the following statement printed on it:

> There is no attempt in either the design or products of this inventory to define the limits of proprietary jurisdiction of any Federal, State, or local government or to establish the geographical scope of the regulatory programs of government agencies.

We believe that with an understanding of the structure and purpose of the classification, as well as the strengths and weaknesses of the inventory products, governmental agencies can use the products as an aid to regulatory decisions.

Principal problem areas

Among the most significant problems encountered during development and application of the USFWS wetland classification were (1) the definition of wetland, (2) the definition of classification taxa; (3) the lack of basic ecological data to support the classification; and (4) the limitations of remote sensing. Each of these problems is discussed below.

Definition of wetland
The definition of wetland is the most basic step in wetland classification, inventory, and management. Unfortunately, attempts to define wetland and to distinguish it from upland (nonwetland) are beset by a number of conceptual, as well as practical, problems. First, the moisture gradient in nature is continuous, and obvious breaks along that continuum are seldom observed, except at abrupt changes in topographic slope. Second, the diversity of hydrologic conditions over a geographic area as large as the United States is so great that it is practically impossible to write a concise definition to cover all situations. Third, while it is difficult to write a sound conceptual definition of wetland, it is even more difficult to develop reasonable, practicable rules for application of this definition in the field. Fourth, although the definition of wetland is a scientific matter, it also may be highly controversial because of land use or regulatory implications.

Hydrology is universally regarded as the most basic feature of wetlands (Gosselink & Turner 1978; Carter *et al.* 1979; Mitsch & Gosselink 1993), but it is also the most difficult parameter to describe accurately. Due to the great variations in physiography and climate across the United States, the hydrologic regimes of wetlands vary widely. Even at the same site, water levels and soil moisture frequently vary markedly among years and among seasons within the same year. 'Wetland hydrology' is easy to recognize in the interior of those wetlands where surface water is present all year, but not so easy near the wetland-upland edge, where boundaries must be drawn for purposes of inventory and land use regulation. To even approximate the longterm, prevailing water regime at a given site, groundwater levels or soil moisture would have to be monitored continuously over many years. Such long-term data are not now available – nor are they ever likely to be – except at isolated research sites. For these reasons, it is virtually impossible to write a hydrologic definition for wetland that is comprehensive, quantitative, and yet practical.

The USFWS wetland definition (Cowardin *et al.* 1979:3) states that wetland can be identified by any one of three features: a predominance of hydrophytes; a predominance of hydric soils; or, in areas where true soils and rooted plants do not exist, a substrate that is saturated with water or covered by shallow water at some time during the growing season of each year. When the authors wrote this definition, they assumed that most wetlands would be identified by their vegetation or their soils, the most obvious indicators of wetland hydrology. We labored long and hard over the wording of the third, or hydrologic, part of the wetland definition (see section above on 'Definition of Wetland and Deepwater Habitat' for exact wording). None of the authors was entirely pleased with the results, but after many months of discussions and review by wetland scientists and managers, we agreed that further modification was not warranted. The authors were pressured by some to quantify the hydrologic aspect of the wetland definition, and to state the minimum number of days of surface inundation or soil saturation required for an area to qualify as wetland, but we declined to do so. We cited as reasons the wide range of hydrologic conditions in the Nation's wetlands and the lack of long-term hydrologic data.

The authors of the USFWS wetland classification maintained that it is neither reasonable nor practicable to establish a quantitative hydrologic criterion for field identification of wetlands. We still believe that, in the great majority of cases, wetlands should be identified by vegetation and soils. We argue that hydrology should be used only where soil and vegetation criteria cannot reasonably be applied, such as in highly disturbed wetlands, and that any hydrologic criteria devised for those special circumstances must be consistent with the hydrologic criteria established for identification of hydric soils (Soil Conservation Service 1991).

Definitions of 'hydrophyte' and 'hydric soil' typically are circular, hydrophytes often being defined as plants that grow in wet soils and hydric soils being defined as soils that support wetland plants. To avoid this problem, the authors asked the USFWS to create a list of those plants that occur in U.S. wetlands (Reed 1988); the list then serves to define 'hydrophyte'. Similarly, they coined the term 'hydric soil' and asked the US Soil Conservation Service to define the term and to identify those specific soils nationwide that met the definition (Soil Conservation Service 1991). Development of these lists has been a prodigious task, but surprisingly successful. In each case, additions to the list required the unanimous agreement of National (and regional, in the case of plants) review teams, composed of wetland scientists from the major federal agencies concerned with wetlands (USFWS, US Soil Conservation Service, US Environmental Protection Agency, and US Army Corps of Engineers); University scientists also served on the National Technical Committee for Hydric Soils. The lists have been widely used by state and federal agency personnel, as well as academicians and private environmental consultants.

When Cowardin *et al.* (1979) wrote the definition of wetland, we also assumed that, in most vegetated wetlands, there would be a strong correlation between vegetation that was predominantly hydrophytic and soils that were predominantly hydric. The USFWS commissioned scientists at North Carolina State University (Wentworth and Johnson 1986) to develop numerical methods for identifying a 'predominance of hydrophytes', using the wetland plant list and indicator categories (Table 1). In 1985, the USFWS then launched a nationwide research effort to test the extent of agreement between the hydric status of soils and the presence of hydrophytic vegetation. These studies found that the agreement between hydric soils and wetland plants was close, with few exceptions (Scott *et al.* 1989; Segelquist *et al.* 1990). Despite this apparent success in making the conceptual definition of wetland operational, significant problems remain. One controversy centers on which of the categories of plants that occur in US wetlands (Reed 1988) are truly hydrophytes or 'wetland plants'. Tiner (1991), for example, has argued that not only Facultative plants, but Facultative Upland plants as well, are true hydrophytes under some circumstances. Others have suggested that only Obligate Wetland and Facultative Wetland plants should qualify as hydrophytes.

The USFWS wetland definition was proposed as an ecological, not a regulatory, definition, as already stated. Although some technical problems still remain, particularly in development of field methods for wetland identification, we believe that the scientific basis for that definition is even stronger today than when it was published in 1979. In our view, the definition of wetland is a scientific issue; the scope and intensity of wetland protection are policy issues. Although policy is of critical importance, we do not believe that it should overrule science in scientific matters.

Definition of classification taxa
By its very nature, habitat classification is problematic because it is an attempt to place artificial boundaries on natural continua. Wetland classification is a prime example. One of our greatest sources of frustration was our inability to clearly define what we perceived as real boundaries between distinct categories of wetlands or wetland characteristics. In some cases, the problem was caused by a lack of basic ecological data; in other cases, we were constrained by the limitations of remote sensing. The definitions of Systems and Water Regime Modifiers gave us the most difficulty.

The authors of the classification and many reviewers would have preferred – at least initially – to see river floodplains placed in the Riverine System, but they soon discovered that it simply was not possible to consistently locate the landward edge of floodplains, either in the field or on high-altitude aerial photographs. This was especially obvious in densely forested terrain where topographic relief was negligible, such as much of the southeastern United States and the Great Lake States. In some areas, rivers course through large emergent wetlands with interspersed lakes, making distinction between river floodplain and lakeshore wetlands almost impossible. Consequently, the authors altered the traditional concept of riverine, limiting it to the river channel (including nonvegetated classes, Aquatic Beds, and Nonpersistent Emergent Wetlands). Persistent wetland vegetation, such as Forested, Scrub-Shrub, and Emergent Wetlands of the adjacent floodplain, was classified as Palustrine. To be consistent in concept and in practice, they drew the boundary of the Lacustrine System at the edge of upland or at the edge of persistent (i.e., Palustrine) wetland vegetation.

The separation of Systems on the basis of persistence of vegetation, while it has some merit, becomes awkward where islands of persistent vegetation occur within river channels or lake basins. NWI staff apprised the authors of the problems created by basing one of the uppermost distinctions in the classification hierarchy (i.e., distinction between Systems) on a feature that cannot always be determined through remote sensing. Using salinity (0.5 ppt) to distinguish coastal from inland Systems has created similar problems. The classification forces the use of data sources other than remote sensing imagery.

Perhaps the most imprecise facet of the USFWS classification is the definition of Water Regime Modifiers. We readily admit that these modifiers are only crude approximations of some of the more common hydrologic situations observed in nature. Some Water Regime Modifiers, such as irregularly flooded, seasonally flooded, and saturated, are too broad to be useful in many cases. Complex water regimes, such as those associated with seiches in lakes, cannot be adequately described by the present water regime modifiers. Additions and modifications are needed. The authors were asked by some reviewers to quantify, in days or weeks, the duration of flooding or soil saturation for each Modifier, but they argued that was inappropriate in light of the lack of data on which to base such figures. Moreover, it would be impractical to have to verify that any such quantitative criteria were met prior to classification of specific sites.

The authors limited consideration of water regime in nontidal areas to the growing season, arguing that the presence of water at that time of year was most significant to vegetation. However, water levels during the dormant season may be of critical importance for certain wetland functions such as floodwater retention and wintering waterfowl habitat. In some regions, the growing season runs virtually year-round. We are confident that the current Water Regime Modifiers can be improved, but the extent of refinement ultimately will be determined by practicality and the availability of basic ecological data.

Lack of basic ecological data
One common approach to classification is to assemble the elements to be classified, describe their attributes, and then group them, often in a hierarchical structure. This approach may employ complex statistical methodology (Anderberg 1973). Developing the classification for the NWI was, in some respects, the

reverse of this process. First the taxonomy was created, based on published descriptions, experience, and tradition; the result was like a large ordered set of pigeonholes. Next, descriptions of the taxa were written in an attempt to assure that an element would be placed in the right pigeonhole. Finally, the user gathered the elements, applied the definitions, and attempted to place each element in a pigeonhole. The last step was equivalent to the NWI mapmaker adding the identifier to a polygon on a map.

This reverse classification procedure had a number of shortcomings, but it was the only choice, because of the lack of supporting data. Cowardin recalls that when he was just starting work on the classification he made a trip to Alaska and met with Dr Bonita Wieland (University of Alaska, Fairbanks). He was asking questions about the vegetation, soils, and hydrology of Alaskan wetlands. Dr Wieland replied, 'You realize that nobody has ever been in many of the wetlands in Alaska'. Even in the lower 48 states there are many wetlands where the basic data necessary for accurate classification are unavailable. The authors also frequently discussed whether a taxon that was being included in the classification actually existed. They had such a discussion about Unconsolidated Shores with an Organic substrate. In 1985 Golet observed this type of wetland in western Alaska, and a photograph of it (Plate 47) was included in the first reprinting of the classification.

Despite the lack of data, the authors found that the classification system often worked well in areas where they had little experience. We are not suggesting that classification be delayed until all wetlands have been described in detail, but we do want to stress that the process of classification does not produce any data. Nevertheless, misclassification does produce errors in wetland inventory data. It is important that basic ecological studies of wetlands be accelerated. As new data become available, refinements in the USFWS classification will be possible.

Limitations of remote sensing

Wetland management decisions are often based on data presented on maps or on data bases derived from maps. Maps are one product of the NWI. Map-based data have the advantage of being readily available. Some decisions cannot be deferred until intensive ground studies are conducted. The trend toward the use of mapbased data will accelerate in the future because of the increasing availability of geographic information systems and computer equipment that allow rapid access to spatial data and complex analyses, which were not practical in the past. Map data, however, have limitations and present problems for wetland classification.

Mapping conventions cannot include all of the detail available in the USFWS classification. Classifications used for mapping are scale-specific (Kuchler 1988). There is a limit to the size of a mapping unit that can practically be placed on a map and to the size of a water body or stand of vegetation that can be interpreted from a photograph. The standard NWI maps have a scale of 1:24 000 and most are prepared from high-altitude (1:58 000 or smaller scale) color-infrared photographs. The inventory staff prepared a standardized set of scale-specific mapping conventions based on the classification. Even at this scale, which is large in comparison to data derived from satellites such as LANDSAT (Jacobson *et al.* 1987), some entire wetland basins and many wetland areas around the margins of basins are not detectable or mappable.

There are also limits to what can be interpreted from a single aerial photograph or remote sensing image. Wetland dynamics present a difficult problem for remote sensing because surface water may or may not be present on a wetland, and the dominant vegetation can vary in both species composition and life form within and among years. Thus a high degree of ecological expertise and familiarity with local areas are required of the photointerpreter. Interpretation of water regimes and sometimes classes in the USFWS classification often requires more than one set of photographs, or supplemental data.

The ordering of the taxa in the classification hierarchy is ecologically based, instead of being geared primarily to the use of remote sensing as are the classes in the national land use classification of Anderson *et al.* (1976). The need to consistently identify Systems has forced the NWI to use supplementary sources of data such as maps of tidal influence and soils, as well as ground study of sample sites. Some of the components of the classification such as water chemistry, though critically important in wetland ecology, cannot be determined from photographs and have been omitted from NWI maps. The Class level in the hierarchy, based on life form of vegetation and substrate form and texture, should be the most easily photointerpreted taxon. Even here, however, there are difficult interpretation problems, such as distinguishing Emergent Wetland from Scrub-Shrub Wetland in Alaskan tundra (Cowardin *et al.* 1979: Plate 75).

Future directions

Anticipated changes in the classification

Since the publication of the classification in 1979, NWI staff have been compiling a list of problem areas for which revisions of the classification are needed. In addition, new users, new data, and new methodologies for describing wetlands have become available. There appears to be a need for revision. However, the inventory is nearing completion, and products are being converted to expensive digital form at an increasing rate, thus imposing severe limits on the nature and extent of revision that is practical. We suspect that any revisions will be in the form of clarification of definitions of the taxa and possibly additions to the hierarchical structure. A major redesign of the classification would invalidate comparisons with past mapping and databases.

Explosion of remote sensing and GIS technology

There is already an increasing need for rapid development of up-to-date wetland maps. New remote sensing products, especially those coming from satellites, will be developed. Techniques involving multispectral scanners have been used to determine water quality (Dekker *et al.* 1991) and water depth (Lyon *et al.* 1992). These advances will increase the demand for classifications that are applicable to computer processing. It is doubtful that the USFWS classification can be completely compatible with automated procedures, but it can provide the basis for mapping conventions developed for special inventories. These types of inventories should be viewed as complementary to the NWI, rather than competitive with it. The purposes and limitations of both kinds of classifications and data sets must be understood (Koeln *et al.* 1988).

Wetland functional assessment

As realization of the importance of wetlands to society has grown, regulations that protect wetlands and restrict human activities in wetlands have increased as well. The USFWS wetland classification and the inventory based on it were not intended for regulatory purposes. Agencies such as the US Army Corps of Engineers and the US Environmental Protection Agency must make regulatory decisions. As a result, there has been increased interest in wetland values to society (Mitsch & Gosselink 1993, Chapter 15). One way of evaluating wetlands is to determine their functions and to determine how human activities affect these func-

tions or vice versa (Sather & Smith 1984). Wetland function has become an increasingly important component of recently proposed classifications (Clairain 1985). Classification and inventory of wetland functions are difficult to achieve, however, considering the current state of our knowledge. Functions of certain wetland types are frequently unknown. Therefore, detailed studies of individual wetlands may be necessary before functional classification can proceed. The problem is similar to the problem of wetland definition based on hydrology, which frequently is not measurable and must be inferred from other attributes. In some cases, function can be broadly inferred from the taxa in the USFWS classification, but in many cases such inference will have to wait for additional basic wetland research.

Global wetland classification

Potential benefits

A standard, global wetland classification would do much to advance the causes of wetland science and management worldwide. One benefit would simply be a standard terminology for communication about wetlands. This, in itself, would permit greater understanding of wetland diversity and functions. Standard terminology also would allow more effective comparisons and integration of research results from different parts of the world. Global classification would facilitate the development of coordinated research efforts for particular, widely distributed, wetland ecosystems, resulting in more rapid advances in our knowledge and less duplication of effort. Finally, global classification would provide the basis for a global wetland inventory, periodic monitoring of wetland extent and condition, improved management of migratory wildlife habitats on an international basis, and more accurate assessments of the global significance of wetlands to humankind.

Some important considerations

We view the following as key considerations in the development of a global wetland classification:

1. No single system can accurately portray the diversity of wetland conditions worldwide. Some important ecological information inevitably will be lost through classification. Despite this fact, a sincere effort should be made to incorporate as much of this diversity as possible into the basic classifica-

tion structure, so that the system truly represents the global wetland resource. Regional, national or continental modifiers might be provided at the lowest levels of the classification hierarchy to assure that unique, but geographically restricted, wetland types are covered.

2. Classification taxa must be clearly defined and readily distinguishable, both in the field and on a variety of remote sensing products. A 'classification by components', such as the USFWS system, would appear to be the most objective approach. Geographically specific terms such as bayou, reedswamp, arroyo, fen, and billabong should be avoided.

3. Insofar as possible, the classification taxa should be functionally relevant, but due to the dearth of information on the relationship between wetland attributes and functions, it seems unwise to pursue a functional classification of wetlands at this time.

4. The classification should be based in ecology, not regulatory concerns, and value-related biases should be avoided. Currently at least, the perceived values of wetlands to society vary widely around the world as a result of wide-ranging socioeconomic conditions, political ideologies, and cultural traditions. International cooperation in efforts such as wetland classification and inventory may eventually lead to more uniform, global policies for wetland management, but that goal cannot logically be addressed through the classification and inventory process.

5. Careful consideration must be given to the development of a wetland definition that will encompass the diversity of wetland conditions worldwide, and, at the same time, be operationally effective at any field location.

6. Special efforts should be made to guarantee compatibility among any global wetland classification and existing national or international systems. The correspondence between global classification taxa and more colloquial terms (e.g., bog, marsh, reedswamp) should be addressed as well, possibly as an appendix or supplement to the classification.

Conclusions

We have reviewed the history of development of the USFWS wetland classification and sketched its structure. Despite undisputed problems, we believe that the classification has met its stated objectives. In the Unit-

ed States we found that there is extreme variation in wetlands, and the variation will be far greater on an international scale. The authors of the U.S. classification also found that each had personal biases based on experience. Ideas that worked well in familiar territory were found to be unworkable elsewhere. An international effort will require a diversity of geographical and technical expertise.

Our experience in the United States suggests that development of a classification should follow five steps in chronological order:

1. The purpose for, and ultimate users of, the classification must be clearly identified.

2. Data sources and methodology for data gathering must be reviewed. This step will identify constraints that will affect the structure of the classification.

3. A classification structure should be drafted. Our experience suggests that there are advantages to a hierarchical structure and that, as much as possible, the structure should follow established, and accepted, concepts. Drafting the structure is, perhaps, the easiest and least time-consuming step.

4. Detailed definitions for wetland and for each taxon of the classification must be written. This essential step is the most difficult and controversial.

5. The system must be tested on a wide array of different wetland classes over a broad geographic area.

Hopefully, discussions generated at this symposium will lead to an international system of classification and inventory that can be used to protect wetland ecosystems. The task will be immense, but it is time to begin.

Acknowledgments

We acknowledge the major contribution of our coauthors of the classification, V. Carter and E. T. LaRoe. Without the efforts of J. H. Montanari, W. O. Wilen, and J. H. Sather, as well as the entire NWI staff, the classification would never have been completed. We hope that this historical review is accurate in the minds of the many individuals involved in the development and testing of the classification. V. Carter, J. H. Montanari, J. H. Sather, L. L. Strong, J. E. Austin, J. T. Lokemoen, and D. H. Johnson kindly reviewed earlier drafts of this paper.

152

References

Anderberg, M. R. Cluster analysis for applications. Academic Press, New York. 359 pp.

Anderson, J. R., Hardy, E. E., Roach, J. T. & Witmer, R. E. 1976. A land use and land cover classification system for use with remote sensor data. US Geological Survey Professional Paper 964. 28 pp.

Bailey, R. G. 1976. Ecoregions of the United States. US Forest Service, Ogden, UT. (map only; scale 1:7 500 000).

Bailey, R. G. 1978. Ecoregions of the United States. US Forest Service, Intermountain Region, Ogden, UT. 77 pp.

Carter, V., Bedinger, M. S., Novitzki, R. P. & Wilen, W. O. 1979. Water resources and wetlands. Pages 344–376 in P. E. Greeson, J. R. Clark and J. E. Clark, eds. Wetland functions and values: the state of our understanding. American Water Resources Association, Minneapolis, MN.

Clairain, E. J., Jr. 1985. National wetland functions and values study plan. Transactions North American Wildlife and Natural Resources Conference 30: 485–494.

Cowardin, L. M. & Carter, V. 1975. Tentative classification for wetlands of the United States. US Fish and Wildlife Service Office of Biological Services Washington, DC Rep. 43 pp. (mimeo)

Cowardin, L. M., Carter, V., Golet, F. C. & LaRoe, E. T. 1976. Interim Classification of wetlands and aquatic habitats of the United States. Addendum to J. H. Sather, ed. Proceedings of the National Wetland Classification and Inventory Workshop. 1975. US Fish and Wildlife Service. FWS/OBS 76/09.

Cowardin, L. M., Carter, V., Golet, F. C. & LaRoe, E. T. 1979. Classification of wetlands and deepwater habitats of the United States. US Fish and Wildlife Service FWS/OBS 79/31. 103 pp.

Dekker, A. G., Malthus, T. J. & Seyhan, E. 1991. Quantitative modeling of inland water quality for high-resolution MSS systems. IEEE Transactions Geoscience and Remote Sensing 29: 89–95.

Golet, F. C. & Larson, J. S. 1974. Classification of freshwater wetlands in the glaciated Northeast. US Fish and Wildlife Service Resource Publication. 116. 56 pp.

Gosselink, J. G. & Turner, R. E. 1978. The role of hydrology in fresh-water wetland ecosystems. Pages 63–78 in R. E. Good, D. F. Whigham, and R. L. Simpson, eds. Freshwater wetlands: ecological processes and management potential. Academic Press, New York.

Jacobson, J. E., Ritter, R. A. & Koeln, G. T. 1987. Accuracy of Thematic Mapper derived wetlands as based on National Wetland Inventory data. Pages 109–118 in American Society Photogrametry and Remote Sensing Technical Papers, 1987 ASPRS-ACSM Fall Convention, Reno, NV.

Jeglum, J. K., Boissoneau, A. N. & Haavisto, V. F. 1974. Toward a wetland classification for Ontario. Canadian Forest Service Information Report O-X-215. 54 pp.

Koeln, G. T., Jacobson, J. E., Wesley, D. E. & Rempel, R. S. 1988. Wetland inventories derived from LANDSAT data for waterfowl management planning. Transactions North American Wildlife and Natural Resources Conference 53: 303–310.

Kuchler, A. W. 1988. Aspects of maps. Handbook of vegetation science, Vol. 10: 97–1040.

Lyon, J. G., Lunetta, R. S. & Williams, D. C. 1992. Airborne multispectral scanner data for evaluating bottom sediment types and water depths of the St. Marys River, MI. Photogramtric Engineering and Remote Sensing. 58: 951–956.

Martin, A. C., Hotchkiss, N., Uhler, F. M. & Bourn, W. S. 1953. Classification of wetlands of the United States. US Fish and Wildlife Service Special Scientific Report Wildlife. 20. 14 pp.

Millar, J. B. 1976. Wetland classification in western Canada: a guide to marshes and shallow open water wetlands in the grasslands and parklands of the Prairie Provinces. Canadian Wildlife Service Report Series 37. 38 pp.

Mitsch, W. J. & Gosselink, J. G. 1993. Wetlands, 2nd ed. Van Nostrand Reinhold, New York, 539 pp.

Odum, H. T., Copeland, B. J. & McMahan, E. A. (eds) 1974. Coastal ecological systems of the United States. The Conservation Foundation, Washington, DC. 4 Vol.

Reed, P. B. 1988. National list of plant species that occur in wetlands: national summary. US Fish and Wildlife Service Biological Report. 88(24). 244 pp.

Sather, J. H. 1976. Proceedings of the National Wetland Classification and Inventory Workshop. US Fish and Wildlife Service, Washington, DC 248 pp.

Sather, J. H. & Smith, R. D. 1984. An overview of major wetland functions and values. US Fish and Wildlife Service Office of Biological Services. FWS/OBS 84/18. 68 pp.

Scott, M. L., Slauson, W. L., Segelquist, C. A. & Auble, G. T. 1989. Correspondence between vegetation and soils in wetlands and nearby uplands. Wetlands 9: 41–60.

Sculthorpe, C. D. 1967. The biology of aquatic vascular plants. Edward Arnold Ltd., London. 610 pp.

Segelquist, C. A., Slauson, W. L., Scott, M. L. & Auble, G. T. 1990. Synthesis of soil-plant correspondence data from twelve wetland studies throughout the United States. US Fish and Wildlife Service Biological Report. 90(19). 24 pp.

Shaw, S. P., Fredine, C. G. 1956. Wetlands of the United States. US Fish and Wildlife Service Circular 39. 67 pp.

Soil Conservation Service. 1982. Hydric soils of the United States. US Department of Agriculture National Bulletin. 430-2-7. (January 4, 1982).

Soil Conservation Service. 1985. Hydric soils of the United States. US Department of Agriculture and National Technical Committee for Hydric Soils, Washington, DC.

Soil Conservation Service. 1987. Hydric soils of the United States. US Department of Agriculture, in cooperation with the National Technical Committee for Hydric Soils, Washington, DC.

Soil Conservation Service. 1991. Hydric soils of the United States. US Department of Agriculture Miscellaneous. Publication. 1491, in cooperation with the National Technical Committee for Hydric Soils. Washington, DC.

Soil Survey Staff. 1975. Soil taxonomy: a basic system of soil classification for making and interpreting soil surveys. US Soil Conservation Service Agricultural Handbook 436. 754 pp.

Stewart, R. E. & Kantrud, H. A. 1971. Classification of natural ponds and lakes in the glaciated prairie region. US Fish and Wildlife Service Resource Publication. 92. 57 pp.

Tiner, R. W. 1991. The concept of a hydrophyte for wetland identification. BioSience 41: 236–247.

Welch, P. S. 1952. Limnology, 2nd ed. McGraw-Hill, New York. 538 pp.

Wentworth, T. R. & Johnson, G. P. 1986. Use of vegetation in the designation of wetlands. US Fish and Wildlife Service, National. Wetland Inventory, Washington, DC. 107 pp.

Wilen, B. & Bates, M. K. 1995. The United States national wetlands inventory. Vegetation (in press).

Zhadin, V. I. & Gerd, S. V. 1963. Fauna and flora of the rivers, lakes and reservoirs of the USSR. Oldbourne Press, London. 626 pp.

Zoltai, S. C., Pollett, F. C., Jeglum, J. K. & Adams, G. D. 1975. Developing a wetland classification for Canada. Proceedings North American Forest Soils Conference 4: 497–511.

Vegetatio **118**: 153–169, 1995.

The US Fish and Wildlife Service's National Wetlands Inventory Project

B.O. Wilen & M.K. Bates
National Wetlands Inventory Project, US Fish & Wildlife Service, Room 400 – Arlington Square, Department of the Interior, 1849 C Street NW, Washington, DC 20240, USA

Key words: Digitizing, Photo-interpretation, Status, Trends, Wetland classification, Wetland mapping

Abstract

In 1974, the US Fish and Wildlife Service directed its Office of Biological Services to design and conduct an inventory of the Nation's wetlands. The mandate was to develop and disseminate a technically sound, comprehensive data base concerning the characteristics and extent of the Nation's wetlands. The purpose of this data base is to foster wise use of the Nation's wetlands and to expedite decisions that may affect this important resource. To accomplish this, state-of-the-art principles and methodologies pertaining to all aspects of wetland inventory were assimilated and developed by the newly formed project. By 1979, when the National Wetlands Inventory (NWI) Project became operational, it was clear that two very different kinds of information were needed. First, detailed wetland maps were needed for site-specific decisions. Second, national statistics developed through statistical sampling on the current status and trends of wetlands were needed in order to provide information to support the development or alteration of Federal programs and policies. The NWI has produced wetland maps (scale = 1:24 000) for 74% of the conterminous United States. It has also produced wetland maps (scale = 1:63 360) for 24% of Alaska. Nearly 9000 of these wetland maps, representing 16.7% of the continental United States, have been computerized (digitized). In addition to maps, the NWI has produced other valuable wetland products. These include a statistically-based report on the status and trends of wetlands that details gains and losses in United States wetlands that have occurred from the mid-1970's to the mid-1980's. Other wetland products include a list of wetland (hydric) soils, a national list of wetland plant species, wetland reports for certain individual States such as New Jersey and Florida, and a wetland values data base.

Introduction

Wetlands provide a variety functions and values. Many people use wetlands for recreational activities, ranging from canoeing to bird watching. A wetland's natural beauty and solitude can be experienced in these unique natural settings. Wetlands also play an integral role in maintaining the quality of human life via material contributions to the national economy (through food supply; water quality improvement; flood control; and fish, wildlife, and plant resources) and thus to the health, safety, recreation, and economic well-being of all United States citizens.

Wetlands also act as natural filtration systems which have the capacity to purify the water that flows through them. The sediments in the wetland act as nutrient sinks, absorbing the nutrients released by plant decomposition. As water flows through the wetland system, the plants, animals, and sediments absorb, assimilate or change the chemical form of many of the contaminants and heavy metals introduced by human activities in the watershed. In addition, significant amounts of suspended sediments are removed from the water as it flows slowly through the wetland.

Wetlands have the ability to slow the flow of water and to store large amounts of water in organic deposits and basins. This includes such functions as erosion and flood control, flow stabilization, discharge of ground water to the surface, and recharging of underground aquifers. Wetlands perform food chain support by producing tremendous amounts of detritus which is consumed by many of the organisms which inhabit wetland

154

ecosystems. Thus, detritus forms the base of a complex food web which cycles energy and nutrients within the wetland environment, and also exports nutrients into adjacent areas.

During the 1780's the conterminous United States contained an estimated 221 million acres of wetlands. Over a 200-year period, wetlands have been drained, dredged, filled, leveled and flooded. Twenty-two States have lost 50 percent or more of their original wetlands since the 1780's (Dahl 1990). Wetlands represent only 5.0 percent of the land area in the conterminous United States. These wetlands provide a wide variety of habitats for some very unique and diverse plant and animal communities. With the current emphasis by many government agencies of preserving biodiversity, it is crucial that these wetland areas be protected. The US Fish and Wildlife Service (FWS) has always recognized the importance of wetlands to waterfowl and other migratory birds, in part because 10–12 million ducks breed annually in the United States, and millions more overwinter here. Consequently, the FWS has a direct interest in protecting wetlands, especially wetlands where waterfowl breed and overwinter.

In 1954, the FWS conducted a nationwide wetlands survey covering roughly 40 percent of the conterminous United States and focusing on important waterfowl wetlands. Although this survey was not a comprehensive wetlands inventory by today's standards, it was instrumental in stimulating public interest in the conservation of waterfowl wetlands. These findings were published in a well-known FWS report – 'Wetlands of the United States', commonly referred to as Circular 39 (Shaw & Fredine 1956).

Since this survey wetlands have undergone many changes, both natural and human-induced. These changes, coupled with our increased understanding of wetland values, led the FWS to establish the National Wetlands Inventory (NWI) Project. The NWI goal is to generate and disseminate scientific information on the characteristics and extent of the Nation's wetlands, in order to foster wise use of the Nation's wetlands and provide data for making quick and accurate resource decisions. Decision makers are not able to make informed decisions about wetlands without knowing how many wetlands, and of what type, are where.

The Emergency Wetlands Resources Act (Act) of 1986 directs the Secretary to the US Department of the Interior, through the Director of the FWS, to produce at 10-year intervals, reports to update and improve the information contained in the report entitled 'Status and

Trends of Wetlands and Deepwater Habitats in the Conterminous United States, 1950's to 1970's' (Frayer *et al.* 1983). The first update of this report was produced in 1991 and was entitled 'Wetlands Status and Trends in the Conterminous United States, Mid-1970's to Mid-1980's (Dahl & Johnson 1991). The next update is due in the year 2000. This Act also requires the FWS to produce, by September 30, 1998, National Wetlands Inventory maps for the remainder of the contiguous United States and, as soon as practicable after 1998, wetland maps for Alaska and noncontiguous portions of the United States. In 1989 the Act was amended to require an assessment of the estimated total number of acres of wetland habitat as of the 1780's in the areas that now comprise each State, an assessment of the estimated total number of acres of wetlands in each State as of the 1980's, and the percentage of the loss of wetlands in each State between the 1780's and the 1980's. This requirement was met by the publication of 'Wetlands Losses in the United States, 1780's to 1980's' (Dahl 1990). NWI mapping mandates under the Emergency Wetlands Resources Act have recently been amended under the Wild Bird Conservation Act of 1992. This Act requires the FWS to: produce NWI maps for Alaska and other noncontiguous portions of the United States by September 30, 2000; produce a digital wetlands data base for the United States by September 30, 2004 based on the final NWI maps and; to archive and make available for dissemination digitized wetlands maps and data as such maps and data become available.

Two different kinds of information are mandated by this legislation: (1) detailed wetland maps; and (2) status and trends reports. Detailed wetland maps are needed for assessing the effects of site-specific projects. These maps serve a purpose similar to the US Department of Agriculture (USDA) Soil Conservation Service's soil survey maps, the US National Oceanic and Atmospheric Administration's coastal geodetic survey maps, and the US Geological Survey's (USGS) topographic maps. Detailed wetland maps are used by local, State and Federal agencies – as well as by private industry and organizations – for many purposes, including comprehensive resource management plans, environmental impact assessments, facility and corridor siting, oil spill contingency plans, natural resource inventories, and habitat surveys. National estimates of the current status and trends (i.e., losses and gains) of wetlands, developed through statistical sampling will be used to evaluate the effectiveness of existing Federal programs and policies, identify national or regional problems and increase general public awareness of wetlands.

National Wetlands Inventory pre-operational phase

Before actually beginning wetland mapping in 1979, the NWI Project reviewed existing State and local wetland inventories and existing classification schemes to determine the best way to inventory wetlands. Researchers determined that a remote sensing technique would be the best method to inventory wetlands. The first step of the pre-operational phase was to review existing wetland inventories. The NWI consulted with Federal and State agencies to learn where and when wetland surveys had previously been completed, what inventory techniques were employed, where to obtain copies of any wetland maps that may have been produced, and the status of State wetlands protection. Only a handful of States had inventoried their wetlands, and most of these had only mapped coastal wetlands. This information was published in a 1976 FWS report – 'Existing State and Local Wetlands Surveys (1965–1975)' (US Department of the Interior 1976).

Before the inventory could begin, NWI researchers had to decide how to classify wetlands. In 1975 the FWS brought together 15 of the Nation's top wetland scientists to evaluate the usefulness of existing wetland classification schemes for the NWI. These scientists determined that none of the existing systems could be used or modified for the NWI and that a new classification system should be developed. The FWS's wetlands classification system (Cowardin *et al.* 1979) was developed by a team of four wetland ecologists, one each from the FWS, the USGS, the US National Oceanic and Atmospheric Administration, and the University of Rhode Island, with the assistance of local, State and Federal agencies as well as many private groups and individuals. The new system went though four major revisions and extensive field testing before to its official adoption by the FWS on October 1, 1980. This classification system describes ecological units having certain common natural attributes, arranges these units in a system that aids resource management decisions, furnishes units for inventory and mapping, and provides uniformity in wetland concepts and terminology throughout the United States. Although it is not an evaluation system, it does provide information upon which evaluations can be made.

Wetlands are extremely diverse and complex. The FWS classification system defines the limits of wetlands according to ecological characteristics and not according to administrative or regulatory programs. In general terms, wetlands are defined as lands where saturation with water is the dominant factor determining the nature of soil development and the types of plant and animal communities living in the soil and on its surface. This includes open water and deep water areas. Under the FWS classification system, wetlands are lands transitional between terrestrial and aquatic systems where the water table is usually at or near the surface or the land is covered by shallow water. For purposes of this classification, wetlands must have one or more of the following three attributes: (1) at least periodically, the land supports predominantly hydrophytes; (2) the substrate is predominantly undrained hydric soil; and (3) the substrate is nonsoil and is saturated with water or covered by shallow water at some time during the growing season of each year (Cowardin *et al.* 1979).

The Cowardin *et al.* system presents a method for grouping ecologically similar wetlands. The system is hierarchical, with wetlands divided among five major Systems at the broadest level: Marine, Estuarine, Riverine, Lacustrine and Palustrine. Each System is further subdivided into Subsystems that reflect hydrologic conditions, such as subtidal vs. intertidal in the Marine and Estuarine Systems. Below Subsystem is the Class level, which describes the appearance of the wetland in terms of vegetation (e.g. emergent wetland, aquatic bed, forested wetland) or substrate if vegetation is inconspicuous or absent (e.g. unconsolidated shore, rocky shore, streambed). Each Class is further divided into Subclasses which are used to describe finer differences in life forms and are named on the basis of the predominant life form (e.g. broadleaved deciduous, moss, floating vascular) or the substrate (e.g. mud, bedrock, rubble). The classification also includes modifiers to describe hydrology (water regime), water chemistry (pH, salinity and halinity) and special modifiers relating to human activities (e.g. impounded, partly drained, farmed, artificial).

Below the Class level, the classification system is open-ended. The Dominance Type is the taxonomic category subordinate to Subclass. Dominance Types are determined on the basis of dominant plant species, dominant sedentary or sessile animal species, or dominant plant and animal species. Cowardin *et al.* (1979) only provides examples of the many dominance types possible. Users of this classification system may identify these dominance types and use them as part of the hierarchical classification system. It is also probable that as the system is used in more detail to meet the user's site-specific needs, the need for additional Subclasses and special modifiers will become clear.

The Cowardin *et al.* wetland classification system has been adopted by many national and international organizations. The States of Illinois, Michigan, Minnesota, Oregon and Vermont have passed State wetlands legislation that relies heavily on NWI wetland information for implementation. The work of NWI was the first phase of a long-range State wetland plan for Illinois. Most States in the Northeast US use NWI wetland information as the primary source for their wetland regulatory guidance policies. The State of Maryland is using the Cowardin *et al.* system and NWI specifications to produce color infrared 1:12 000 scale orthophoto maps.

When the first International Wetlands Conference met in New Delhi, India, on 10–17 September 1980, conference participants passed a motion to adopt the Cowardin *et al.* classification system (Gopal *et al.* 1982). The basic concept and hierarchy of the Cowardin *et al.* system has been adopted for use in Central America, Brazil, Poland, India, Greece, and Russia. The wetland classification system used by the Convention on Wetlands of International Importance (the Ramsar Convention) is based upon Cowardin *et al.* In addition, the International Waterfowl and Wetlands Research Bureau uses the same wetland classification system as the Ramsar Convention. International interest in the Cowardin *et al.* system is still active. For example, the NWI gave a 1 week seminar in Mexico on the use of Cowardin *et al.* Next year, representatives from Hungary will be coming to the United States to visit the NWI offices and to learn about the Cowardin *et al.* system.

The main advantage of the Cowardin *et al.* system is its versatility. Since the Cowardin *et al.* system is hierarchical, a country has a variety of choices when deciding upon which level of wetland classification to pursue. For example, a country interested in classifying their wetlands on a very broad level would select Cowardin *et al.*'s five major Systems as the basis for that country's wetland classification program. This general inventory could be done on existing aerial photography and transferred to existing base maps using inexpensive transfer equipment. A major limitation for the international use of Cowardin *et al.* is that it has not been translated into other languages.

Selecting a remote sensing tool

Because of the magnitude of performing an inventory covering the entire geographic area of the United States, remote sensing was the obvious choice of techniques for inventory of the Nation's wetlands. The basic choice was between high-altitude photography and satellite imagery (Landsat). After comparing Landsat's capabilities with the FWS's and other agencies' needs for wetland information, it was evident that Landsat could not provide the needed data for classification detail and wetland determinations within the desired accuracy requirements. Therefore the inventory is being conducted using mid- and high-altitude color infrared aerial photography.

The NWI Project has continued testing of satellite technologies. In conjunction with the National Aeronautic and Space Administration's (NASA) Jet Propulsion Laboratory, NWI conducted a year-long test of the multispectral scanner to detect and map wetlands in Alaska. With Ducks Unlimited, NWI also tested Thematic Mapper data, as well as data from the French satellite SPOT. A year-long test is now being conducted by the Earth Observation Satellite Company to test the feasibility of using Thematic Mapper satellite data to detect wetlands, map wetlands or update existing wetland maps. None of these tests has provided any hope that present satellite configurations can provide the needed data for classification detail and wetness determinations within desired accuracy requirements of the NWI Project and its State and Federal cooperators. The Federal Geographic Data Committee (FGDC) is an interagency effort that 'promotes the coordinated development, use, sharing, and dissemination of geographic data'. The Wetlands Subcommittee of the FGDC published a report called 'Application of Satellite Data for Mapping and Monitoring Wetlands' which supports the use of aerial photography instead of satellites for obtaining accuracy in wetland mapping (Federal Geographic Data Committee 1992).

National Wetlands Inventory operational phase

The FWS employs a small, full time staff of 40 persons that include biologists, photointerpreters, cartographers, computer technicians and computer analysts who are assembled into three basic groups: NWI Project Leader, Central Control Group, and Regional Wetland Coordinators. The NWI Project Leader and his Assistant work out of the Washington, DC office and coordinate the budget, annual work plans and strategic planning. The budget for NWI is $8 million annually, with $5 million spent on wetland mapping and the remaining $3 million spent on wetland status and trends work. The NWI Central Control Group in St. Petersburg, Florida is the focal point for all operational activities of the NWI. It acquires all materials necessary for performing the inventory, provides technical assistance and work materials to the Regional Wetland Coordinators, and produces the wetlands maps. A private service support contractor is responsible for map production, and provides needed personnel (about 140 technicians and professionals).

Regional Wetland Coordinators and their Assistants at FWS's seven Regional Offices are responsible for the inventory of wetlands within their regions and ensuring that all NWI products meet regional needs. They manage contracts for wetland photo-interpretation, coordinate interagency review of draft maps, secure cooperative funding from other agencies, produce regional wetland reports, and provide training in the use of products.

Photo-interpretation and field work are performed by approximately 150 contract personnel hired by FWS. These contractors photo-interpret wetlands with stereoscopes, and, in addition, review soil maps, conduct field checks, and examine existing information on an area's wetlands to ensure accurate identification of wetlands.

The operational phase of the NWI, initiated on 1 October 1979, involves two main efforts: (1) wetlands mapping, and (2) wetlands status and trends analysis. In addition to the wetlands maps and the trends reports (produced through statistical analysis), NWI has produced other products that compliment the mapping effort, including the 'National List of Plant Species That Occur in Wetlands' (Reed 1988), numerous wetland reports, and textual and geographic computerized data bases. NWI has also contributed to a list of hydric soils (USDA, Soil Conservation Service 1991).

The primary map product of the NWI is large-scale (1:24 000) maps that show the location, shape, and characteristics of wetlands and deepwater habitats on USGS base topographic maps. These detailed maps are excellent for site-specific project evaluation (Figs 1 and 2).

To produce a final map, NWI undertakes the following steps: (1) Review of aerial photography to identify obvious wetland types and problematic areas; (2) selection of sites for possible field-checking and layout of a route for a field trip; (3) preliminary field investigations and collection of site specific data resolving photo-interpretation questions; (4) review of field sites on aerial photos in stereo; (5) stereoscopic photo-interpretation of high-altitude photographs, delineation of wetland boundaries, classification of each polygon, and review of existing wetland information; (6) follow-up field trip if necessary; (7) regional and national consistency quality control of interpreted photos; (8) draft map production; (9) interagency review of draft map, conduct field checking; (10) preparation of edited draft map; and (11) final map production. The cost of map production varies with each map based on the amount of wetlands present in a given area. Costs can be as low as $200 to a high of $1000 for a single 7.5 minute quadrangle covering approximately 50 square miles. Swartwout (1982) and Crowley et al. (1988) evaluated NWI maps and determined that the maps were 95 and 91 percent accurate, respectively. Accuracy determinations included errors of omission and commission. This high accuracy was achieved because the NWI technique involves a combination of field studies, photo-interpretation, use of existing information and interagency review of draft maps.

The NWI has produced wetland maps for 74% of the conterminous United States and 24% of Alaska (Figs 3 and 4). Mapping priorities are based principally on the needs of the FWS and other Federal and State agencies. They include the coastal zone (including the coastline of the Great Lakes), prairie wetlands, playa lakes, floodplains of major rivers and other areas that reflect the goals of the joint US-Canadian North American Waterfowl Management Plan. The actual priority of mapping depends on the availability of funds and the existence of high-quality aerial photography. Obtaining acceptable photographs for the Prairie Potholes region of the US was particularly difficult because of the need to capture optimum water conditions. Consequently, NWI established a special agreement with NASA to obtain that photography. The NWI produces wetland maps at the rate of 5% of the conterminous United States and 2% of Alaska annually. This is the equivalent of 3200 1:24 000-scale quads a year in the conterminous United States and 60 1:63 360-scale quads in Alaska.

The FWS has established a 3-tiered distribution system for NWI maps composed of State-run distribution centers, regional centers, and a national toll free number 1-800-USAMAPS. The State-run tier consists of 29 State-run centers covering 36 States. The second

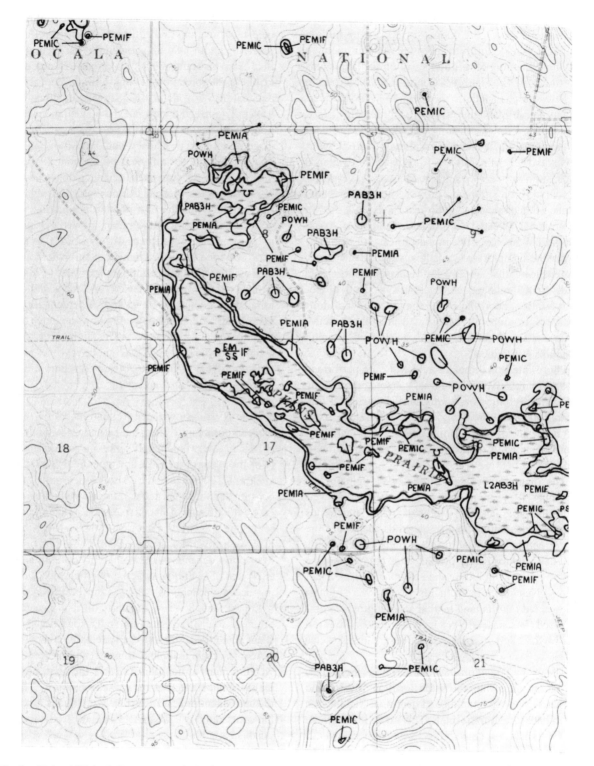

Fig. 1. National Wetlands Inventory map in the State of Florida: close-up of a delineated wetland. Alpha-numeric designations represent wetland classification codes as described in Cowardin *et al.* (1979). Example of alpha-numeric code: PEM1C. P = palustrine; EM = emergent; 1 = persistent; C = seasonally flooded. PAB3H: P = palustrine; AB = aquatic bed; 3 = rooted vascular; H = permanently flooded.

tier consists of regional centers. Information on NWI wetland map availability may be obtained and maps can be ordered through the 6 USGS's Earth Science Information Center regional offices. All these Earth Science Information Center offices have an on-line computer link into the NWI's office in St. Petersburg, Florida to allow greater efficiency of the map ordering process. The third tier is the toll free number from which the user can obtain information on map availability and ordering information. More than 1 560 250 copies of draft and final wetlands maps have been distributed by the NWI. This figure does not include the secondary distribution made through the State-run distribution centers covering Alabama, Arizona, Arkansas, Connecticut, Delaware, Florida, Georgia, Guam, Hawaii, Illinois, Iowa, Kentucky, Louisiana, Maine, Maryland, Massachusetts, Minnesota, Mississippi, Missouri, Nebraska, New Hampshire, New Jersey, New Mexico, New York, North Carolina, Ohio, Oklahoma, Oregon, Pennsylvania, Rhode Island, South Carolina, Tennessee, Texas, Vermont, Washington, West Virginia, and Wyoming.

National Wetlands Inventory digital data base
The NWI is constructing a georeferenced wetland data base using geographic information system (GIS) technologies. Digitizing is done in arc-node format with attributes assigned to the left, center and right sides of each arc. Wetland attributes are coded according to Cowardin *et al.* (1979). As digitization occurs, points are converted to latitude/longitude coordinates. As a result, all map data are stored in a common, ground-based geographic reference system.

To date, almost 9000 NWI maps, representing 16.7% of the continental United States, have been digitized (Fig. 5). Statewide data bases have been built for New Jersey, Delaware, Maryland, Illinois, Washington, and Indiana and are in progress for and Virginia, Minnesota, and South Carolina. NWI digital data also are available for portions of the 25 other States. The graphic map products can be combined with other GIS layered information such as soils and land-use planning, and transportation routes. These digital data are being used for such applications as resource management planning, impact assessment, facility siting, wetland trend analysis and information retrieval. Copies of data base files can be purchased at cost from the NWI Office in St. Petersburg, Florida at telephone (813) 893-3624. The data are provided on magnetic tape in Map Overlay and Statistical System (MOSS) export,

Digital Line Graph 3 (DLG3) optional, and Digital Exchange File (DXF), International Graphic Exchange Standard (IGES), or Geographic Resources Analysis Support System (GRASS) formats. Other digital products available at cost include acreage statistics by quadrangle, county, or study area, and color-coded wetland maps.

Map and digital data: Users and uses
The number of users has grown steadily since the maps were first introduced. Requests are common from individuals, private organizations, industry, consulting firms, developers, agencies from all levels of government (municipal, town, county, State, Federal), and educational/research groups (universities, colleges). User surveys have documented over 100 different uses of the wetland maps. Resource managers in the FWS and in the States are provided with information on wetland location and type, which is essential to effective habitat management and acquisition of important wetland areas. These areas are needed to perpetuate waterfowl populations and other migratory bird populations as called for in the North American Waterfowl Management Plan.

Regulatory agencies use the maps to help in advanced identification, determining wetland values, and mitigation requirements. For example, the USDA uses the maps as a major tool in the identification of wetlands for the administration of the 'Swampbuster' provisions of the 1985 Food Security Act. Copies of more than 74 260 draft and final NWI maps have been sent to the Soil Conservation Service's county offices as of July 1993. Private sector planners use the maps to determine the location and nature of wetlands to aid in framing alternative plans to meet regulatory requirements. These maps are instrumental in preventing problems that arise because the maps eliminate confusion over whether an area is a wetland. They are also instrumental because they provide facts that allow sound business decisions to be made quickly, accurately, and efficiently.

Map of the nation's wetlands
The National Wetlands Inventory has produced a 3.5 × 5.5 feet. color wall map that shows the relative location and abundance wetlands present in the conterminous United States, Hawaii and Puerto Rico. The map is called 'Wetland Resources of the United States' (Dahl 1991) and is at a scale of 1 inch equals 50 miles. The purpose of this color map is to increase the pub-

WETLANDS AND DEEPWATER HABITATS CLASSIFICATION

SYSTEM M — MARINE

SUBSYSTEM 1 — SUBTIDAL

CLASS: RB — ROCK BOTTOM | UB — UNCONSOLIDATED BOTTOM | AB — AQUATIC BED | RF — REEF | OW — OPEN WATER/ Unknown Bottom

Subclass:
- RB — ROCK BOTTOM: 1 Bedrock; 2 Rubble
- UB — UNCONSOLIDATED BOTTOM: 1 Cobble-Gravel; 2 Sand; 3 Mud; 4 Organic
- AB — AQUATIC BED: 1 Algal; 3 Rooted Vascular; 5 Unknown Submergent
- RF — REEF: 1 Coral; 3 Worm

SUBSYSTEM 2 — INTERTIDAL

CLASS: AB — AQUATIC BED | RF — REEF | RS — ROCKY SHORE | US — UNCONSOLIDATED SHORE

Subclass:
- AB — AQUATIC BED: 1 Algal; 3 Rooted Vascular; 5 Unknown Submergent
- RF — REEF: 1 Coral; 3 Worm
- RS — ROCKY SHORE: 1 Bedrock; 2 Rubble
- US — UNCONSOLIDATED SHORE: 1 Cobble-Gravel; 2 Sand; 3 Mud; 4 Organic

SYSTEM E — ESTUARINE

SUBSYSTEM 1 — SUBTIDAL

CLASS: RB — ROCK BOTTOM | UB — UNCONSOLIDATED BOTTOM | AB — AQUATIC BED | RF — REEF | OW — OPEN WATER/ Unknown Bottom

Subclass:
- RB — ROCK BOTTOM: 1 Bedrock; 2 Rubble
- UB — UNCONSOLIDATED BOTTOM: 1 Cobble-Gravel; 2 Sand; 3 Mud; 4 Organic
- AB — AQUATIC BED: 1 Algal; 3 Rooted Vascular; 4 Floating Vascular; 5 Unknown Submergent; 6 Unknown Surface
- RF — REEF: 2 Mollusc; 3 Worm

SUBSYSTEM 2 — INTERTIDAL

CLASS: AB — AQUATIC BED | RF — REEF | SB — STREAMBED | RS — ROCKY SHORE | US — UNCONSOLIDATED SHORE | EM — EMERGENT | SS — SCRUB-SHRUB | FO — FORESTED

Subclass:
- AB — AQUATIC BED: 1 Algal; 3 Rooted Vascular; 4 Floating Vascular; 5 Unknown Submergent; 6 Unknown Surface
- RF — REEF: 2 Mollusc; 3 Worm
- SB — STREAMBED: 1 Cobble-Gravel; 2 Sand; 3 Mud; 4 Organic
- RS — ROCKY SHORE: 1 Bedrock; 2 Rubble
- US — UNCONSOLIDATED SHORE: 1 Cobble-Gravel; 2 Sand; 3 Mud; 4 Organic
- EM — EMERGENT: 1 Persistent; 2 Nonpersistent
- SS — SCRUB-SHRUB: 1 Broad-Leaved Deciduous; 2 Needle-Leaved Deciduous; 3 Broad-Leaved Evergreen; 4 Needle-Leaved Evergreen; 5 Dead; 6 Deciduous; 7 Evergreen
- FO — FORESTED: 1 Broad-Leaved Deciduous; 2 Needle-Leaved Deciduous; 3 Broad-Leaved Evergreen; 4 Needle-Leaved Evergreen; 5 Dead; 6 Deciduous; 7 Evergreen

SYSTEM R — RIVERINE

SUBSYSTEM 1 — TIDAL

CLASS: RB — ROCK BOTTOM | UB — UNCONSOLIDATED BOTTOM | AB — AQUATIC BED | *SB — STREAMBED | RS — ROCKY SHORE | US — UNCONSOLIDATED SHORE | **EM — EMERGENT | OW — OPEN WATER/ Unknown Bottom

Subclass:
- RB — ROCK BOTTOM: 1 Bedrock; 2 Rubble
- UB — UNCONSOLIDATED BOTTOM: 1 Cobble-Gravel; 2 Sand; 3 Mud; 4 Organic
- AB — AQUATIC BED: 1 Algal; 2 Aquatic Moss; 3 Rooted Vascular; 4 Floating Vascular; 5 Unknown Submergent; 6 Unknown Surface
- SB — STREAMBED: 1 Bedrock; 2 Rubble; 3 Cobble-Gravel; 4 Sand; 5 Mud; 6 Organic; 7 Vegetated
- RS — ROCKY SHORE: 1 Bedrock; 2 Rubble
- US — UNCONSOLIDATED SHORE: 1 Cobble-Gravel; 2 Sand; 3 Mud; 4 Organic; 5 Vegetated
- EM — EMERGENT: 2 Nonpersistent

SUBSYSTEM 2 — LOWER PERENNIAL / 3 — UPPER PERENNIAL / 4 — INTERMITTENT / 5 — UNKNOWN PERENNIAL

*STREAMBED is limited to TIDAL and INTERMITTENT SUBSYSTEMS, and comprises the only CLASS in the INTERMITTENT SUBSYSTEM.

**EMERGENT is limited to TIDAL and LOWER PERENNIAL SUBSYSTEMS.

Classification of Wetlands and Deepwater Habitats of the United States
Cowardin ET AL. 1979 as modified for National Wetland Inventory Mapping Convention

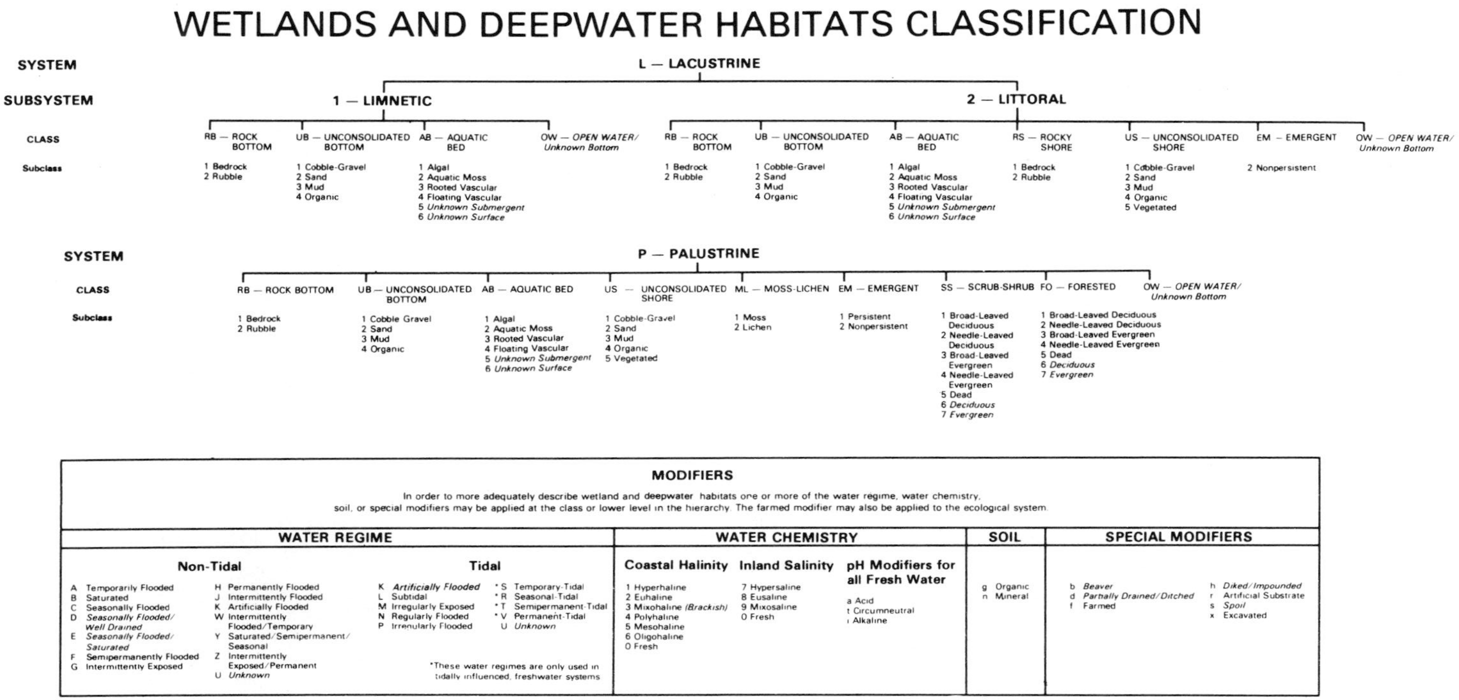

161

WETLANDS AND DEEPWATER HABITATS CLASSIFICATION

L — LACUSTRINE

1 — LIMNETIC

SYSTEM

SUBSYSTEM

CLASS

RB — ROCK BOTTOM	UB — UNCONSOLIDATED BOTTOM	AB — AQUATIC BED	OW — *OPEN WATER/ Unknown Bottom*

Subclass

RB — ROCK BOTTOM
1 Bedrock
2 Rubble

UB — UNCONSOLIDATED BOTTOM
1 Cobble-Gravel
2 Sand
3 Mud
4 Organic

AB — AQUATIC BED
1 Algal
2 Aquatic Moss
3 Rooted Vascular
4 Floating Vascular
5 *Unknown Submergent*
6 *Unknown Surface*

2 — LITTORAL

RB — ROCK BOTTOM	UB — UNCONSOLIDATED BOTTOM	AB — AQUATIC BED	RS — ROCKY SHORE	US — UNCONSOLIDATED SHORE	EM — EMERGENT	OW — *OPEN WATER/ Unknown Bottom*

RB — ROCK BOTTOM
1 Bedrock
2 Rubble

UB — UNCONSOLIDATED BOTTOM
1 Cobble-Gravel
2 Sand
3 Mud
4 Organic

AB — AQUATIC BED
1 Algal
2 Aquatic Moss
3 Rooted Vascular
4 Floating Vascular
5 *Unknown Submergent*
6 *Unknown Surface*

RS — ROCKY SHORE
1 Bedrock
2 Rubble

US — UNCONSOLIDATED SHORE
1 Cobble-Gravel
2 Sand
3 Mud
4 Organic
5 Vegetated

EM — EMERGENT
2 Nonpersistent

P — PALUSTRINE

SYSTEM

CLASS

RB — ROCK BOTTOM	UB — UNCONSOLIDATED BOTTOM	AB — AQUATIC BED	US — UNCONSOLIDATED SHORE	EM — EMERGENT	ML — MOSS-LICHEN	SS — SCRUB-SHRUB	FO — FORESTED	OW — *OPEN WATER/ Unknown Bottom*

Subclass

RB — ROCK BOTTOM
1 Bedrock
2 Rubble

UB — UNCONSOLIDATED BOTTOM
1 Cobble-Gravel
2 Sand
3 Mud
4 Organic

AB — AQUATIC BED
1 Algal
2 Aquatic Moss
3 Rooted Vascular
4 Floating Vascular
5 *Unknown Submergent*
6 *Unknown Surface*

US — UNCONSOLIDATED SHORE
1 Cobble-Gravel
2 Sand
3 Mud
4 Organic
5 Vegetated

EM — EMERGENT
1 Persistent
2 Nonpersistent

ML — MOSS-LICHEN
1 Moss
2 Lichen

SS — SCRUB-SHRUB
1 Broad-Leaved Deciduous
2 Needle-Leaved Deciduous
3 Broad-Leaved Evergreen
4 Needle-Leaved Evergreen
5 Dead
6 *Deciduous*
7 *Evergreen*

FO — FORESTED
1 Broad-Leaved Deciduous
2 Needle-Leaved Deciduous
3 Broad-Leaved Evergreen
4 Needle-Leaved Evergreen
5 Dead
6 *Deciduous*
7 *Evergreen*

MODIFIERS

In order to more adequately describe wetland and deepwater habitats one or more of the water regime, water chemistry, soil, or special modifiers may be applied at the class or lower level in the hierarchy. The farmed modifier may also be applied to the ecological system.

WATER REGIME

Non-Tidal

A Temporarily Flooded
B Saturated
C Seasonally Flooded
D *Seasonally Flooded/ Well Drained*
E *Seasonally Flooded/ Saturated*
F Semipermanently Flooded
G Intermittently Exposed
H Permanently Flooded
J Intermittently Flooded
K Artificially Flooded
W *Intermittently Flooded/Temporary*
Y *Saturated/Semipermanent/ Seasonal*
Z *Intermittently Exposed/Permanent*
U *Unknown*

Tidal

K *Artificially Flooded*
L Subtidal
M Irregularly Exposed
N Regularly Flooded
P Irregularly Flooded
S Temporary-Tidal
R Seasonal-Tidal
T Semipermanent-Tidal
V Permanent-Tidal
U *Unknown*

*These water regimes are only used in tidally influenced freshwater systems

WATER CHEMISTRY

Coastal Halinity
1 Hyperhaline
2 Euhaline
3 Mixohaline (Brackish)
4 Polyhaline
5 Mesohaline
6 Oligohaline
0 Fresh

Inland Salinity
7 Hypersaline
8 Eusaline
9 Mixosaline
0 Fresh

pH Modifiers for all Fresh Water
a Acid
t Circumneutral
i Alkaline

SOIL
g Organic
n Mineral

SPECIAL MODIFIERS
b *Beaver*
d *Partially Drained/Ditched*
f Farmed
h *Diked/Impounded*
r Artificial Substrate
s *Spoil*
x Excavated

Fig. 2. National Wetlands Inventory map legend

STATUS OF NATIONAL WETLANDS INVENTORY

OCTOBER 1992

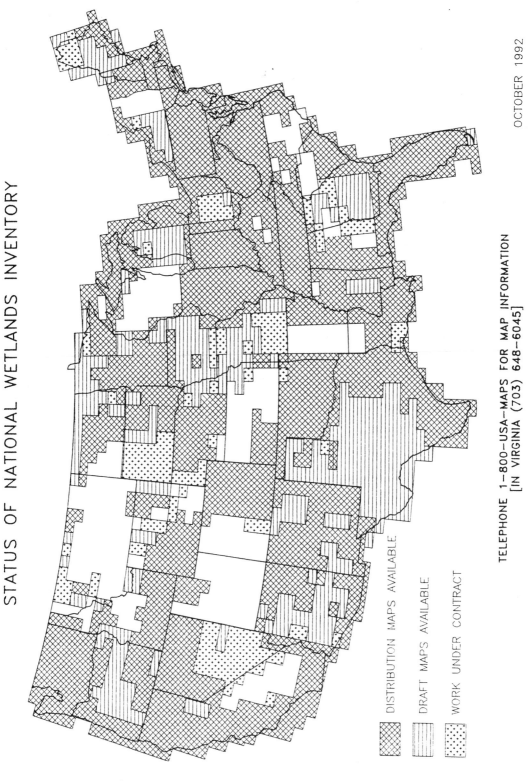

TELEPHONE 1-800-USA-MAPS FOR MAP INFORMATION
[IN VIRGINIA (703) 648-6045]

DISTRIBUTION MAPS AVAILABLE

DRAFT MAPS AVAILABLE

WORK UNDER CONTRACT

Fig. 3. Status of the National Wetlands Inventory: conterminous United States

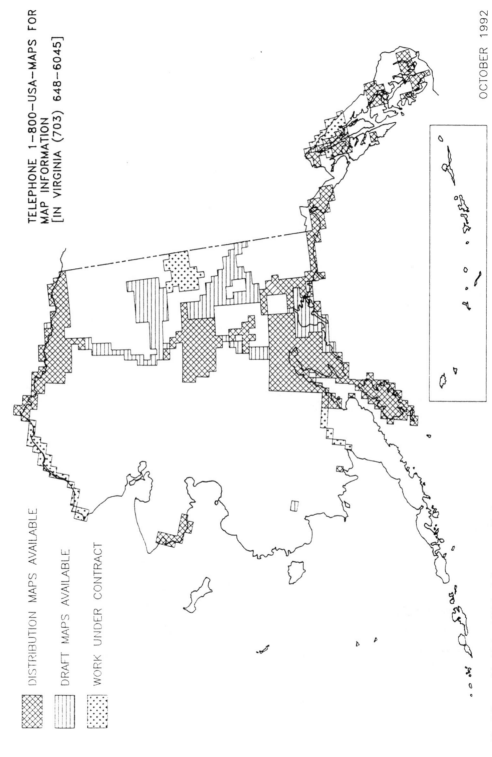

STATUS OF NATIONAL WETLANDS INVENTORY IN ALASKA

TELEPHONE 1—800—USA—MAPS FOR
MAP INFORMATION
[IN VIRGINIA (703) 648—6045]

OCTOBER 1992

DISTRIBUTION MAPS AVAILABLE

DRAFT MAPS AVAILABLE

WORK UNDER CONTRACT

Fig. 4. Status of the National Wetlands Inventory: Alaska

164

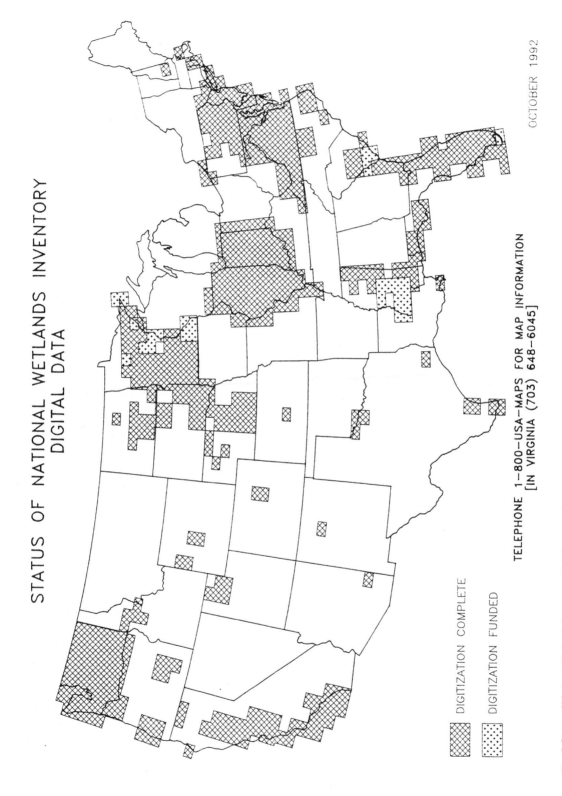

Fig. 5. Status of National Wetlands Inventory: digital map products

lic's understanding of our nation's wetlands. The map was compiled in cooperation with the Water Resources Division of the USGS. A companion map of Alaska's wetland resources has also been published. It is called 'Wetlands Resource Map of Alaska' (Hall 1991). This map is also a color wall sized map measures approximately 3 × 4 feet and shows Alaska's wetlands at a scale of 1 inch equals 40 miles. Copies of both the United States and Alaska maps are available from the USGS's Earth Science Information Centers. For further ordering information, call 1-800-USA-MAPS.

National wetland status and trends study

The national wetlands status and trends analysis study originated from the need for national estimates on the present extent of our Nation's wetland resources in the conterminous United States, and on corresponding losses and gains over the past 20 years. A statistical survey of United States wetlands in the mid-1950's and mid-1970's was conducted through conventional air photo-interpretation techniques. The status of wetlands in the mid-1950's and mid-1970's was determined, and estimates of losses and gains during that interval were computed. The national sampling grid consists of a stratified random sample of 3635 permanent, 4-square-mile plots distributed within strata being formed by State boundaries, and the 35 physical subdivisions described by Hammond (1965). Additional strata were added to include: (1) a coastal zone stratum encompassing estuarine wetlands and, (2) the area immediately adjacent to the Great Lakes. Sample units were allocated to strata in proportion to the expected amount of wetland and deepwater habitat acreage estimated as determined by the earlier work of Shaw and Fredine (Shaw & Fredine 1956). The results of this study were published in four major reports – 'Status and Trends of Wetlands and Deepwater Habitats in the Conterminous United States, 1950's to 1970's' (Frayer *et al.* 1983); 'Wetlands of the United States: Current Status and Recent Trends' (Tiner 1984); 'Wetlands: Their Use and Regulation' (US Congress, Office of Technology Assessment 1984); and 'The Impact of Federal Programs on Wetlands: Volume I. The Lower Mississippi Alluvial Plain and the Prairie Pothole Region' (Goldstein 1988).

Results of the mid-1970's to mid-1980's study

The following information on study results is taken from Dahl and Johnson (1991). In the mid-1970's, there were an estimated 105.9 million acres of wetlands in the conterminous United States. In the mid-1980's, an estimated 103.3 million acres of wetlands remained. These data indicate a net loss of 2.6 million acres over the nine-year study period. The study design recognized that aerial photography is not available in each successive year for the same plot or necessarily in the same year for all plots. For these reasons, estimates of average annual rates of wetland loss have not been developed by this study. One possible way of calculating an average annual net loss of wetlands for the study period would be to use the wetland acreage estimate for the mid-1980's (1983) minus the acreage estimate for the mid-1970's (1974) and divide by the nine-year study period. Using this method, the average annual loss of wetlands for this period would be approximately 290 thousand acres per year. By comparison, during the mid-1950's to mid-1970's study period, the average annual net loss was 458 thousand acres per year.

Of the remaining wetland acreage in the conterminous United States, 97.8 million acres or 95.0 percent were freshwater wetlands. Another 5.5 million acres (5.0 percent) were estuarine wetlands. The acreage of deepwater habitats was also included in this study. There were an estimated 63.0 million acres of deepwater habitat in the lacustrine (lake) and riverine systems in the mid-1980's. This represents an increase of 271.2 thousand acres from the mid-1970's estimate and was primarily due to the construction of reservoirs and lakes in the States of Alabama, Florida, Georgia, Mississippi, and South Carolina. If wetlands and deepwater acres were combined, about 9.3 percent of the land area in the conterminous United States is made up of these areas.

Wetland losses from the mid-1970's to the mid-1980's were more evenly distributed between agricultural land use (54 percent) and 'other' land use (41 percent) than losses from the 1950's to the 1970's. A substantial portion of the lands classified as 'other' were wetlands that had been cleared and drained, but not yet put to an identifiable use. Conversion to urban land uses (5 percent) were responsible for net loss of an estimated 59.9 thousand acres of palustrine forested wetlands, 37.5 thousand acres of palustrine emergent wetlands, and 21.0 thousand acres of palustrine scrub-shrub wetlands, about 5.0 percent of the total wetlands loss. Overall, wetland acreage in the mid-1980's constituted 5.0 percent of the land area of the conterminous United States.

Comparison with the mid-1950's to mid-1970's shows that the acreage of wetlands continued to decline, at about two-thirds of the loss rate measured from the 1950's to the 1970's. During the study period covering the mid-1950's to mid-1970's, a net loss of 9 million acres occurred. There is a substantial decrease in rates of wetland loss documented previously from the mid-1950's to the mid-1970's in which agricultural conversion represented 87.0 percent of all wetland losses, 'other' development caused 5 percent of the losses, and urban development accounted for 8 percent of the losses.

The acreage of estuarine wetlands declined 1.0 percent between the mid-1970's and the mid-1980's. Losses in the estuarine system were evidenced by the decrease in estuarine vegetated wetlands, which declined by 71.0 thousand acres. The majority of these losses occurred in the Gulf Coast States, and most of the loss (about 57 percent) was due to change from emergent wetlands to open salt water (bays). Estuarine nonvegetated wetlands increased by an estimated 11.6 thousand acres from the mid-1970's to the mid-1980's. Similarly, between the mid-1950's to mid-1970's estuarine wetland losses were heaviest in the Gulf States of Louisiana, Florida, and Texas. During that time most of Louisiana's coastal marsh losses were attributed to submergence by coastal waters. In other areas, urban development was the major direct human-induced cause of coastal wetland loss.

By the 1980's, there were significant differences in the status of freshwater and estuarine wetlands based on vegetative cover type: 73.1 percent of all coastal wetlands were estuarine emergent whereas inland 52.9 percent of freshwater wetlands were forested. Freshwater emergent marshes and shrubs made up 25.1 and 15.7 percent of all freshwater wetlands, respectively. Freshwater wetlands experienced 98.0 percent of the losses that occurred during the study period. By the mid-1980's, an estimated 97.8 million acres of freshwater wetlands and 5.5 million acres of estuarine wetlands remained.

Palustrine forested wetlands suffered the biggest loss during the study period. An estimated 3.4 million acres were converted, primarily in the southern portion of the country, representing an annual net loss of 378.2 thousand acres. Over 2.1 million acres of these wetlands were converted to non-wetland land uses, including about 1.0 million acres that were lost to agri-

culture. Most of the remaining acreage was converted from forested wetland to other wetland categories.

Palustrine emergent wetlands increased by 220.2 thousand acres during the nine-year study period, despite significant losses. About 375.2 thousand acres of emergent wetlands were converted to agricultural land uses, 151.2 thousand acres were converted to 'other' land uses, and 37.5 thousand acres were converted to urban land uses. An additional 49.1 thousand acres of emergent wetlands were converted to non-vegetated wetlands. At the same time, 722.2 thousand acres of forested wetlands and 68.6 thousand acres of scrub-shrub wetlands were converted to emergent wetlands, more than offsetting the losses in emergent wetland acreage.

About 249.0 thousand acres of palustrine scrub-shrub wetlands were converted to agricultural land uses and 265.0 thousand acres were converted to 'other' land uses. These losses were partially offset by the conversion of 482.8 thousand acres of forested wetlands to scrub-shrub wetlands, resulting in a net loss of 161.1 thousand acres of scrub-shrub wetlands. During the mid-1950's to mid-1970's scrub-shrub wetlands were hardest hit in North Carolina, where pocosins in wetlands were being converted to cropland, pine plantations, or mined for peat.

Palustrine nonvegetated wetlands increased by 794.0 thousand acres. There were 6.1 million acres of palustrine nonvegetated wetlands in the mid-1980's. Gains in this wetlands category, which were well distributed throughout the conterminous United States, totalled 792.4 thousand acres. Almost all of this increase occurred in palustrine unconsolidated bottoms (primarily ponds) and mainly resulted from ponds built on former upland areas.

Other national wetlands inventory products

Hydric (wetland), soils list
Hydric soils are defined by soil saturation for a significant period or by frequent flooding for long periods during the growing season. To clarify the meaning of 'hydric soils', the NWI, in cooperation with the USDA Soil Conservation Service, developed the first list the Nation's hydric soils. Since then, the Soil Conservation Service has chaired of the Interagency National Technical Committee for Hydric Soils. The 'National List of Hydric Soils of the United States' (USDA, Soil Conservation Service 1991) is available from the Soil

Conservation Service. This soils list is useful for making wetland determinations in the field, or in the office through use of soil survey maps.

List of plants that occur in wetlands
The FWS published the 'National List of Plants Species that Occur in Wetlands: 1988 National Summary' (Reed 1988). The plants in the list are divided into four indicator categories based on plant's frequency of occurrence in wetlands: (1) obligate – always found in wetlands more than 99% of the time); (2) facultative wet – usually found in wetlands (66–99% of the time); (3) facultative – sometimes found in wetlands (33–66%); and (4) facultative upland – seldom found in wetlands (less than 33%). This list is available from the Superintendent of Documents, US Government Printing Office, Washington, DC, 20402, telephone (202) 783-3238. When ordering use Stock Number 024-010-00682-0. Thirteen regional subdivisions of the national wetland plant list as well as individual State lists are available from the National Technical Information Service, 5285 Port Royal Rd., Springfield, Virginia, 22161, telephone (703) 487-4650.

Two wetland plant list data bases have been developed based on the National List of Plant Species that Occur in Wetlands. The first is the wetland plant list data base is a listing of plants associated with wetlands, as defined by the FWS's wetland definition and classification system (Cowardin *et al.* 1979). It lists scientific and common names of plants, distribution, and regional wetland indicator status of almost 6700 species. It can be accessed by plant name, region, State, and wetland indicator status. The data base is updated as additional information is received. Regional and State subdivisions of the wetland plant list data base are available on floppy disks in ASCII format for use on IBM XT/AT-compatible computers running the equivalent of MS-DOS 2.0 or higher. Contact BIO-DATA, Inc., 13950 West 20th Ave., Golden, Colorado, 80401, telephone (303) 278-1046.

The second is the wetland plant species data base which is comprised of two parts. The first part, PLANTS, contains detailed taxonomic, distributional and habitat information on more than 6200 wetland plants found in the United States and its territories. The second part, BOOKS, contains bibliographic citations for more than 280 sources such as floras, checklists, and botanical manuals used to compile PLANTS.

Wetland reports
Two basic wetland reports are developed by NWI: map reports and State wetland reports. The map reports briefly outline NWI procedures and findings (e.g., list of wetland plant communities, photo-interpretation problems). Map reports are available for all mapped area. By contrast, the State wetland report is a comprehensive publication on the results of wetlands inventory in a given State. It is prepared upon completion of the wetlands acreage summary in a State. The State report includes wetland statistics and detailed discussions of NWI techniques, wetland plant communities, hydric soils and wetland values. To date, State reports have been produced for New Jersey, Delaware, Rhode Island, Connecticut, Maryland, Pennsylvania and Florida. NWI expects to prepare reports for Hawaii, Washington, Indiana, Illinois and Alaska when statistics become available.

Wetland values data base
The Wetland Values Data Base is a bibliographic listing of nearly 15 000 scientific articles about the functions and values of wetlands. It is intended to support the Fish and Wildlife Service's efforts to identify and map wetlands. Field names include author, year, hydrologic unit number (USGS/Water Resources Council geographic area descriptor), land surface form, location, State, US Army Corps of Engineers District Codes, wetland classification (Cowardin *et al.* 1979), ecoregion codes (Bailey 1980), and subject keywords. For further information or to request a search of the data base contact: Wetland Values Data Base Administrator, US Fish & Wildlife Service, National Wetlands Inventory, 9720 Executive Center Dr. Monroe Bld. Suite 101, St. Petersburg, Florida, 33702-2440, telephone (813) 893-3865.

Acknowledgments

We wish to thank all the staff and contract personnel of the NWI for their work over the last 14 years. The accomplishments presented in this paper could not have been achieved without the financial support of many Federal, State and local cooperators. The accuracy of the maps, in good part is the result of the voluntary map review by many Federal, State, local and private sector agencies and organizations, as well as persons such as George Fore of Texas.

References

Bailey, R. G. 1980. Description of the Ecoregions of the United States. US Department of Agriculture, Washington, DC, Miscellaneous Publication No. 1391. 77 pp.

Cowardin, L. M., Carter, V., Golet, F. C. & LaRoe, E. T. 1979. Classification of Wetlands and Deepwater Habitats of the United States. US Fish and Wildlife Service, Washington, DC, FWS/OBS–79/31. 130 pp.

Crowley, S., O'Brien, C. & Shea, S. 1988. Results of the Wetland Study and the 1988 Draft Wetland Rules. Report by the Agency of Natural Resources Divisions of Water Quality, Waterbury, Vermont. 33 pp.

Dahl, T. E. 1990. Wetlands Losses in the United States 1780's to 1980's. US Fish and Wildlife Service, National Wetlands Inventory Project, Washington, DC 21 pp.

Dahl, T. E. 1991. Wetlands Resources of the United States. US Fish and Wildlife Service, National Wetlands Inventory Project, Washington, DC, 3.5 × 5.5 ft. color wall map.

Dahl, T. E. & Johnson, C. E. 1991. Wetlands Status and Trends in the Conterminous United States, Mid-1970's to Mid-1980's. US Fish and Wildlife Service, National Wetlands Inventory Project, Washington, DC 28 pp.

Federal Geographic Data Committee. 1992. Application of Satellite Data for Mapping and Monitoring Wetlands – Fact Finding Report; Technical Report 1. Wetlands Subcommittee, FGDC, Washington, DC 32 pp. plus Appendices.

Frayer, W. E., Monahan, T. J., Bowden, D. C. & Graybill, F. A. 1983. Status and Trends of Wetlands and Deepwater Habitats in the Conterminous United States, 1950's to 1970's. Department of Forest and Wood Sciences, Colorado State University, Ft. Collins, Colorado. 32 pp.

Goldstein, J. H. 1988. The Impact of Federal Programs on Wetlands – Volume I: The Lower Mississippi Alluvial Plain and the Prairie Pothole Region. Report to Congress by the Secretary of the Interior. US Department of the Interior, Washington, DC 114 pp.

Gopal, B., Turner, R. E., Wetzel, R. G. & Whigham, D. F. 1982. Wetlands Ecology and Management. Proceedings of the First International Wetlands Conference (September 10–17, 1980; New Delhi, India). National Institute of Ecology and International Scientific Publications. Jaipur, India. 514 pp.

Hall, J. V. 1991. Wetland Resources of Alaska. US Fish and Wildlife Service, National Wetlands Inventory Project, Anchorage, Alaska. 3 × 4 ft. color wall map.

Hammond, E. H. 1965. Physical subdivisions of the United States. pp. 61–64. In: The National Atlas of the United States of America. 1970. US Geological Survey, Washington, DC 417 pp.

Reed, P. B. Jr. 1988. National List of Plant Species that Occur in Wetlands: National Summary. US Fish and Wildlife Service, National Wetlands Inventory Project, Washington, DC, Biological Report 88 (24). 244 pp.

Shaw, S. P. & Fredine, C. G. 1956. Wetlands of the United States. US Fish and Wildlife Service, Washington, DC Circular 39. 67 pp.

Swartwout, D. J. 1982. An evaluation of National Wetlands Inventory in Massachusetts. Unpublished master's thesis, University of Massachusetts, Amherst, Massachusetts. 123 pp.

Tiner, R. W. Jr. 1984. Wetlands of the United States: Current Status and Recent Trends US Fish and Wildlife Service, National Wetlands Inventory Project, Newton Corner, Massachusetts. 59 pp.

US Congress, Office of Technology Assessment, 1984. Wetlands: Their Use and Regulation. Office of Technology Assessment, Washington, DC, OTA-0-026. 208 pp.

US Department of Agriculture, Soil Conservation Service. 1991. Hydric Soils of the United States (3rd ed.). In cooperation with the National Technical Committee for Hydric Soils. US Department of Agriculture, Soil Conservation Service, Washington, DC Miscellaneous Publication No. 1491, pages not numbered.

US Department of Interior, Fish and Wildlife Service. 1976. Existing State and Local Wetlands Surveys 1965–75. 2 Volumes: Vol. I – map atlas; Vol. II – narrative profile of each inventory. US Fish and Wildlife Service, Office of Biological Services, Washington, DC

Other publications of interest

Frayer, W. E., Peters, D. D. & Pywell, H. R. 1989. Wetlands of the California Central Valley, Status and Trends, 1939 to Mid-1980's. US Fish and Wildlife Service, National Wetlands Inventory Project, Portland, Oregon. 28 pp.

Frayer, W. E. & Hefner, J. M. 1991. Florida Wetlands: Status and Trends, 1970's to 1980's. US Fish and Wildlife Service, National Wetlands Inventory Project, Atlanta, Georgia. 31 pp.

Metzler, K. & Tiner, Jr., R. W. 1991. Wetlands of Connecticut. State Geological & Natural History Survey of Connecticut, Dept. of Environmental Protection, Hartford Conn. and US Fish and Wildlife Service, National Wetlands Inventory Project, Newton Corner, Massachusetts. Cooperative publication. 115 pp.

Peters, D. D. & Browers, H. W. 1991. Atlas of National Wetlands Inventory for Marin County, California. US Fish and Wildlife Service, National Wetlands Inventory Project, Portland, Oregon. 30 pp.

Tiner, R. W. Jr. 1985. Wetlands of Delaware. US Fish and Wildlife Service, National Wetlands Inventory Project, Newton Corner, Massachusetts and Delaware Dept. of Natural Resources & Environmental Control – Wetlands Section, Dover Delaware. Cooperative publication. 77 pp.

Tiner, R. W. Jr. 1985. Wetlands of New Jersey. US Fish and Wildlife Service, National Wetlands Inventory Project, Newton Corner, Massachusetts. 117 pp.

Tiner, R. W. Jr. 1989. Wetlands of Rhode Island. US Fish and Wildlife Service, National Wetlands Inventory Project, Newton Corner, Massachusetts. 71 pp. plus Appendix.

Tiner, R. W. Jr. 1990. Pennsylvania's Wetlands: Current Status and Recent Trends. US Fish and Wildlife Service, National Wetlands Inventory Project, Newton Corner, Massachusetts. 104 pp.

Tiner, R. W. Jr. 1990. Preliminary NWI wetland acreage reports for Massachusetts (1989) Connecticut (1988), and Vermont (1987). US Fish and Wildlife Service, National Wetlands Inventory Project, Newton Corner, Massachusetts. 104 pp.

Tiner, R. W. Jr. 1992. Field Guide to Coastal Wetland Plants of the Southeastern United States. University of Massachusetts Press, Amherst, Massachusetts. 285 pp.

Tiner, R. W. Jr. & Burke, D. G. 1992. Wetlands of Maryland. Maryland Dept. of Natural Resources – Water Resources Administration, Annapolis, Maryland and US Fish and Wildlife Service, National Wetlands Inventory Project, Newton Corner, Massachusetts. Cooperative publication. (in progress)

Tiner, R. W. Jr. & Finn, J. T. 1986. Status and Recent Trends of Wetlands in Five Mid-Atlantic States: Delaware, Maryland, Pennsylvania, Virginia, and West Virginia. US Fish and Wildlife Service, National Wetlands Inventory Project, Newton Corner, Massachusetts and US Environmental Protection Agency, Philadelphia, Pennsylvania. Cooperative publication. 40 pp.

Tiner, R. W. Jr., Stone, J. & Gookin, J. 1989. Current Status and Recent Trends in Wetlands in Central Connecticut. US Fish and Wildlife Service, National Wetlands Inventory Project, Newton Corner, Massachusetts. 9 pp.

Tiner, R. W. Jr. & Zinni, W. 1988. Recent Wetland Trends in Southeastern Massachusetts. US Fish and Wildlife Service, National Wetlands Inventory Project, Newton Corner, Massachusetts. 9 pp.

US Department of Interior, Fish and Wildlife Service. 1990. Cartographic Conventions for the National Wetlands Inventory. US Fish and Wildlife Service, National Wetlands Inventory Project, St. Petersburg, Florida. 73 pp.

US Department of the Interior, Fish and Wildlife Service. 1990. Photointerpretation Conventions for the National Wetlands Inventory. US Fish and Wildlife Service, National Wetlands Inventory Project, St. Petersburg, Florida. 45 pp. plus Appendices.

US Department of the Interior, Fish and Wildlife Service. 1992. Digitizing Conventions for the National Wetlands Inventory. US Fish and Wildlife Service, National Wetlands Inventory Project, St. Petersburg, Florida. 21 pp. plus attachments.

US Department of the Interior, Fish and Wildlife Service. 1993. Continuous Wetland Trend Analysis Project Specifications (Manual Interpretation and Measurement Procedures). US Fish and Wildlife Service, National Wetlands Inventory Project, St. Petersburg, Florida. 45 pp.

Wilen, B. O. & Tiner, R. W. 1993, Wetlands of the United States. pp. 515–636. In: Wetlands of the World: Inventory, ecology and management (Volume I), D. F. Whigham et al. (eds.) 1993. Kluwer Academic Publishers, Dordrecht, The Netherlands. 768 pp.

Vegetatio **118:** 171–184, 1995.
© 1995 *Kluwer Academic Publishers. Printed in Belgium.*

EMAP-Wetlands: A sampling design with global application *

R.P. Novitzki
Man Tech Environmental Technology, Inc., US EPA Environmental Research Laboratory, 200 SW 35th Street, Corvallis, OR 97330, USA

Key words: EMAP-wetlands, Monitoring, Probability sample design, Wetland classification, Wetlands

Abstract

The U.S. Environmental Protection Agency (EPA) initiated the Environmental Monitoring and Assessment Program (EMAP) in 1988. The wetland component (EMAP-Wetlands) is designed to provide quantitative assessments of the current status and long-term trends in the ecological condition of wetland resources. EMAP-Wetlands will develop a wetland monitoring network and will identify and evaluate indicators that describe and quantify wetland condition. The EMAP-Wetlands network will represent a probability sample of the total wetland resource. The EMAP sample is based on a triangular grid of approximately 12,600 sample points in the conterminous U.S. The triangular grid adequately samples wetland resources that are common and uniformly distributed in a region, such as the prairie pothole wetlands of the Midwest. However, the design is flexible and allows the base grid density to be increased to adequately sample wetland resources, such as the coastal wetlands of the Gulf of Mexico, which are distributed linearly along the coast. The Gulf sample network required a 49-fold increase in base grid density. EMAP-Wetlands aggregates the 56 U.S. Fish and Wildlife Service's (FWS) National Wetland Inventory (NWI) categories (Cowardin *et al.* 1979) into 12 functionally similar groups (Leibowitz *et al.* 1991). Both the EMAP sample design and aggregated wetland classes are suitable for global inventory and assessment of wetlands.

Introduction

This paper provides a brief introduction to the EMAP-Wetlands Program and describes the basic EMAP sampling design as applied to the Nation's wetland resources. It provides a brief discussion of the EMAP-Wetland classes (aggregated from NWI classes) and the EMAP probability sample design, and demonstrates application of the EMAP sample design to developing a wetland sampling/monitoring network for Gulf coast salt marshes and Midwest prairie pothole wetlands.

The U.S. Environmental Protection Agency (EPA) initiated the Environmental Monitoring and Assess-

ment Program (EMAP) in 1988. The program is designed to provide information on the current status and long-term trends in the extent and condition of the Nation's ecological resources. The program includes seven broad resource categories: near-coastal waters, the Great Lakes, surface waters, wetlands, forests, arid lands, and agroecosystems. A coordinated monitoring network and series of indicator measurements to assess the condition of the resource are being developed independently for each category. However, a major strength of the program is that ultimately the assessments of individual resources will be combined into landscape-level assessments of ecological resources.

The overall goal of EMAP-Wetlands is to provide a quantitative assessment of the current status and long-term trends in wetland condition. EMAP will coordinate this activity with the U.S. Fish and Wildlife Service (FWS) National Wetland Inventory (NWI) program to produce joint reports assessing the extent and condition of the Nation's wetlands. To accomplish that

---
* The research described in this report has been funded by the U.S. Environmental Protection Agency. This document has been prepared at the EPA Environmental Research Laboratory in Corvallis, OR, through contract No. 68-C8-0006 to Man Tech Environmental Technology, Inc. This paper has been subjected to the Agency's peer and administrative review and approved for publication. Mention of trade names or commercial products does not constitute endorsement or recommendation for use.

goal it is necessary to define the resource, select indicators of resource condition, and establish a monitoring network. Condition will be defined by a series of measurements or observations (indicators) of wetland characteristics, used individually or in combination (e.g., indices), and compared to ranges of similar measures/indices defined by reference wetlands in the region. The assessments of wetland condition in a region will be available to the scientific community and wetland regulators/managers as soon after collection as practical. This information will be valuable for assessing the impact on wetland resources of the "no net loss" policy, the Conservation Reserve Program, regulatory decisions on wetland mitigation or mitigation banking, and other environmental policy decisions that may impact wetland condition. The information will also be valuable for assessing the success of wetland creation, restoration, or mitigation projects by providing a basis for comparing project site characteristics to those of the wetland population in the region.

EMAP-Wetlands uses the wetland classification developed by Cowardin *et al.* (1979), aggregated to reduce the number of categories (see discussion in CLASSIFICATION section). The EMAP sample design and classification approach adequately defines the distribution and extent of all but rare wetland resources (those representing less than one percent of the total resource). The number and acreage of common wetland classes predicted by the EMAP sample of 1/16th of the land surface (see discussion in SAMPLING DESIGN section) compares closely to the total wetland resource as mapped by NWI (Liebowitz et al. 1993). Hence, the EMAP approach could be applied to describe the distribution and extent of wetlands in countries or regions where complete mapping and inventory of the resource is impractical. In the United States, NWI is mandated by Congress to completely map all wetland resources; hence the EMAP probability sample approach is not needed to define the Nation's wetland resources.

EMAP-Wetlands will conduct research to develop and evaluate indicators of wetland condition. Indicators will be identified in workshops attended by local wetland experts. The experts will also suggest criteria for identifying wetlands in good condition and in degraded condition. Pilot studies will be designed to identify wetland sites in good and degraded condition and evaluate which proposed indicators successfully discriminate between the extremes of condition. Leibowitz *et al.* (1991) provide a general list of potential indicators, and those selected for evaluation in current pilot studies underway in the Gulf of Mexico and prairie pothole region of the Midwest are listed in Tables 1A and 1B. A discussion of EMAP-Wetland's indicator development strategy is provided by Squires (1993).

Potential indicators that successfully discriminate between good and degraded wetlands will be tested in regional demonstration studies designed to develop an approach for assessing wetland condition (for one wetland class) in the region. A detailed discussion of the EMAP-Wetland's approach for assessing wetland condition is provided by Novitzki (1994). The regional demonstration studies will be conducted at sites representing a random sample of the population of the wetland resource in the region. A major contribution of the regional demonstration studies will be the development of a wetland sampling/monitoring network, as described in this paper.

EMAP-Wetlands plans to assess the condition of estuarine emergents (e.g., salt marshes), palustrine emergents (e.g., prairie potholes), and palustrine forested wetlands (e.g., bottomland hardwoods) nationally by the year 2004. These three classes include approximately 79 percent of the wetland resources of the Nation (Dahl & Johnson 1991). EMAP currently expects that approximately 200 samples will be required for each wetland class for each assessment region. Assuming that each of the three classes occur in four regions, the national wetland sample frame will include 2400 sample sites. EMAP further has decided to visit one-forth of the sample sites each year (see discussion in WETLAND SAMPLING NETWORK section) so each year as many as 600 sites may be visited. In order to make this a practical program, EMAP seeks to identify and evaluate indicators that integrate over time and that can be measured during a single one-day visit to each sample site.

The information from the sample network will be combined to assess the condition of the wetland resources in each region annually. These assessments will be structured so that they can be used to answer specific assessment questions, such as 'What portion of prairie wetlands support waterfowl production similar to those of reference wetlands in the region?' A preliminary list of assessment questions and the associated indicators is presented in Tables 2A and 2B. A detailed discussion linking wetland values, assessment questions, and associated indicators will be included in reports presenting conceptual models for each wetland class.

Table 1A. EMAP-Wetlands proposed conceptual linkages of values, sub-values and indicators for estuarine emergent wetlands.

WETLAND VALUES	SUB-VALUES	INDICATORS
Biological integrity	Plant diversity (community composition)	# of native, rare, threatened, endangered and nuisance species; total # of species
	Animal diversity (community composition)	# of native, rare, threatened, endangered and nuisance species, total # of species
	Plant abundance	% cover/stem density, stem height and width index, spectral reflectance
	Animal abundance	biomass or # of each species
Harvestable productivity	Plant abundance	% cover, stem height and width index, spectral reflectance
	Animal abundance	biomass or # of each species, wetland extent
Hydrology	Shoreline protection	hydraulic conductivity, % cover/stem density
	Water regime	wetland area, tidal amplitude, range, water depth, Eh, sulfides, soil salinity
	Salinity	soil salinity
Water quality improvement	Sediment retention	accretion rate, bulk density, % organic matter, tissue and soil analysis for contaminants
	Nutrient processing	% cover/stem density, dead vegetation, sulfides, plant tissue nutrient analysis, C/N Ratio, redox-Eh

Table 2A. Examples of EMAP-Wetlands assessment questions and associated indicators for estuarine emergent wetlands.

What portion of salt marshes exhibit plant and animal communities similar to reference wetlands, as indicated by the total number of species, relative abundance, and number of native, threatened, endangered, and nuisance species.

What portion of salt marshes provide a degree of shoreline protection similar to reference wetlands, as indicated by plant species and density, wetland area, and tidal range.

What portion of salt marshes accumulate sediments in a manner similar to reference wetlands, as indicated by accretion rates, percentage of organic matter, and bulk density.

What portion of salt marshes provide habitat for fish and wildlife similar to that of reference wetlands, as indicated by shellfish production, finfish production, and invertebrate species and abundance (food).

Classification

EMAP-Wetlands will build upon and complement existing programs to the extent possible. FWS is mandated to inventory the Nation's wetland resources and to report on status and trends in wetland acreage. EMAP-Wetlands will report on status and trends of wetland condition. One of the primary indicators of

Table 1B. EMAP-Wetlands proposed conceptual linkages of values, sub-values and indicators for palustrine emergent wetlands.

WETLAND INTEGRITY	SUB-VALUES	INDICATORS
Biological integrity	Plant diversity (community composition)	# of native, rare, threatened, endangered and nuisance species; total # of species
	Animal diversity (community composition)	# of native, rare, threatened, endangered and nuisance species; total # of species
	Plant abundance	% cover/stem density
	Animal abundance	breeding bird surveys
	Macroinvertebrates	biomass for each species
Harvestable productivity	Wildlife production	breeding dabbling ducks, % cover, wetland extent
	Plant abundance	% cover/stem density
	Animal abundance	wetland extent
Hydrology	Water storage	water depth, range in water level, basin morphology, # and size of wetland basins, # of drained wetlands
	Water regime	water level fluctuation, classification of wetlands, drainage ditches (# & length)
Water quality improvement	Sediment retention	sediment deposition, particle size, number, size and classification of wetland basins
	Nutrient processing	% cover/stem density, depth of fibric litter, water & soil chemistry (N, P, Ca, pH, NaCl)

Table 2B. Examples of EMAP-Wetlands assessment questions and associated indicators for palustrine emergent wetlands.

What portion of the prairie wetlands exhibit plant and animal communities similar to reference wetlands, as indicated by the total number of species, relative abundance, and number of native, threatened, endangered, and nuisance species.

What portion of the prairie wetlands perform hydrologic functions similar to those performed by reference wetlands, as indicated by water level fluctuation and basin morphology (in individual wetland basins) and the number, size and area of wetland basins; the mix of wetland classes; and the number of drained wetlands (within wetland complexes).

What portion of the prairie wetlands support waterfowl production levels similar to those of reference wetlands, as indicated by numbers of breeding waterfowl, invertebrate species and abundance (food), and plant species and abundance (food and habitat).

What portion of the prairie wetlands retain sediments similar to reference wetlands, as indicated by sediment accretion rates, soil particle size, and basin morphology.

wetland condition is expected to be wetland extent (Leibowitz *et al.* 1991) so EMAP expects to integrate NWI wetland acreage data – especially acreage change reported by NWI's Status and Trends program – into the assessment of wetland condition. EMAP-Wetlands will use the NWI definition and classification system

(Cowardin *et al.* 1979) and the most recent NWI maps or digital data to define the location, distribution, and extent to the wetland resource.

The NWI classification recognizes five systems, eight subsystems, and eleven classes (Fig. 1). Systems are based primarily on broad hydrogeomrphic characteristics of the wetland site (e.g., marine, estuarine). Subsystems are based on more specific hydrologic chracteristics (e.g., subtidal, intertidal). Classes are based primarily on substrate and vegetative lifeform (e.g., rock bottom, emergent). In practice, applying these three levels results in 56 wetland categories (Fig. 1). These categories may be divided into subclasses on the basis of other vegetation characteristics (e.g., needle-leaved deciduous) or substrate characteristics (e.g., cobbles and gravel) and may be further refined by adding modifiers to describe water regime, water chemistry, soils, or other site characteristics.

EMAP-Wetlands, with NWI assistance, has grouped NWI classes into functionally distinct EMAP classes having similar characteristics, settings, and functions (Table 3). EMAP-Wetlands classes include only the NWI Estuarine and Palustrine Systems – the marine, Riverine, and Lacustrine Systems are included in the EMAP-Near Coastal, Great Lakes, and Surface Waters resource groups. EMAP-Wetlands will coordinate activities with other EMAP groups so that compatible approaches for describing ecosystem condition are developed and that indicators can be used by any resource group to describe condition from their perspective. This is especially important, because wetlands are changeable ecosystems, and during extremely wet periods they may be more like lakes than wetlands, while in extremely dry periods they may be more like uplands.

Recent efforts by EPA, the Corps of Engineers, and other federal and state agencies to categorize or otherwise understand wetland functions and condition have focused on basic hydrogeomorphic settings of wetlands (Brinson 1993). These efforts seek to group wetlands by similar wetland function or site characteristics, similar to the goals of EMAP-Wetlands. To make the EMAP-Wetlands program compatible with such efforts, *yet also* compatible with the current NWI classification system, NWI and EMAP-Wetlands staff have developed a location modifier to be used with the NWI classification. The modifier identifies palustrine wetlands that adjoin either a lake (L modifier) or river (R modifier) separately from palustrine wetlands that adjoin neither. These are approximately equivalent to Brinson's 'fringe', 'riverine', and 'basin' categories

(L modifier, R modifier, and no modifier, respectively).

Sampling design

EMAP sought a design that adequately sampled the United States, but that also could be applied globally. The design is based on a probability sample, so that measurements or observations from the sample can be used to make statements about the entire resource population, with known confidence. The design is based on a uniform grid that preserves spatial balance and aerial distribution so that the sample sites represent/preserve the spatial properties of the entire resource population. The design also maintains an equal area projection on the earth's surface so that the probability of a grid point falling in a resource of interest is proportional to the size of the resource, and the proportionality is the same everywhere.

EMAP is evaluating different techniques for projecting geometric boundaries upon the surface of the globe. One has been identified that provides good representation in the conterminous U.S. and neighboring regions, including Alaska, and that can be adapted for use for other parts of the world. It is recognizable as the most common construction of the surface of a soccer ball (Fig. 2). It is based on the semi-regular geometric solid named the truncated icosahedron, which is a modification of the regular icosahedron, the 20-sided regular platonic solid. The truncated icosahedron has 20 hexagonal faces and 12 pentagonal faces.

One hexagonal face, when appropriately centered on the U.S., covers the land area and part of the adjacent continental shelf of the conterminous U.S., southern Canada, and northern Mexico (Fig. 3). This orientation meets the needs of EMAP, but provides poor representation for a number of areas on earth. However, it is possible to center hexagonal faces over these areas, relatively independently of each other, to provide the needed representation.

The hexagon centered on the U.S. measures approximately 2,599 km on a side. The geometry of the hexagon allows construction of a regular triangular grid comprised of 27,937 points in the full hexagon, and approximately 12,600 points in the conterminous U.S. (Fig. 3). These points are approximately 27.1 km apart. The entire grid was given a small random shift to provide a randomized sample while retaining the uniform grid spacing and spatial balance.

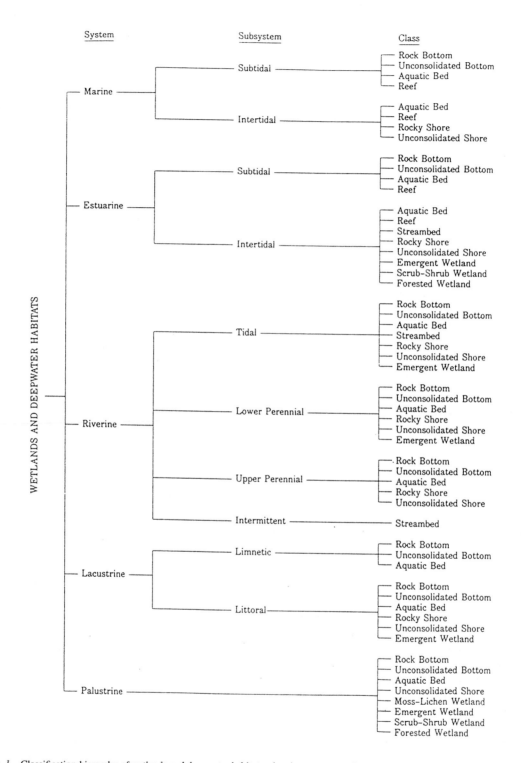

Fig. 1. Classification hierarchy of wetlands and deepwater habitats, showing systems, subsystems, and classes (Cowardin *et al.* 1979).

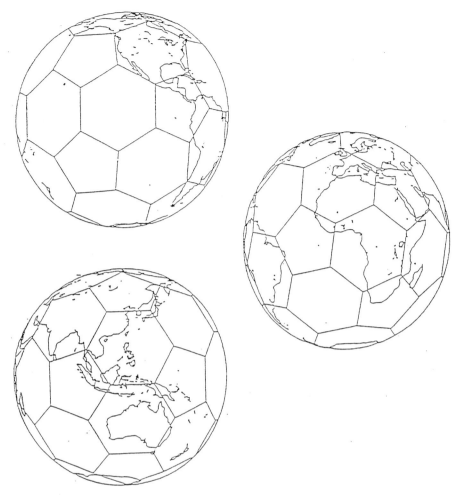

Fig. 2. The familiar soccer ball pattern (truncated icosahedron) projected onto the earth (Overton *et al.* 1990).

EMAP will describe the resources in an area centered around each of these grid points at two levels (tiers). The first (Tier 1) will be based on existing information or information collected using remote sensing. The area associated with each grid point is bounded by a hexagon with an area of 634.5 km^2 (referred to as the 640-hex – Fig. 4). However, EMAP proposes to characterize only 1/16th of the land surface i.e., the area within a hexagon of approximately 39.7 km^2 (referred to as the 40-hex) centered around the grid point. The relationship of the 40-hex to the 640-hex is shown in Figure 5. The characterization will primarily be based on aerial photography, Landsat images, maps, or other existing data. These classified areas constitute the Tier 1 probability sample of the resources of the U.S. From these descriptions regional estimates of areal extent of all landscape entities can be generated, as well as regional estimates of numbers of entities or discrete objects such as lakes, stream reaches, or wetlands. A strength of the EMAP design is that the same Tier 1 sample may be used later to sample other resource classes or attributes. The Tier 2 sample is a subset of the Tier 1 sample, selected randomly from the Tier 1 sample, which will be sampled on the ground.

For certain resources that are rare (occur in relatively few of the 40-hexes), highly localized (e.g., redwood forests), or linear (e.g., wetlands along streams) the Tier 1 sample may not provide a large enough Tier 2 sample. In these cases, the baseline grid can be augmented to increase the Tier 1 sample size. Because of the hierarchical geometrical properties of triangular grids, the grid can be enhanced by a factor of 3, 4, or 7 times. This process is described in Overton *et al.* (1990).

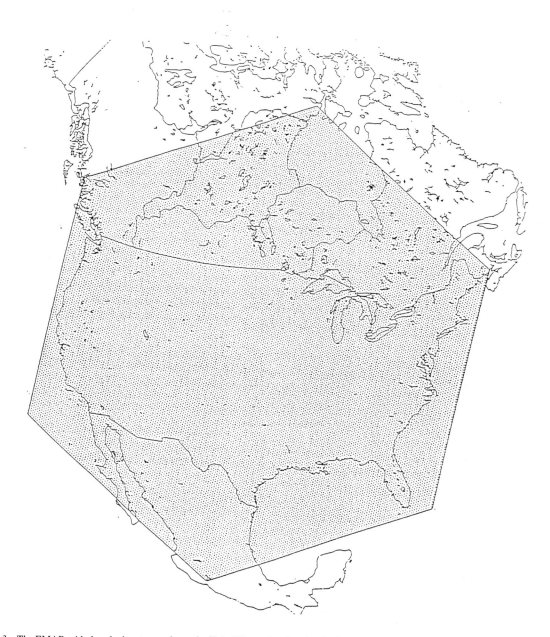

Fig. 3. The EMAP grid placed advantageously on the United States showing the distribution of the 27,937 grid points (Overton *et al.* 1990).

Wetland sampling network

The network of sample sites visited or remotely sensed to obtain measurements of indicators of wetland condition is a subset of the Tier 1 sample. The network sample sites (Tier 2 sample) also are selected by probability methods. Network sample sites will only be selected from the Tier 1 hexes where the resource of concern exists, thereby reducing the area which needs

to be examined in detail. Therefore, even though the EMAP grid is uniform over the entire U.S., the Tier 2 sample network for wetlands only exists where wetlands actually exist. As an initial criterion for describing condition for a wetland type in a region, EMAP proposes to characterize at least 50 sample sites within the region each year (Leibowitz *et al.* 1991).

There are several approaches that can be used to systematically select a Tier 2 sample from the Tier 1

Table 3. EMAP-Wetlands classes (after Cowardin *et al.* 1979).

System	Class	Description
Palustrine	Shallows	Unconsolidated bottom, aquatic bed, or unconsolidated shore
	Emergent	Emergent herbaceous vegetation
	Forested	Trees, living or standing dead
	Scrub-scrub	Woody vegetation
Palustrine L-locator	Emergent	Emergent herbaceous vegetation adjacent to a Lacustrine Wetland or lake
	Forested	Trees adjacent to a Lacustrine Wetland or lake
	Scrub-scrub	Woody vegetation adjacent to a Lacustrine Wetland or lake
Palustrine R-locator	Emergent	Emergent herbaceous vegetation adjacent to a Riverine Wetland or river
	Forested	Trees adjacent to a Riverine Wetland or river
	Scrub-scrub	Woody vegetation adjacent to a Riverine Wetland or river
Estuarine	Emergent	Emergent herbaceous vegetation regularly or irregularly flooded by tides
	Forested/ Scrub/scrub	Trees and woody vegetation regularly or irregularly flooded by tides

Fig. 4. The EMAP sample grid hexes superimposed on the Prairie Pothole Region of the Midwest.

sample. These approaches are discussed and evaluated in Overton *et al.* (1990). As a simple example, for a wetland type uniformly represented in the Tier 1 sample (Fig. 4) for a region, the wetland of that type nearest the center of all the Tier 1 40-hexes (Fig. 5) in which it occurs might be selected as the Tier 2 sample population, and then a network of Tier 2 sample sites selected as a subset of that sample.

Data collected at the Tier 2 sample network will be the basis for reporting on regional status and trends in indicators of ecological response, pollutant exposure, or other attributes, measured during field visits or by remote sensing. However, these two objectives – to describe status and trends – have somewhat conflicting criteria. Status would be best assessed by including different population units in the sample each year, whereas detection of trend requires revisits to selected wetlands sites over time. EMAP currently plans to revisit sites no more frequently than every 4 years, allowing trends to be examined at many more sites than annual revisits would allow. The design represents what appears to be a workable compromise to accomplish both EMAP objectives.

The total number of Tier 1 sample sites is divided into four subsets (the subsets selected systematically

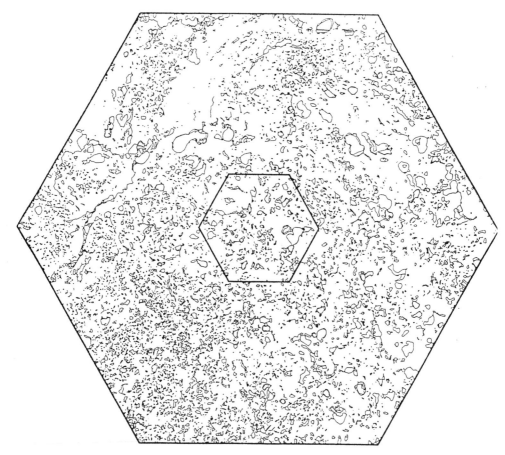

Fig. 5. The relation of the 640-hex to the 40-hex, and the distribution of wetland resources at each scale (Prairie Pothole Region of the Midwest).

to ensure uniform spatial coverage – Fig. 6). Each year, the Tier 2 sample sites are selected from the appropriate Tier 1 sample subset. (The first subset, year 1, was used for 1991, year 2 for 1992, year 3 for 1993, and year 4 for 1994; this pattern will repeat with year 1 used again for 1995, year 2 for 1996, and so on.) The approach for identifying the Tier 1 sample population, and subsequently selecting a subset to constitute a Tier 2 sample and establish a sampling network is best demonstrated by example.

Sample network for gulf coast salt marshes
In 1991 EMAP-Wetlands initiated a pilot study in cooperation with Louisiana State University to identify and evaluate indicators of condition of estuarine emergent wetlands (salt marshes) along the Louisiana coast. Results of the pilot study will be applied in a demonstration study in the Gulf of Mexico (presently planned for 1994) and the sample sites must be selected randomly, following the EMAP sample design.

The coastal salt marshes represent a somewhat unique resource because of their linear distribution along the Atlantic and Gulf coasts. To develop the Tier 1 sample for the Gulf, EMAP-Wetlands staff obtained from the Fish and Wildlife Service the NWI maps for the entire coastal region from the Florida Keys to Brownsville, Texas. Further, the boundary representing the extreme landward extent to salt marsh was from Dahl and Johnson (1991). The EMAP grid was then superimposed on the maps to determine the Tier 1 sample.

It was immediately apparent that the 27.1 km standard grid point spacing was too sparse to generate the minimum 200 Tier 1 sample sites (a minimum of 50 Tier 2 sites required each year). EMAP project and design staff evaluated the salt marsh data along the coast and determined that a 49-fold intensification (Fig. 7) would provide a Tier 1 sample of approximately 500 'hits' – grid points where salt marsh existed at the time of mapping. This number of sites should allow selection of 50 sample sites per year (some will

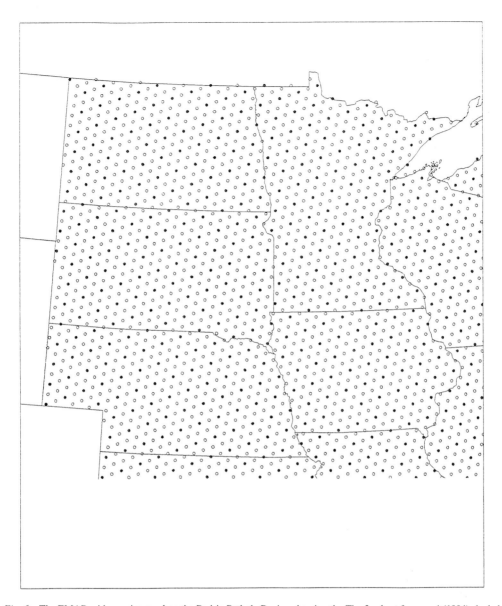

Fig. 6. The EMAP grid superimposed on the Prairie Pothole Region showing the Tier 2 subset for year 4 (1994) shaded.

be rejected because they are no longer salt marsh or because access is denied or logistically infeasible).

The 49-fold intensification results in a grid of points approximately 3.9 km apart. (The hexagons centered around these grid points are approximately 12.9 km^2.) Figure 7 shows the distribution of the Tier 1 sample points for 1993 (year 3 subset). From each grid point a random direction (0–360°) and distance (0–2.23 km) was determined to identify a sample point. If there is salt marsh at that point it was included in the Tier 1 sample. From this population of 'hits' a Tier 2 sample of 50 sites was drawn.

The grid points shown in Figure 7 will not be in the Tier 1 sample again until four years from now, and at that time a random sample point again will be selected, existence of salt marsh will be verified, and the Tier 1 sample points identified. The Tier 2 sample of 50 sites will be drawn from these. This strategy minimizes the likelihood that a specific point would be resampled. Based on field experience and preliminary results of the 1991 pilot study, we decided to minimize revisits to a particular site to avoid impacts of destructive sampling or other disturbances caused by the preceding sample visit.

182

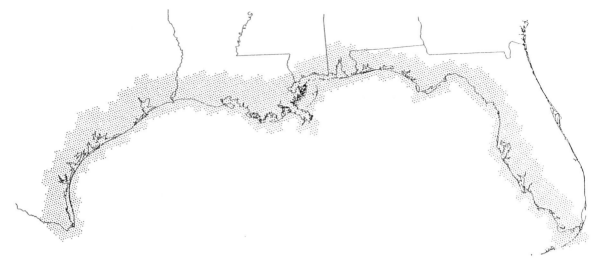

Fig. 7. The EMAP grid intensified 49-fold for the Gulf of Mexico salt marsh study – only one fourth of the grid points (the Tier 2 subset for year 3 – 1993) are shown.

Sample network for prairie potholes

A second example of the sample selection process will be demonstrated when developing a sample network for the prairie pothole region of the Midwest. In 1992 EMAP-Wetlands initiated a pilot study in cooperation with the U.S. Fish and Wildlife Service's Northern Prairie Wildlife Research Center in Jamestown, ND to identify and evaluate indicators of condition of palustrine emergent wetlands (prairie pothole marshes). Results of the pilot study will be incorporated in a demonstration study in the prairie pothole region of the Midwest (presently planned for 1995) and the sample sites must be selected randomly, using the EMAP sample design.

Prairie pothole wetlands are widely distributed and seemingly occupy independent basins, distinctly different from the linear coastal wetland resources. However, in a meeting of prairie wetland experts held at the Jamestown Wildlife Research Center in July 1991, a consensus view suggested that the prairie pothole wetlands are interrelated and function together as a wetland complex, rather than as individual, independent wetland basins. Dr. Lewis M. Cowardin, senior author of the Cowardin *et al.*/NWI wetland classification and principal investigator on the EMAP prairie pothole wetland pilot study, suggests that, based on preliminary study results and earlier research, the appropriate size of the wetland complex for reporting condition may be the EMAP 40-hex.

The EMAP grid superimposed on the prairie pothole region is shown in Figure 6. The shaded hexes represent the Tier 1 sample points (year 4) from which the Tier 2 sample for 1994 will be selected. Figure 5 is an example of the distribution of wetland resources contained in the 640-hexes and 40-hexes in this region. For the prairie pothole wetlands, a 'hit' will be any 40-hex that contains classified wetlands. It is apparent that the standard grid spacing will generate a Tier 1 sample large enough to readily provide a Tier 2 sample of at least 50 sites per year.

Undoubtedly further refinement of this approach will occur as the variability of different indicator measurements is defined during the pilot study. One such refinement is already being considered (L. Cowardin, U.S. Fish and Wildlife Service, Jamestown, ND, pers. comm.). It is apparent that there will be a mix of temporary, seasonal, semi-permanent, and permanent wetlands in a wetland complex. Because some indicators are relevant in some of these wetland types, but not necessarily in another (e.g., observations of amphibian populations may be appropriate in permanent wetlands but not appropriate in temporary wetlands that are dry most of the time) subpopulations of wetlands in the 40-hexes may be established before selecting the Tier 2 sample.

As with salt marshes, there is concern for repeated sampling of the same wetland basin. The concern is greatest for temporary wetlands which are typically small and sensitive to impacts, and least for permanent wetlands which are typically large and less sensitive to impact. (Further, permission to gain access to permanent water bodies appears more readily granted than access to temporary wetlands.) Early examination suggests that there are 3 to 5 times as many temporary and seasonal wetlands as there are semi-permanent and permanent wetlands. If the wetlands contained in the 40-hex are separated into subpopulations comprised of temporary, seasonal, and semi-permanent/permanent wetlands, the Tier 2 sample for any year might be 2 temporary and 2 seasonal wetlands (selected randomly), and all semi-permanent and permanent wetlands contained in the 40-hex. If there are typically at least 10 each of the temporary and seasonal wetlands, and if they were sampled without replacement, this would ensure that each of these sensitive systems would be sampled no more frequently than once in 20 years, whereas the less sensitive systems might be sampled every 4 years.

Discussion

The EMAP sampling design has been developed for global application. Although the EMAP program has applied the design primarily to the conterminous U.S., the design can be applied to the rest of the world with only minor adjustment. It has sufficient flexibility to provide a sample network to characterize common as well as rare and unique resources in different regions. Because the probability based sample design allows description of resources by examining a relatively small part (e.g., 1/16th) of the land surface, the design has great potential to provide resource inventories, with known confidence, at greatly reduced costs.

EMAP-Wetlands will use the NWI classification system for wetlands, although EMAP will aggregate the Cowardin *et al.* (1979) classes into functional groupings. With NWI and EMAP jointly reporting status and trends in wetland acreage and wetland condition, it is obvious that the two programs need to use the same resource classification. It is likely that current efforts (e.g., Brinson 1993) to identify functional groups of wetland classes will be based upon or incorporate the Cowardin system, and consequently will be compatible with EMAP-Wetlands classes as well.

The EMAP design allows wetland condition data (Tier 2 sample) to be reaggregated for re-analysis. This might be desirable if, based on data generated by the program, regional boundaries are modified, or if subpopulations are identified (or recombined). This also assures that the data collected in the early part of the program can be re-analyzed later when new functional

groupings or regionalization schemes may be developed.

The EMAP sample design also provides a technique for transferring detailed wetland process information, derived from long-term research studies, to other sites or entire regions. For example, long-term data on water regime (depth, duration, and periodicity of flooding) in wetlands is quite rare. However, it is widely accepted that water regime is a dominant influence on rate of organic soil accretion (wetter conditions favor accretion) in some wetland types. If the indicator 'rate of organic soil accumulation' is measured at the research site and at EMAP Tier 2 sample sites (revisited every 4 years) inference relative to water regime can be made to the population of wetlands in the region. For example the portions of the regional wetland population that have water regimes similar to, wetter than, and drier than the reference sites can be derived from the relative rates of organic soil accumulation.

Acknowledgements

The information in this document has been funded by the U.S. Environmental Protection Agency under contract 68-C8-006 to Man Tech Environmental Technology, Inc. It has been subjected to Agency review and approved for publication. I wish to thank Dr. Scott Urquhart, Oregon State University, for his substantial assistance in preparing and reviewing this manuscript. I also wish to thank Dr. Arnold van der Valk, Iowa State University and Mr. Thomas E. Dahl, U.S. Fish and Wildlife Service, National Wetland Inventory, for their helpful reviews of the manuscript.

References

Brinson, M. M. 1993. A Hydrogeomorphic Classification for Wetlands. U.S. Army Corps of Engineers, Waterways Experiment Station, Vicksburg, MS, Wetland Research Program Technical Report WRP-DE-4 79 pp..

Cowardin, L. M., Carter, V., Golet, F. C. & LaRoe, E. T. 1979. Classification of Wetlands and Deepwater Habitats of the United States, U.S. Fish and Wildlife Service, Washington, DC. 131 pp.

Dahl, T. E. & Johnson, C. E. 1991. Status and Trends of Wetlands in the Conterminous United States, Mid-1970's to Mid-1980's. U.S. Department of the Interior, Fish and Wildlife Service, Washington, DC. 28 pp.

Leibowitz, N. C., Ernst, T. L., Urquhart, N. S., Stehman, S. & Roose, D. 1993. Evaluation of EMAP-Wetlands Sampling Design Using National Wetlands Inventory Data. EPA/620/R-93/773. U.S. Environmental Protection Agency, Environmental Research Laboratory, Corvallis, OR. 95pp.

184

Leibowitz, N. C., Squires, L. & Baker, J. P. 1991. Research plan for monitoring wetland ecosystems. EPA/600/3-91/010, U.S. Environmental Protection Agency, Corvallis, OR. 157 pp.

Novitzki, R. P. 1994. EMAP-Wetlands: A program for assessing wetland condition, in Mitsch, W. J. and R. E. Turner (editors). Wetlands of the World: Biogeochemistry, Ecological Engineering, Modeling and Management. Elsevier Publishers, Amsterdam p 691–709.

Overton, W. S., White, D. & Stevens, D. L. 1990. Design Report for Environmental Monitoring and Assessment Program (EMAP). EPA/600/3-91/053, U.S. Environmental Protection Agency, Washington, DC. 43 pp.

Squires, L. 1993. A Research Strategy to Develop Ecological Indicators of Wetland Condition, In: Landin, Mary C., Proceedings of the 13th Annual Symposium of the Society of Wetland Scientists, South Central Chapter, Society of Wetland Scientists, Utica, MS, USA. p. 778–791.

Vegetatio **118**: 185–192, 1995.

Wetland classification and inventory: A summary

C.M. Finlayson[1,2] & A.G. van der Valk[3]
[1]*International Waterfowl & Wetland Research Bureau (IWRB), Slimbridge, Glos GL2 7BX, UK*
[2]*Current address: Office of the Supervising Scientist, Locked Bag 2, Post Office, Jabiru, NT 0886, Australia*
[3]*Department of Botany, 141A Bessey Hall, Iowa State University, Ames, IA 50011, USA*

Key words: Classification, Conservation, Inventory, Management, Wetland

Abstract

Regional, national and local wetland classifications have been developed and successfully applied. These have invariably been orientated towards conservation and management goals, and the information used to assess wetland loss or to assign management priorities. Existing national and regional classification systems have not only been useful, but they provide an essential base for developing an international system. At the international level, differences among existing systems in the definition of a wetland and how wetland types are defined assume great importance and need to be resolved. Classification is an essential prerequisite for wetland inventory. A number of international inventories have been undertaken, although these have not generally utilized the available high technology and data storage systems available through remote sensing and geographic information systems. More extensive international inventories will require standardization of techniques for data collection, storage and dissemination. A minimum data set needs to be defined with standards for data accuracy. An international committee under the auspices of an international agency (e.g. IWRB, Ramsar Bureau, IUCN) needs to be established to develop an international classification system and guidelines for carrying out a complete inventory of the world's wetlands.

Introduction

The information collected through wetland inventories is nowadays regarded as a necessary prerequisite for wetland conservation and management at a holistic level, involving planning on a national, regional and international scale (Dugan 1990; Hollis *et al.* 1992; Taylor *et al.* 1995; Hughes 1995; Naranjo 1995; Scott & Jones 1995). An inventory is regarded by Dugan (1990) as the first step in assembling an information base for wetland management. In fact, Contracting Parties to the Ramsar Convention undertake to compile an inventory as part of the process of developing and implementing a national wetland policy for the wise use of all wetlands on their territory.

The term 'wetland' groups together a wide range of habitats that share a number of common features, the most important of which is continuous, seasonal or periodic standing water or saturated soils. Under the widely accepted Ramsar definition of a wetland

(Scott & Jones 1995) some 30 natural wetland types are recognized (Scott 1989a). There is often great confusion and sometimes controversy locally over whether a given type of habitat is or is not a wetland. This can even occur in countries with well established wetland classifications and ongoing national wetland inventories such as the United States (Environmental Defense Fund and World Wildlife Fund 1992). The purpose of wetland classification is to standardize and define the terms being used to describe the various wetland types. At an international level a uniform set of terms is needed (Cowardin & Golet 1995; Scott & Jones 1995; Zoltai & Vitt 1995). However, at a local or national level this may not be necessary (Pressey & Adam 1995), although there would seem to be little argument that the adoption of standardized terms and definitions has definite advantages for comparative and broad planning purposes (Cowardin & Golet 1995; Hughes 1995; Zoltai & Vitt 1995).

Techniques used in wetland inventories vary from basic field and literature surveys (Hughes 1995; Taylor *et al.* 1995; Pressey & Adam 1995) to highly sophisticated technological approaches using aerial photography (Taylor *et al.* 1995; Wilen & Bates 1995; Zoltai & Vitt 1995) and satellite imagery (Menantean 1991; Nakayama 1993; Hess *et al.* 1990). To be effective in promoting the conservation of wetlands these inventories must be available to and understood by all those formulating and implementing wetland management policies (Federal Geographic Data Committee, Wetlands Subcommittee 1994, Naranjo 1995; Pressey & Adam 1995; Wilen & Bates 1995). Thus, they must be framed in a manner suitable for management purposes. Additionally, to remain useful tools for management they need to be regularly reviewed and updated (Naranjo 1995; Scott & Jones 1995; Wilen & Bates 1995). At an international level the inventories need to be comparable and available in commonly used languages (Hughes 1995).

Wetland classification

The starting point for many wetland inventories is the development and adoption of a wetland classification. The wetland literature contains a large number of terms to designate and describe different kinds of wetlands. This has partly come about because wetlands occupy an intermediate position between truly terrestrial and aquatic ecosystems and therefore encompass a diverse array of habitats. This array of habitats is difficult to define and a multiplicity of terms has sprung up in many languages to describe wetland types (Cowardin & Golet 1995; Gopal *et al.* 1990; Pakarinen 1995; Scott & Jones 1995). Unfortunately, this richness of terms can make classification an exceedingly difficult task if uniformity of terms and comparability are major parts of the exercise. If these are not important objectives, then one potential stumbling block to classification is removed (Pressey & Adam 1995). However, Cowardin & Golet (1995) argue that at a national level, in the USA, consistency in concepts and terminology were one of the purposes of the development of the US wetland classification system. Furthermore, Cowardin & Golet (1995) support the adoption of an international classification system that incorporates the diversity of wetland types from around the world. Such a system would avoid the use of geographically specific terms, but could include regional, national or continental modifiers. With this approach, colloquial fears

about adoption of an international system (Pressey & Adam 1995) would be supplanted.

One of the most commonly accepted wetland definitions is that of the Ramsar Convention 'areas of marsh, fen, peatland or water, whether natural or artificial, permanent or temporary, with water that is static or flowing, fresh, brackish or salt, including areas of marine water the depth of which at low tide does not exceed six meter'. This definition has been accepted by more than 70 Contracting Parties to the Convention; however, this does not imply wide acceptance of the definition within these countries (Pressey & Adam 1995; Hughes 1995). Often, narrower definitions that reflect unique national or regional wetland characteristics have been accepted (Cowardin & Golet 1995; Lu 1995; Pakarinen 1995; Pressey & Adam 1995). Furthermore, the six meters depth criterion for marine wetlands causes difficulties of delineation (Lu, 1995), particularly for coral reefs that often extend deeper than six meters (Pressey & Adam, 1995; Scott & Jones 1995). Scott & Jones (1995) point out that this is a legacy of the origins of the Ramsar Convention; for greater acceptance of the conservation benefits of having the Convention such historical and outmoded stumbling blocks should be reconsidered.

Many national wetland classifications already exist and more are being proposed (see Brinson 1993, Cowardin & Golet 1995; Gopal & Sah 1995, Lu 1995; Pakarinen 1995; Pressey & Adam 1995; Semeniuk & Semeniuk 1995; Zoltai & Vitt 1995). These invariably incorporate local terms and definitions that are not necessarily known or accepted elsewhere. For national purposes this may not be a major problem, but for comparisons and management at an international level these differences may present difficulties. However, even at the national level it can be extremely difficult to develop a classification that is acceptable to all wetland scientists and experts (Cowardin & Golet 1995; Lu 1995; Pressey & Adam 1995). Management of wetlands that transcend both inter- and intra-national political boundaries is a difficult enough task (e.g. Hollis 1990) without unnecessary confusion over terms and definitions.

One of the most comprehensive and widely applauded wetland classification systems is that developed for the USA by Cowardin *et al.* (1979). This system is described in depth by Cowardin & Golet (1995). Basically, it divides wetlands into systems, sub-systems, classes and sub-classes, along with a series of water regime, chemistry and soil modifiers. It is hierarchical with the system (marine, estuarine, riverine, lacustrine and palustrine) as the basic unit. The US classification is also accompanied by a list of plants known to occur in wetlands, and also a list of hydric soils.

An alternative approach to classification is that based on the underlying, unifying features of wetlands, i.e. landform and hydrology (Brinson 1993; Semeniuk & Semeniuk 1995). This approach can be used for both coastal and inland wetlands regardless of climate and vegetation types. The hydrogeomorphic classification of wetlands should make it more feasible to develop methods for assessing the physical, chemical and biological functions of wetlands (Brinson 1993).

An international wetland classification has value if it provides readily understood terms, a framework for international legal instruments for wetland conservation, and assists in the dissemination of information (Scott & Jones 1995). Again, under the auspices of the Ramsar Convention an internationally accepted classification system was developed by Scott (1989a). This was based loosely on the USA system (Scott & Jones 1995). In an internationally acceptable classification system the correspondence between globally adopted taxa and more colloquial terms (e.g. bayou, vlei, billabong, jheel, valle etc.) should be addressed, perhaps in an appendix (Cowardin & Golet 1995). However, the use of local terms is generally considered inappropriate at even the national level (e.g. Cowardin & Golet 1995) and for international purposes readily understood descriptors, rather than a multitude of poorly defined and confusing local names are needed.

In developing a classification system Scott & Jones (1995) issue a note of warning concerning the level of sophistication adopted in relation to the amount of information required, particularly where the information is not available. Careful consideration of the need for information and the requirements for management purposes are points strongly made by Pressey & Adam (1995). Furthermore, for large wetland systems a detailed classification of habitats can be extremely cumbersome and possibly irrelevant. By its very nature, wetland classification is beset by problems as it is an attempt to place artificial boundaries on natural continua (Cowardin & Golet 1995; Pressey & Adam 1995). However, the very fact that so many classification systems exist is evidence that some order or standardization of habitat types is required by scientist, managers and policy makers. The ideal classification system would therefore, be everything to everyone; however, due to the very nature of wetlands and

management systems, compromises are necessary and boundaries need to be drawn. Acceptance of such artificial delineation should promote unity of purpose and not serve as dampeners to further innovation, as argued by Pressey & Adam (1995).

Purpose of inventories

Wetland inventories are useful in the first stages of developing effective wetland conservation programs (Novitzki 1995, Pakarinen 1995, Taylor *et al.* 1995; Hughes 1995; Naranjo 1995; Scott & Jones 1995; Wilen & Bates 1995). An inventory can assist in identification of conservation priorities, establish the basis for monitoring the ecological status of wetlands, promote awareness of wetland sites and management issues, and facilitate exchange of information and comparisons between sites and regions (Garcia– Orcoyen *et al.* 1992). As importantly, information gathered for inventories can illustrate the economic value of wetlands (Lu 1995) and provide valuable data for resource utilization decisions (Wilen & Bates 1995).

The usefulness of inventories can quickly diminish if they are not regularly updated (Naranjo 1995; Wilen & Bates 1995; Scott & Jones 1995). To enable rapid updating of inventories the data should be stored in a centralized location and easily accessible through standardized or interchangeable computerized formats. Unfortunately, at an international level the information gathered during broad-scale continental wide inventories remains scattered, making coordinated updating extremely difficult (Hughes 1995; Scott & Jones 1995). The collection of minimum data sets and the utilization of Geographic Information Systems can help overcome such limitations (Pressey & Adam 1995; Wilen & Bates 1995).

Maintenance of the Ramsar site database is one instance where a specific and coordinated approach to updating an international inventory is being undertaken (Scott & Jones 1995). However, this inventory is extremely limited and only covers (in November 1993) 641 of the recognized internationally important wetlands in 80 countries around the world. This database will be expanded as more wetlands are added to the Ramsar list of internationally important wetlands.

Inventories are particularly valuable for assessing wetland loss and degradation (Taylor *et al.* 1995; Hughes 1995; Lu 1995; Wilen & Bates 1995). Information on rates of wetland loss and reasons for this loss can prove invaluable for promoting awareness of issues and developing conservation and restoration programs (Hollis & Jones 1991; Hughes 1995; Wilen & Bates 1995). Once the basic information on wetland occurrence, distribution and status has been collated it is essential that it is utilized as the basis of further conservation effort (Naranjo 1995); otherwise it quickly becomes dated and not seriously regarded by conservation officials. However, even when inventories are available they may only be of limited usefulness (Hughes 1995; Naranjo 1995). This is particularly so where the information is not comprehensive or is restricted in scope and coverage, or is not brought to the attention of governmental officials responsible for setting policies that affect wetlands.

Wetland inventories

A number of international wetland inventories exist, although the first of these, the IBP Project Aqua (Luther & Rzoska 1971) and IUCN Mar list (Olney 1965), were limited in scope and quickly dated. These were followed by inventories emphasizing waterbird habitats in the Western Palearctic (Carp 1980) and Western Europe and North West Africa (Scott 1980). Again, these were not comprehensive and it was not until the publication of Grimmett & Jones (1989) that a thorough inventory of waterbird habitats in all of Europe became available. Fairly extensive inventories are now also available for the Neotropics (Scott & Carbonell 1986), Asia (Scott 1989b), and Africa (Hughes & Hughes 1992; Burgis & Symoens 1987) (see Table 1 and Fig. 1). More recent inventories include those undertaken in Oceania, Australia and New Zealand (Pressey & Adam 1995; Scott & Jones 1995) and others are being proposed for the Commonwealth of Independent States, the Baltic Republics and the Middle East as part of the 1993–96 program for the IWRB.

Thus, at least some information is available for many parts of the world. However, much of this coverage is not comprehensive or needs updating (Lu 1995; Scott & Jones 1995; Naranjo 1995). Unfortunately, many of these international inventories can only be considered as preliminary (Scott & Jones 1995). More thorough and sophisticated techniques are needed to expand the available databases and to make the inventories more useful for conservation management. Further information on wetland functions and data suitable for monitoring ecological change in wetlands is required and needs to be gathered and compiled in a format that allows for ready access and updating

(Hughes 1995; Scott & Jones 1995). Furthermore, the data gathered during inventories needs to be carefully examined to ascertain if it is actually required to further the conservation and wise use of wetlands that are continually being degraded or lost (Hughes 1995; Pressey & Adam 1995; Scott & Jones 1995).

A number of national wetland inventories are now available (Taylor *et al.* 1995; Hughes. 1995; Lu, 1995; Cowardin & Golet 1995; Pressey & Adam 1995; Zoltai & Vitt 1995; Hudec *et al.* 1993). Some of these have been spawned by the international inventories (Lu 1995; Silvius *et al.* 1987). However, despite acceptance of the protocols of the Ramsar Convention and recognition of the values of wetlands many countries have yet to undertake detailed nation-wide inventories (Gopal & Sah 1995; Hughes 1995; Taylor *et al.* 1995).

In southern Africa, inventories are restricted to relatively small regions or are not very detailed (Taylor *et al.*, 1995). In many cases these were compiled from soil and vegetation maps originally designed for other purposes. In northern Africa, several countries now have inventories with information on wetland loss and functions (Atta & Sorensen 1992; Maamouri & Hughes 1992). A number of European countries have produced national inventories and many local regional inventories are also available, although the extent of detail varies enormously (Hughes 1995). Many of the European inventories have concentrated on waterfowl habitat with relatively little attention given to other wetland values. However, a great deal of wetland conservation effort has been initiated as a consequence of inventories based on waterfowl criteria (Hollis *et al.* 1992). In Australia, a large number of localized inventories have been produced and a national inventory project has recently been completed (Usbank & James 1993) and linked to an inventory of wetlands in Oceania (Scott 1993). The most comprehensive attempts at national inventories are those undertaken and being updated in the USA (Wilen & Bates 1995) and Canada (Zoltai & Vitt 1995). This effort has not extended to Central and Southern America, nor to Asia despite the enthusiasm generated by the international inventories for these regions (Scott & Carbonell 1986; Scott 1989b); indicating a fatal flaw in the conservation effort associated with the data gathering or inventory projects (Naranjo 1995).

Table 1. Examples of international and national wetland inventories.

Coverage	Reference
International	
Asia	Scott (1989b)
Neotropics	Scott & Carbonell (1986)
Africa	Hughes & Hughes (1992)/Burgis & Symoens (1987)
Oceania & Australia	Scott (1993)/Usbank & James (1993)
Europe	Grimmett & Jones (1989)
Western Europe/north-west Africa	Scott (1980)
National	
Australia	Usbank & James (1993)
China	Lu (1990)
Czechia	Hudec *et al.* (1993)
Egypt	Atta & Sorensen (1992)
France	Leiderman & Mermet (1991)
Greece	Heliotis (1988)
India	DeRoy & Hussain (1993)
Indonesia	Silvius *et al.* (1987)
Namibia	Simmons *et al.* (1991)
Spain	Montes & Bifani (1989)
Switzerland	Marti (1988)
Tunisia	Maamouri & Hughes (1992)
USA	Tiner (1984)

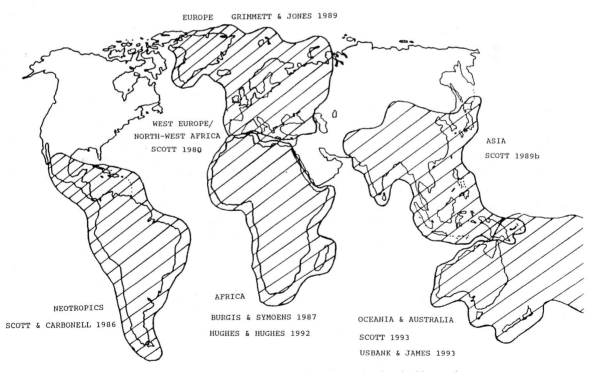

Fig. 1. Areas of the world covered by completed international wetland inventories.

Inventory techniques

Soil, vegetation and topographical maps have proved to be useful sources of information for wetland inventories (Maamouri & Hughes 1992; Pressey & Adam 1993; Psilovikos 1992). However, even where the quality of maps is reasonably reliable they can not always be relied on to provide sufficient information (Taylor *et al.* 1995; Cowardin & Golet 1995). Where maps are available estimates can be obtained by assessing the occurrence of hydromorphic soils or alluvium. In fact, where reconnaissance and literature surveys are utilized for inventory purposes maps are an essential means of locating the wetlands and for assessing the degree of wetland loss. Casado *et al.* (1992) report that during the Spanish wetland inventory around 1500 wetlands were visited; unfortunately this is either not possible nor feasible in many countries. In many instances information and decisions are required before ground studies can be undertaken (Cowardin & Golet 1995).

High altitude aerial photography has been used for wetland inventory purposes in a number of countries (Taylor *et al.* 1995; Psilovikos 1992; Wilen & Bates 1995; Zoltai & Vitt 1995). Problems do occur where the scale varies from region to region and the whole process can be very time consuming and necessitate large numbers of photographs. Despite these problems a large amount of information can be assessed from aerial photography. Interpretation of the data, especially water regimes and vegetation cover can often prove difficult and there is still a need for a high degree of local ecological knowledge (Cowardin & Golet 1995; Pressey & Adam 1995; Zoltai & Vitt 1995).

Orthophotographs which are corrected for height distortion and tilt are being successfully used for wetland inventory in Natal, South Africa (Taylor *et al.* 1995). The photographs are relatively recent and can be used for identification of small wetlands and the data interfaced with other spatial data. As with aerial photographs large scale mapping necessitates the use of large numbers of orthophotographs.

Three satellite sensing techniques are available for wetland inventory purposes. These are the Landsatt Multispectral Scanner (MSS), the Thematic mapper (TM) and the SPOT satellites. All have distinct advantages and disadvantages (Taylor *et al.* 1995). Landsatt MSS data is available, fairly cheaply, for about 20 years in many parts of the world, but the resolution is often not adequate for detecting small linear features. The TM data has only been available in recent years and

is much more expensive than MSS data. SPOT data is also useful for inventory purposes and has been successfully used in parts of southern Africa and is being tried in the Mediterranean Basin (P. Tomas-Vives personal comm.). Overall, satellite sensing gives a wide coverage and is particularly valuable for widescale mapping in remote and inaccessible areas. However, the wide scale of the coverage raises problems with accuracy and detail and cloud cover obstructs data acquisition (Wilen & Bates 1995; Zoltai & Vitt 1995). These problems are currently being addressed with that of cloud cover interference being overcome by the use of radar imagery (Hess *et al.* 1990). When satellite data is collected it is possible to interface with other geographic data storage systems and compile comprehensive data bases that can be interrogated and used to test models at different scales.

Conclusions

There is very little demonstrable acceptance of worldwide classification systems for wetlands, reflecting vast differences in approach and goals (see Pressey & Adams 1995; Scott & Jones,1995). The Ramsar classification system, loosely based on the USA example, is possibly one system that does have the capability of being used internationally, at least for the purposes of maintaining comparability and a common language. At a national and regional level the adoption of this common approach to wetland classification has not been widely accepted, probably due to poor promotion by national conservation authorities and wetland experts familiar with the Convention.

Despite the inherent problems in developing widely accepted classifications, much effort, in response to perceived local needs for classification, has been undertaken. A new and innovative system of classification, using landform and wetness, has been developed at a local level and then expanded to then international by Semeniuk & Semeniuk (1995). However, even this system suffers by introducing yet another set of new terms and definitions; a problem that bedevils classification.

Regardless of the classification system adopted, many wetland inventories have been done. These have utilized a variety of techniques. In this respect, the highly sophisticated USA example of using extensive aerial photography to produce wetland status and trends reports (Wilen & Bates 1995) stands out. At the other end of the scale of technical sophistication, the survey techniques used to produce the first Asian and Neotropical wide (see Scott & Jones 1995) inventories have demonstrated the value of compiling current information; a process also undertaken in Canada (Zoltai & Vitt 1995). However, the sophisticated follow-up effort that occurred in Canada has not been undertaken in the Neotropics, thereby negating some of the conservation value of this inventory (Naranjo, 1995).

From a conservation perspective, the value of an inventory can only be shown by the usefulness of the information for conservation purposes (Naranjo 1995; Pressey & Adam 1995; Scott & Jones 1995). Such shortfalls in the conservation effort stemming from classification and inventory projects could be picked up by the development of a centralized data storage repository under the auspices of international agencies, such as IWRB and the Ramsar Bureau, that are already committed to global wetland conservation programs.

References

Atta, G. A. M. & Sorensen, U. G. 1992. Waterbird populations and the loss and degradation of wetlands in Egypt. pp. 125–129. In: Finlayson, C. M., Hollis, G. E. & Davis, T. J. (eds), Managing Mediterranean wetlands and their birds. IWRB Special Publication No. 20. IWRB, Slimbridge, United Kingdom.

Brinson, M. M. 1993. A hydrogeomorphic classification for wetlands. Wetlands Research Program Technical Report WRP-DE-4. U.S. Army Engineers Experiment Station, Vicksburg, MS, USA.

Burgis, M. J. & Symoens, J. J. (eds) 1987. African wetlands and shallow water bodies. ORSTOM, Paris.

Carp, E. 1980. A directory of Western Palearctic wetlands. UNEP, Nairobi & IUCN, Gland, Switzerland.

Casado, S., Florin, M., Molla, S. & Montes, C. 1992. Current status of Spanish wetlands. pp. 56–58. In: Finlayson, C. M., Hollis, G. E. & Davis, T. J. (eds), Managing Mediterranean wetlands and their birds. IWRB Special Publication No. 20. IWRB, Slimbridge, United Kingdom.

Cowardin, L. M. & Golet, F. C. 1995. U.S. Fish and Wildlife Service 1979 wetland classification – a review. Vegetatio 118 (this issue).

Cowardin, L. M., Carter, V., Golet, F. C. & LaRoe, E. T. 1979. Classification of wetlands and deepwater habitats of the United States. U.S. Fish and Wildlife Service FWS/OBS 79/31.

DeRoy, R. & Hussain, S. A. (eds) 1993. Directory of Indian wetlands. WWF-India, New Delhi and Asian Wetland Bureau, Kuala Lumpur. 240 pp.

Dugan, P. J. 1990. Wetland conservation – a review of current issues and required action. IUCN, Gland, Switzerland.

Environmental Defense Fund and World Wildlife Fund. 1992. How wet is a wetland? Environmental Defense Fund, New York, NY and World Wildlife Fund, Washington, DC, USA.

Federal Geographic Data Committee, Wetlands Subcommittee. 1994. Strategic interagency approach to developing a national digital wetlands data base (second approximation). Federal Geographic Data Committee, Washington, DC, USA.

Garcia-Orcoyen Tormo, C., Vallecillo, C. G. & Valladares, A. M. 1992. How many inland Mediterranean wetlands will there be in the year 2000? pp. 28–31. In: Finlayson, C. M., Hollis, G. E. & Davis, T. J. (eds), Managing Mediterranean wetlands and their birds. IWRB Special Publication No. 20. IWRB, Slimbridge, United Kingdom.

Gopal, B. & Sah, M. 1995. Inventory and classification of wetlands in India. Vegetatio 118 (this issue).

Gopal, B., Kvet, J., Loffler, H., Masing, V. & Patten, B. C. 1990. Definition and classification. pp. 9–15. In: Patten, B. C. (ed.), Wetlands and shallow continental water bodies. SPB Academic Publishing bv, The Hague.

Grimmett, R. F. A. & Jones, T. A. 1989. Important bird areas in Europe. ICBP Technical Publication No. 9, ICBP, Cambridge.

Heliotis, F. D. 1988. An inventory and review of the wetland resources of Greece. Wetlands 8, 1–18.

Hess, L. H., Melack, J. M. & Simonett, D. S. 1990. Radar detection of flooding beneath the forest canopy. International Journal of Remote Sensing 11: 1313–1325.

Hollis, G. E. 1990. Wetland systems: science, management, control and controllers. pp. 20–28. In: Kusler, J. A. & Day, S. (eds), Wetlands and river corridor management. Association of Wetland Managers, Inc, New York.

Hollis, G. E. & Jones, T. A. 1991. Europe and the Mediterranean basin. In: Finlayson, M. & Moser, M. (eds), Wetlands. Facts on File, Oxford.

Hollis, G. E., Patterson, J. H., Papayannis, T. & Finlayson, C. M. 1992. Sustaining wetlands: policies, programmes and partnerships. pp. 281–285. In: Finlayson, C. M., Hollis, G. E. & Davis, T. J. (eds), Managing Mediterranean wetlands and their birds. IWRB Special Publication No. 20. IWRB, Slimbridge, United Kingdom.

Hudec, K., Husak, S., Janda, J. & Pellantova, J. 1993. Survey of aquatic and wetland biotopes of the Czech Republic. Czech Ramsar Committee, Trebon, Czech Republic. 32 pp.

Hughes, J. M. R. 1995. The current status of European wetland inventories and classifications. Vegetatio 118 (this issue).

Hughes, R. H. & Hughes, S. 1992. A directory of African wetlands. UNEP, Nairobi & IUCN, Gland Switzerland/WCMC, Cambridge.

Leiderman, E. & Mermet, L. 1991. Mise en Place d'un Observatoire de Zones Humides. Identification de Zones Humides d'Importance Majeure au Plan National. Ministère de l'Environnement – DNP, AIDA & SPRN unpublished report.

Lu, J. 1990. Wetlands in China (in Chinese). East China Normal University Press, Shanghai, China. 177 pp.

Lu, J. 1995. Ecological significance and classification of Chinese wetlands. Vegetatio 118 (this issue).

Luther, H. & Rzoska, J. 1971. Project Aqua: a source book of inland waters proposed for conservation. IBP Handbook No. 21, Blackwell Scientific Publications, Oxford.

Maamouri, F. & Hughes, J. 1992. Prospects for wetlands and waterfowl in Tunisia. pp. 47–52. In: Finlayson, C. M., Hollis, G. E. & Davis, T. J. (eds), Managing Mediterranean wetlands and their birds. IWRB Special Publication No. 20. IWRB, Slimbridge, United Kingdom.

Marti, C. 1988. Zones d'Importance Internationale pour les oiseaux d'eau en Suisse; Cartes commentées pour la première révision de l'inventaire, 1987. Station Ornithologique Suisse, Sempach.

Menantean, L. 1991. Zones humides du littoral de la Communauté Européenne vues de l'espace.

Montes, C. & Bifani, P. 1989. An ecological and economic analysis of the current status of Spanish wetlands. Report to OECD, Paris.

192

Nakayama, N. 1993. Monitoring Asian wetlands and lake basins using remote sensing techniques. pp. 39–42 In: Moser, M. E., Prentice, R. C. & van Vessem, J. (eds), Waterfowl and wetland conservation in the 1990s. IWRB Special Publication No. 26.

Naranjo, L. G. 1995. An evaluation of the first inventory of South America wetlands. Vegetatio 118 (this issue).

Novitzki, R. P. 1995. EMAP-Wetlands: a sampling design with global application. Vegetatio 118 (this issue).

Olney, P. (ed.) 1965. Project MAR: list of European and North African wetlands of international importance. IUCN New Series, IUCN, Morges, Switzerland.

Pakarinen, P. 1995. Classification of boreal mires in Finland and Scandinavia – a review. Vegetatio 118 (this issue).

Pressey, R. L. & Adam, P. 1995. A review of wetland inventory and classification in Australia. Vegetatio 118 (this issue).

Psilovikos, A. A. 1992. Prospects for wetlands and waterfowl in Greece. pp. 53–55. In: Finlayson, C. M., Hollis, G. E. & Davis, T. J. (eds), Managing Mediterranean wetlands and their birds. IWRB Special Publication No. 20. IWRB, Slimbridge, United Kingdom.

Scott, D. A. 1980. A preliminary inventory of wetlands of international importance for waterfowl in west Europe and northwest Africa. IWRB Special Publication No. 2, IWRB, Slimbridge, United Kingdom.

Scott, D. A. 1989a. Design of wetland data sheets for database on Ramsar sites. Photocopied report to Ramsar Bureau, Gland, Switzerland.

Scott, D. A. 1989b. A directory of Asian wetlands. IUCN, Gland, Switzerland and Cambridge, United Kingdom.

Scott, D. A. 1993. A directory of wetlands in Oceania. IWRB, Slimbridge, UK and AWB, Kuala Lumpur, Malaysia. 444 pp.

Scott, D. A. & Carbonell, M. 1986. A directory of Neotropical wetlands, IUCN, Gland, Switzerland.

Scott, D. A. & Jones, T. A. 1995. Classification and inventory of wetlands: a global overview. Vegetatio 118 (this issue).

Semeniuk, C. A. & Semeniuk, V. 1995. A geomorphic approach to global wetland classification for inland wetlands. Vegetatio 118 (this issue).

Silvius, M. J., Djuharsa, E., Taufik, A. W., Steeman, A. P. J. M. & Berczy, E. T. 1987. The Indonesian wetland inventory – a compilation of information on wetlands in Indonesia. PHPA-AWB/Interwader Indonesia & EDWIN, Netherlands.

Simmons, R. E., Brown, C. J. & Griffin, M. (eds) 1991. The status and conservation of wetlands in Namibia. Special Edition Madoqua 17, 254 pp.

Taylor, A. R. D., Howard, G. W. & Begg, G. W. 1995. Developing wetland inventories in southern Africa: a review. Vegetatio 118 (this issue).

Tiner, R. W. Jr. 1974. Wetlands of the United States: current status and recent trends. US Fish and Wildlife Service, National Wetlands Inventory Project, Newton Corner, Massachusetts. 59 pp.

Usbank, S. & James, R. 1993. A directory of important wetlands in Australia. Australian Nature Conservation Agency, Canberra. 687 pp.

Wilen, B. O. & Bates, M. K. 1995. The U.S. Fish and Wildlife Service's national wetland inventory project. Vegetatio 118 (this issue).

Zoltai, S. C. & Vitt, D. H. 1995. Canadian wetlands: environmental gradients and classification. Vegetatio 118 (this issue).